The Finite Element Method

GW00399984

The Finite Element Method

An Introduction with Partial Differential Equations

Second Edition

A. J. DAVIES

Professor of Mathematics
University of Hertfordshire

UNIVERSITY PRESS

OXFORD
UNIVERSITY PRESS

Great Clarendon Street, Oxford OX2 6DP

Oxford University Press is a department of the University of Oxford.
It furthers the University's objective of excellence in research, scholarship,
and education by publishing worldwide in

Oxford New York

Auckland Cape Town Dar es Salaam Hong Kong Karachi
Kuala Lumpur Madrid Melbourne Mexico City Nairobi
New Delhi Shanghai Taipei Toronto

With offices in

Argentina Austria Brazil Chile Czech Republic France Greece
Guatemala Hungary Italy Japan Poland Portugal Singapore
South Korea Switzerland Thailand Turkey Ukraine Vietnam

Oxford is a registered trade mark of Oxford University Press
in the UK and in certain other countries

Published in the United States
by Oxford University Press Inc., New York

© A. J. Davies 2011

The moral rights of the author have been asserted
Database right Oxford University Press (maker)

First Edition published 1980
Second Edition published 2011

All rights reserved. No part of this publication may be reproduced,
stored in a retrieval system, or transmitted, in any form or by any means,
without the prior permission in writing of Oxford University Press,
or as expressly permitted by law, or under terms agreed with the appropriate
reprographics rights organization. Enquiries concerning reproduction
outside the scope of the above should be sent to the Rights Department,
Oxford University Press, at the address above

You must not circulate this book in any other binding or cover
and you must impose the same condition on any acquirer

British Library Cataloguing in Publication Data

Data available

Library of Congress Cataloging in Publication Data

Data available

Typeset by SPI Publisher Services, Pondicherry, India
Printed in Great Britain by
Ashford Colour Press Ltd, Gosport, Hampshire

ISBN 978–0–19–960913–0

1 3 5 7 9 10 8 6 4 2

Preface

In the first paragraph of the preface to the first edition in 1980 I wrote:

It is not easy, for the newcomer to the subject, to get into the current finite element literature. The purpose of this book is to offer an introductory approach, after which the well-known texts should be easily accessible.

Writing now, in 2010, I feel that this is still largely the case. However, while the 1980 text was probably the only introductory text at that time, it is not the case now. I refer the interested reader to the references.

In this second edition, I have maintained the general ethos of the first. It is primarily a text for mathematicians, scientists and engineers who have no previous experience of finite elements. It has been written as an undergraduate text but will also be useful to postgraduates. It is also suitable for anybody already using large finite element or CAD/CAM packages and who would like to understand a little more of what is going on. The main aim is to provide an introduction to the finite element solution of problems posed as partial differential equations. It is self-contained in that it requires no previous knowledge of the subject. Familiarity with the mathematics normally covered by the end of the second year of undergraduate courses in mathematics, physical science or engineering is all that is assumed. In particular, matrix algebra and vector calculus are used extensively throughout; the necessary theorems from vector calculus are collected together in Appendix B.

The reader familiar with the first edition will notice some significant changes. I now present the method as a numerical technique for the solution of partial differential equations, comparable with the finite difference method. This is in contrast to the first edition, in which the technique was developed as an extension of the ideas of structural analysis. The only thing that remains of this approach is the terminology, for example 'stiffness matrix', since this is still in common parlance. The reader familiar with the first edition will notice a change in notation which reflects the move away from the structural background. There is also a change of order of chapters: the introduction to finite elements is now via weighted residual methods, variational methods being delayed until later. I have taken the opportunity to introduce a completely new chapter on boundary element methods. At the time of the first edition, such methods were in their infancy, but now they have reached such a stage of development that it is natural to include them; this chapter is by no means exhaustive and is very much an introduction. I have also included a brief chapter on computational aspects. This is also an introduction; the topic is far too large to treat in any depth. Again the interested reader can follow up the references. In the first edition, many of the examples and exercises were based on problems in journal papers of around that time. I have kept the original references in this second edition.

In Chapter 1, I have written an updated historical introduction and included many new references. Chapter 2 provides a background in weighted residual and variational methods. Chapter 3 describes the finite element method for Poisson's equation, concentrating on linear elements. Higher-order elements and the isoparametric concept are introduced in Chapter 4.

Chapter 5 sets the finite element method in a variational context and introduces time-dependent and non-linear problems. Chapter 6 is almost identical with Chapter 7 of the first edition, the only change being the notation. Chapter 7 is the new chapter on the boundary element method, and Chapter 8, the final chapter, addresses the computational aspects. I have also expanded the appendices by including, in Appendix A, a brief description of some of the partial differential equation models in the physical sciences which are amenable to solution by the finite element method.

I have not changed the general approach of the first edition. At the end of each chapter is a set of exercises with detailed solutions. They serve two purposes: (i) to give the reader the opportunity to practice the techniques, and (ii) to develop the theory a little further where this does not require any new concepts; for example, the finite element solution of eigenvalue problems is considered in Exercise 3.24. Also, some of the basic theory of Chapter 6 is left to the exercises. Consequently, certain results of importance are to be found in the exercises and their solutions.

An important development over the past thirty years has been the wide availability of computational aids such as spreadsheets and computer algebra packages. In this edition, I have included examples of how a spreadsheet could be used to develop more sophisticated solutions compared with the 'hand' calculations in the first edition. Obviously, I would encourage readers to use whichever packages they have on their own personal computers.

Well, this second edition has been a long time coming; I've been working on it for quite some time. It has been confined to very concentrated two-week spells over the Easter periods for the past six years, these periods being spent with Margaret and Arthur, *Les Meuniers*, at their home in the Lot, south-west France. The environment there is ideal for the sort of focused work needed to produce this second edition. I am grateful for their friendship and, of course, their hospitality. The production of this edition would have been impossible without the help of Dr Diane Crann, my wife. I am very grateful for her expertise in turning my sometimes illegible handwritten script into OUP LaTeX house style.

A.J.D.

Lacombrade
Sabadel Latronquière
Lot
August 2010

Contents

1 Historical introduction

The fundamental idea of the finite element method is the replacement of continuous functions by piecewise appproximations, usually polynomials.

Although the finite element method itself is relatively new, its development and success expanding with the arrival and rapid growth of the digital computer, the idea of piecewise approximation is far from new. Indeed, the early geometers used 'finite elements' to determine an approximate value of π. They did this by bounding a quadrant of a circle with inscribed and circumscribed polygons, the straight-line segments being the finite element approximations to an arc of the circle. In this way, they were able to obtain extremely accurate estimates. Upper and lower bounds were obtained, and by taking an increasing number of elements, monotonic convergence to the exact solution would be expected. These phenomena are also possible in modern applications of the finite element method. One remark regarding ancient finite elements: Archimedes used these ideas to determine areas of plane figures and volumes of solids, although of course he did not have a precise concept of a limiting procedure. Indeed, it was only this fact which prevented him from discovering the integral calculus some two thousand years before Newton and Leibniz. The interesting point here is that whilst many problems of applied mathematics are posed in terms of differential equations, the finite element solution of such equations uses ideas which are in fact much older than those used to set up the equations initially.

The modern use of finite elements really started in the field of structural engineering. Probably the first attempts were by Hrennikoff (1941) and McHenry (1943), who developed analogies between actual discrete elements, for example bars and beams, and the corresponding portions of a continuous solid. These methods belonged to a class of semi-analytic techniques which were used in the 1940s for aircraft structural design. Matrix methods for the solution of such problems were developed at this time, and it is interesting to note that the work of Argyris (1955), in an engineering context, introduced a minimization process which is also the basis of the mathematical underpinning of the finite element method. With the development of high-speed, jet-powered aircraft, these semi-analytic methods were soon found to be inadequate and the quest began for a more reliable approach. A direct approach, based on the principle of virtual work, was given by Argyris (1955), and in a series of papers he and his colleagues developed this work to solve very complex problems using computational techniques (Argyris and Kelsey 1960). At about the same time,

Turner *et al.* (1956) presented the element stiffness matrix, based on displacement assumptions, for a triangular element, together with the direct stiffness method for assembling the elements. The term 'finite element' was introduced by Clough (1960) in a paper describing applications in plane elasticity.

The engineers had put the finite element method on the map as a practical technique for solving their elasticity problems, and although a rigorous mathematical basis had not been developed, the next few years saw an expansion of the method to solve a large variety of structural problems. Solutions of three-dimensional problems required only simple extensions to the basic two-dimensional theory (Argyris 1964). The obvious problem to consider after plane problems was that of plate bending; here, researchers found their first real difficulties and the early attempts were not altogether successful. It was not until some time later that the problems of compatibility were resolved (Bazely *et al.* 1965).

One area of application of plate elements was that of modelling thin shells, and some success was achieved (Clough and Johnson 1968). However, the representation of a thin shell by a polyhedral surface of flat plates can cause serious problems in the presence of pronounced bending, and it soon became clear that shell elements themselves were necessary.

Plate elements presented difficulties to researchers, but these were small compared with the problems associated with shell elements. The first actual shell elements developed were axisymmetric elements (Grafton and Strome 1963), and these were followed by a whole sequence of cylindrical and other shell elements (Gallagher 1969). Such elements are still being developed, and it is probably fair to say that this is the only area of linear analysis that still has potential for further work in the context of finite elements.

The workers in the early 1960s soon turned their attention towards the solution of non-linear problems. Turner *et al.* (1960) showed how to use an incremental technique to solve geometrically non-linear problems, i.e. problems in which the strains remain small but displacements are large. Stability analysis also comes into this category and was discussed by Martin (1965). Plasticity problems, involving non-linear material behaviour, were modelled at this time (Gallagher *et al.* 1962) and the method was also applied to the solution of problems in viscoelasticity (Zienkiewicz *et al.* 1968).

Besides the static analysis described above, dynamic problems were also being tackled, and Archer (1963) introduced the concept of the consistent mass matrix. Both vibration problems (Zienkiewicz *et al.* 1966) and transient problems (Koenig and Davids 1969) were considered. Thus the period from its conception in the early 1950s to the late 1960s saw the method being applied extensively by the engineering community. With the successes of these practical applications in the structural field, it was open for engineers in other disciplines (Silvester and

Ferrari 1983) to get hold of the finite element method. An obvious candidate was fluid mechanics.

Potential flow (Doctors 1970) and Stokes flow were easy to develop (Atkinson *et al.* 1970), and it wasn't long before the appearance of a textbook on the finite element method in viscous flow problems (Connor and Brebbia 1976). However, the more general form of the Navier–Stokes equations was much more difficult, the convection terms yield non-self-adjoint operators and, consequently, there are no obvious variational principles. The method was extended further when it was seen to fit in with the method of *weighted residuals* (Szabó and Lee 1969). This then allowed the solution of such problems posed as partial differential equation boundary-value problems. The method had been well known for some time; Crandall (1956) had used the term to classify a variety of numerical approximation techniques, although Galerkin (1915) was the first to use the method. Probably the first finite element solution of the Navier–Stokes equations was given by Taylor and Hood (1973). However, problems that had been encountered using finite differences (Spalding 1972) were apparent in the finite element approach, and the so-called *up-wind* approach was brought into a finite element context (Zienkiewicz, Heinrich *et al.* 1977). Also, the so-called *finite-volume* approach was developed (Jameson and Mavriplis 1986), which has the important physical property that certain conservation laws are maintained. The scene was now set for rapid developments in fluid mechanics and other areas such as heat and mass transfer (Mohr 1992), for diffusion–convection problems, and for other coupled problems (Elliott and Larsson 1995).

As far as this historical introduction is concerned, this is where we shall leave the contributions from the engineering community. There are excellent accounts of applications from the mid 1970s onwards in the texts by Zienkiewicz and Taylor (2000a,b). Let us return to the early days of the developments: at the same time as the engineers were pushing forward with the practical aspects of the method, similar work was being carried out by applied mathematicians, each group apparently unaware of the work of the other. Courant (1943) gave a solution to the torsion problem, using piecewise linear approximations over a triangular mesh, formulating the problem from the principle of minimum potential energy. Zienkiewicz (1995) noted that Courant had already developed some of the ideas in the 1920s without taking them further. Similar papers followed by Polya (1952) and Weinberger (1956). Greenstadt (1959) presented the idea of considering a continuous region as an assembly of several discrete parts and making assumptions about the variables in each region, variational principles being used to find values for these variables. We note here also that the work of Schoenberg (1946) was very much in the spirit of finite elements, since the piecewise polynomial approximation led to the development of the theory of splines.

Similar work was being carried out in the physics community. In the late 1940s, Prager and Synge (1947) developed a geometric approach to a variational principle in elasticity which led to the so-called *hypercircle method*, which is also in the spirit of the finite element method. The method is discussed in detail in the book by Synge (1957). A three-dimensional problem in electrostatics was solved, using linear tetrahedral elements, by McMahon (1953).

It was some time before Birkhoff *et al.* (1968) and Zlamal (1968) published a convergence proof and error bounds in the applied mathematics literature. However, the first convergence proof in the engineering literature had already been given by Melosh (1963), who used the principle of minimum potential energy, and this work was extended by Jones (1964) using Reissner's variational principle. Once it was realized that the method could be interpreted in terms of variational methods, the mathematicians and engineers were brought together and many extensions of the method to new areas soon followed. In particular, it was realized that the concept of piecewise polynomial approximation offered a simple and efficient procedure for the application of the classical Rayleigh–Ritz method. The principles could be clearly seen in the much earlier work of Lord Rayleigh (Strutt 1870) and Ritz (1909). From a physical point of view, it meant that problems outside the structural area could be solved using standard structural packages by associating suitable meanings to the terms in the corresponding variational principles. This was just what was done by Zienkiewicz and Cheung (1965) in the application of the finite element method to the solution of Poisson's equation and by Doctors (1970) in the application to potential flow. Similarly, transient heat conduction problems were considered by Wilson and Nickell (1966).

The mathematical basis of the method then started in earnest, and it is well beyond the scope of this text to do more than indicate where the interested reader may wish to start. Error estimation is clearly an important aspect of any numerical approximation, and the first developments were by Babuška (1971, 1973) and Babuška and Rheinboldt (1978, 1979), who showed how to estimate errors and how convergence was ensured by suitable mesh refinement. The basis was then set for the possibility of adaptive mesh refinement, in which meshes are automatically refined in response to knowledge of computed solutions. Mesh generation and adaption is an area in which much work is still needed; for a recent account, see Zienkiewicz *et al.* (2005). Ciarlet (1978) provided the first of what would usually be described as a 'mathematical' account of the finite element method, and the text has since been extended and updated by Ciarlet and Lions (1991). The reader interested in becoming familiar with current mathematical approaches to the method should consult Brenner and Scott (1994) or the very readable text by Axelsson and Barker (2001).

At about the same time as Courant was working on variational methods for elliptic problems, Trefftz (1926) developed a technique in which a partial differential equation, defined over a region, becomes an integral equation over the boundary of that region. The immediate advantage is in the reduction of the dimension of the problem. Such integral techniques have been known since the late nineteenth century; the theorems of Green (1828) are the bedrock of the solution of potential problems, the term *potential* first being coined by Green in his seminal paper. These techniques have been the basis of the formulation of potential theory and elasticity by, amongst others, Fredholm (1903) and Kellog (1929).

It was with developments in computing and numerical procedures that the technique became attractive to physicists and engineers in the 1960s (Hess and Smith 1964), and the ideas developed at that time were collected together in a single text (Jaswon and Symm 1977). A very good overview of the early development of boundary elements was given by Becker (1992). It is interesting to note here the work of Rizzo (1967), who applied the ideas for potential problems to use boundary elements to solve problems in elasticity, in contrast with Zienkiewicz and Cheung (1965), who used codes for structural analysis to obtain the finite element solution of potential problems.

The first boundary element textbook was written by Brebbia (1978), and since then there has been a variety of similar texts, each with the intention of making the technique accessible to those who would wish to develop their own code. See, for example, Gipson (1987), Becker (1992) and París and Cañas (1997). The text by Hall (1993) is particularly useful to those for whom boundary elements are a completely new idea. Finally, the two-volume set by Aliabadi (2002) and Wrobel (2002) provides a similar state-of-the-art work on boundary elements, as does the three-volume set by Zienkiewicz and Taylor (2000a,b) and Zienkiewicz *et al.* (2005) for finite elements.

There is always the question 'Which is better, the finite element method or the boundary element method?' See Chapter 8, where we discuss the merits of each case. It is usually accepted that boundary elements are more appropriate for infinite regions. However, in a recent text, Wolf and Song (1997) set out a finite element procedure to cope with unbounded regions. Zienkiewicz, Kelly *et al.* (1977) proposed a coupling of the two methods to get the best out of each: finite elements in regions of material non-linearity and boundary elements for unbounded regions.

Recently, further developments in so-called *mesh-free methods* have been proposed (Goldberg and Chen 1997, Liu 2003); included is the *method of fundamental solutions* (Goldberg and Chen 1999), which has its origins in the work on potential problems by Kupradze (1965).

Currently, the terms 'finite element method', 'boundary element method', 'mesh-free method' etc. are used, and they are all really variations on a more general weighted residual theme. Zienkiewicz (1995) suggested that a more appropriate generic name would be the *generalized Galerkin method* (Fletcher

1984). For further details of background and history, see the following: for finite elements, Zienkiewicz (1995) and Fish and Belytschko (2007), who gave a very good account of the commercial development of finite elements; and for boundary elements, see Becker (1992) and Cheng and Cheng (2005). There is now a very large body of work in the finite element method field; a quick Internet search on the words 'finite element' and 'boundary element' yielded more than 20 million pages. The website run at the University of Ohio gives details of more than 600 finite element books.

The finite element method has now reached a very sophisticated level of development, so much so that it is applied routinely in a wide variety of application areas. We mention here just two of them. (i) Biomedical engineering: Zienkiewicz (1977) performed stress analysis calculations for human femur transplants. Recently, Phillips (2009) has extended these ideas significantly and, instead of modelling just the bone, he has made a complete finite element analysis of the bone and the associated muscles. (ii) Financial engineering: in the early days of the development of finite elements, the study of financial systems would have seemed to have been outside the scope of the method. However, Black–Scholes models (Wilmott *et al.* 1995), describing a variety of option-pricing schemes, have been set in a finite element context by, amongst others, Topper (2005) and Tao Jiang *et al.* (2009).

For a general guide to current research from both an engineering and a mathematical perspective, the reader is referred to the sets of conference proceedings MAFELAP (From 1973 to 2010), edited by Whiteman, and BEM (From 1979 to 2010), edited by Brebbia.

Finally, if there is one person whose work forms a basis for both the finite element method and the boundary element method then it is George Green. Green's theorem underpins both methods, and, of course, fundamental solutions are themselves Green's functions.

2 Weighted residual and variational methods

2.1 Classification of differential operators

The quantities of interest in many areas of applied mathematics are often to be found as the solution of certain partial differential equations, together with prescribed boundary and/or initial conditions.

The nature of the solution of a partial differential equation depends on the form that the equation takes. All linear, and quasi-linear, second-order equations are classified as *elliptic, hyperbolic* or *parabolic*. In each of these categories there are equations which model certain physical phenomena. The classification is determined by the coefficients of the highest partial derivatives which occur in the equation.

In this chapter, we shall consider functions which depend on two independent variables only, so that the resulting algebra does not obscure the underlying ideas.

Consider the second-order partial differential equation

$$(2.1) \qquad\qquad \mathcal{L}u = f,$$

where $\mathcal{L}$ is the operator defined by

$$\mathcal{L}u \equiv a\frac{\partial^2 u}{\partial x^2} + b\frac{\partial^2 u}{\partial x \partial y} + c\frac{\partial^2 u}{\partial^2 y^2} + F\left(x, y, u, \frac{\partial u}{\partial x}, \frac{\partial u}{\partial y}\right).$$

a, b and c are, in general, functions of x and y; they may also depend on u itself and its derivatives, in which case the equation is *non-linear*. Non-linear equations are, in general, far more difficult to deal with than linear equations, and they will not be discussed here. They will, however, be considered briefly in Section 5.3.

Equation (2.1) is said to be:

- elliptic if $b^2 < 4ac$;
- hyperbolic if $b^2 > 4ac$;
- parabolic if $b^2 = 4ac$.

Unlike ordinary differential equations, it is not usually profitable to investigate the solution of a partial differential equation in isolation from the associated boundary and/or initial conditions. Indeed, it will always be an equation together with prescribed conditions which forms a mathematical model of a particular situation.

In general, elliptic equations are associated with steady-state phenomena and require a knowledge of values of the unknown function, or its derivative, on the boundary of the region of interest. Thus *Poisson's equation*,

$$-\nabla^2 u = \frac{\rho}{\epsilon},$$

gives a model which describes the variation of the electrostatic potential in a medium with permittivity ϵ and in which there is a charge distribution ρ per unit volume. In the case $\rho \equiv 0$ we have *Laplace's equation*,

$$\nabla^2 u = 0.$$

In order that the solution is unique, it is necessary to know the potential or charge distribution on the surrounding boundary. This is a pure boundary-value problem.

Hyperbolic equations are, in general, associated with propagation problems and require the specification of certain initial values and/or possible boundary values as well.

Thus, the *wave equation*,

$$\frac{\partial^2 u}{\partial x^2} = \frac{1}{c^2}\frac{\partial^2 u}{\partial t^2},$$

gives the small transverse displacement, $u(x,t)$, of a uniform vibrating string; waves are propagated along the string with speed c. In such problems it is usually required to know the displacement, or its derivative, at the ends, together with the initial displacement and velocity distribution. This is an initial boundary-value problem.

Finally, parabolic equations model problems in which the quantity of interest varies slowly in comparison with the random motions which produce these variations. As is the case with hyperbolic equations, they are associated with initial-value problems. Thus the *heat equation*, or *diffusion equation*,

$$\frac{\partial^2 u}{\partial x^2} = \frac{1}{\alpha}\frac{\partial u}{\partial t},$$

describes the temperature variation, $u(x,t)$, in a thin rod which has a uniform thermal diffusivity α. The temperature changes in the material are produced by the motion of individual molecules and occur slowly in comparison with these molecular motions. Either the temperature or its derivative is usually given at the ends of the rod, together with an initial temperature distribution. This again is an initial boundary-value problem. The three equations occur in many areas of applied mathematics, engineering and science. In Appendix A we provide a table of application areas, so that the interested reader may be able to associate the equations with appropriate applications.

In this chapter we shall consider the approximate methods on which our finite element techniques, described in Chapters 3, 4, 5 and 6, will be based. It is in the area of elliptic partial differential equations that finite element methods have been used most extensively, since the differential operators involved belong to the important class of positive definite operators. However, finite elements are widely used for the solution of hyperbolic and parabolic equations, and all three categories will be discussed; elliptic equations in Chapters 3 and 4, and hyperbolic and parabolic equations in Chapter 5.

2.2   Self-adjoint positive definite operators

Suppose that the function u satisfies eqn (2.1) in a two-dimensional region D bounded by a closed curve C, i.e.

$$\mathcal{L}u = f,$$

where $f(x, y)$ is a given function of position. Suppose also that u satisfies certain given homogeneous conditions on the boundary C. Usually these conditions are of the following types:

(2.2)   *Dirichlet* boundary condition: $u = 0$;

(2.3)   *Neumann* condition: $\dfrac{\partial u}{\partial n} = 0$;

(2.4)   *Robin* condition: $\dfrac{\partial u}{\partial n} + \sigma(s)u = 0$.

Here s is the arc length measured along C from some fixed point on C, and $\partial/\partial n$ represents differentiation along the outward normal to the boundary. Note that the Neumann condition may be obtained from the Robin condition by setting $\sigma \equiv 0$.

A problem is said to be *properly posed*, in the sense of Hadamard (1923), if and only if the following conditions hold:

1. A solution exists.
2. The solution is unique.
3. The solution depends continuously on the data.

The third condition is equivalent to saying that small changes in the data lead to small changes in the solution (Renardy and Rogers 1993). If at least one of these conditions does not hold, then the problem is said to be *poorly posed* or *ill-posed*.

For an elliptic operator $\mathcal{L}$, the problem is properly posed only when one of these conditions holds at each point on the boundary.

The numerical methods which we shall discuss involve processes which change our partial differential equation into a system of linear algebraic equations. Two important properties of $\mathcal{L}$ lead to particularly useful properties of the system matrix.

1. The operator $\mathcal{L}$ is said to be *self-adjoint* if and only if the expression

$$\iint_D v\mathcal{L}u \, dx \, dy - \iint_D u\mathcal{L}v \, dx \, dy$$

is a function only of u, v and their derivatives evaluated on the boundary. In particular, for homogeneous boundary conditions, $\mathcal{L}$ is self-adjoint if and only if

$$\iint_D v\mathcal{L}u \, dx \, dy = \iint_D u\mathcal{L}v \, dx \, dy.$$

2. The operator $\mathcal{L}$ is said to be *positive definite* if and only if, for all functions u,

$$\iint_D u\mathcal{L}u \, dx \, dy \geq 0,$$

equality occurring if and only if $u \equiv 0$.

In both definitions, it is assumed that u and v satisfy suitable differentiability conditions in order that the operations exist.

Example 2.1 Suppose that $\mathcal{L} = -\nabla^2$, so that eqn (2.1) becomes Poisson's equation; then

$$\iint_D v\mathcal{L}u \, dx \, dy - \iint_D u\mathcal{L}v \, dx \, dy = \iint_D u \, \nabla^2 v \, dx \, dy - \iint_D v \, \nabla^2 u \, dx \, dy$$

$$= \oint_C \left(u\frac{\partial v}{\partial n} - v\frac{\partial u}{\partial n} \right) ds$$

by virtue of the second form of Green's theorem. Thus the operator $-\nabla^2$ is self-adjoint.

Also, $v \, \nabla^2 u = \text{div}\,(v \, \text{grad}\,u) - \text{grad}\,u \cdot \text{grad}\,v$, and thus

$$\iint_D u\,(-\nabla^2 u) \, dx \, dy = -\iint_D \text{div}\,(u \, \text{grad}\,u) \, dx \, dy + \iint_D |\,\text{grad}\,u\,|^2 \, dx \, dy$$

$$= - \oint_C u \frac{\partial u}{\partial n} ds + \iint_D |\operatorname{grad} u|^2 \, dx \, dy,$$

using the *divergence theorem*; see Appendix B.

Thus if u satisfies either of the boundary conditions (2.2) or (2.3) it follows that $-\nabla^2$ is positive definite. For the Robin boundary condition (2.4), the boundary integral becomes

$$\oint_C \sigma u^2 \, ds$$

and hence $-\nabla^2$ is positive definite provided that $\sigma > 0$.

Example 2.2 Suppose now that $\mathcal{L}u = -\operatorname{div}(k \operatorname{grad} u)$, where $k(x, y)$ is a scalar function of position, and suppose also that the problem is isotropic. Anisotropy can be taken into account by replacing the scalar k by a tensor represented by

$$(2.5) \qquad \qquad \boldsymbol{\kappa} = \begin{bmatrix} \kappa_{xx} & \kappa_{xy} \\ \kappa_{yx} & \kappa_{yy} \end{bmatrix};$$

see eqn (2.68).

$$\iint_D v\mathcal{L}u \, dx \, dy - \iint_D u\mathcal{L}v \, dx \, dy$$

$$= \iint_D u \operatorname{div}(k \operatorname{grad} v) \, dx \, dy - \iint_D v \operatorname{div}(k \operatorname{grad} u) \, dx \, dy$$

$$= \iint_D \{\operatorname{div}(uk \operatorname{grad} v) - \operatorname{grad} u \cdot k \operatorname{grad} v\} \, dx \, dy$$

$$- \iint_D \{\operatorname{div}(vk \operatorname{grad} u) - \operatorname{grad} v \cdot k \operatorname{grad} u\} \, dx \, dy$$

$$= \oint_C k \left(u \frac{\partial v}{\partial u} - v \frac{\partial u}{\partial n} \right) ds$$

using the divergence theorem.

Hence $\mathcal{L}$ is self-adjoint. Also,

$$\iint_D u\mathcal{L}u \, dx \, dy = - \oint_C ku \frac{\partial u}{\partial n} ds + \iint k \, |\operatorname{grad} u|^2 \, dx \, dy.$$

If u satisfies eqn (2.2) or eqn (2.3), then the boundary integral vanishes and $\mathcal{L}$ is positive definite if $k > 0$. If, however, u satisfies eqn (2.4), then the boundary integral is negative if $\sigma > 0$, so that again $\mathcal{L}$ is positive definite if $k > 0$.

When one is modelling physical phenomena, an important property that the model must possess is that it has a unique solution. If the physical system is modelled by eqn (2.1) and $\mathcal{L}$ is linear and positive definite, then the solution is unique. The proof is as follows.

Suppose that u_1 and u_2 are two solutions of eqn (2.1). Let

$$v = u_1 - u_2,$$

so that

$$\mathcal{L}v = \mathcal{L}u_1 - \mathcal{L}u_2 = 0.$$

Hence

$$\iint_D v\mathcal{L}v\, dx\, dy = 0.$$

Now $\mathcal{L}$ is positive definite, so that $v \equiv 0$. Thus $u_1 = u_2$ and the solution is unique.

2.3   Weighted residual methods

Consider the boundary-value problem

(2.6) $$\mathcal{L}u = f \quad \text{in} \quad D$$

subject to the non-homogeneous Dirichlet boundary condition

$$u = g(s)$$

on some part C_1 of the boundary, and the non-homogeneous Robin condition

$$\frac{\partial u}{\partial n} + \sigma(s)u = h(s)$$

on the remainder C_2.

An approximate solution $\tilde{u}$ will not, in general, satisfy eqn (2.6) exactly, and associated with such an approximate solution is the residual defined by

$$r(\tilde{u}) = \mathcal{L}\tilde{u} - f.$$

If the exact solution is u_0, then

$$r(u_0) \equiv 0.$$

We choose a set of basis functions $\{v_i : i = 0, \ldots, n\}$, and make an approximation of the following form:

$$(2.7) \qquad \tilde{u}_n = \sum_{i=0}^{n} c_i v_i.$$

In the weighted residual method, the unknown parameters c_i are chosen to minimize the residual $r(\tilde{u})$ in some sense. Different methods of minimizing the residual yield different approximate solutions.

All the methods we shall consider result in a system of equations of the form

$$\mathbf{Ac} = \mathbf{h}$$

for the unknowns c_i. The different methods yield different matrices $\mathbf{A}$ and $\mathbf{h}$.

The methods presented in this section are illustrated in Examples 2.3–2.6, in which all calculations are done by hand. A comparison of these solutions with the exact solution is shown in Table 2.1 and Fig. 2.1. We shall consider the simple two-point boundary-value problem

$$(2.8) \qquad \begin{aligned} -u'' &= x^2, \qquad 0 < x < 1, \\ u(0) &= u(1) = 0. \end{aligned}$$

This problem has the exact solution

$$u_0(x) = \frac{x}{12}\left(1 - x^3\right).$$

Example 2.3 Firstly, consider the collocation method, in which the trial function (2.7) is chosen to satisfy the boundary conditions.

N.B. We choose homogeneous Dirichlet boundary conditions. One-dimensional problems with non-homogeneous conditions of the form $u(0) = a$, $u(1) = b$ may be transformed to a problem with homogeneous conditions by the change of dependent variable $w(x) = u(x) - ((1 - x)a + xb)$.

The parameters c_i are then found by forcing $\tilde{u}_n$ to satisfy the differential equation at a given set of n points, i.e. at these points the residual vanishes.

Table 2.1 Approximate solutions ($\times 10^2$) to the boundary-value problem of Examples 2.3–2.6

x	0	0.2	0.4	0.6	0.8	1
Collocation	0	1.422	2.933	3.733	3.022	0
Overdetermined collocation	0	1.533	3.100	3.900	3.133	0
Least squares	0	1.867	3.600	4.400	3.467	0
Galerkin	0	1.600	3.200	4.000	3.200	0
Exact	0	1.653	3.120	3.920	3.253	0

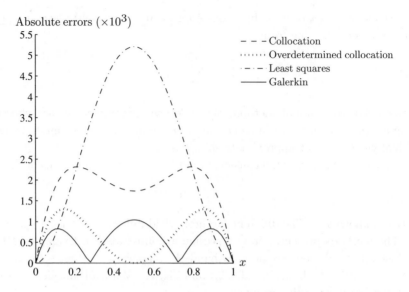

Fig. 2.1 Comparison of the errors in the four approximate solutions in Examples 2.3–2.6.

The residual is

$$r\left(\tilde{u}_n\right) = -\left(\tilde{u}_n'' + x^2\right).$$

A cubic approximation which satisfies the boundary conditions is

$$\tilde{u}_1(x) = x(1-x)(c_0 + c_1 x),$$

so that

$$r\left(\tilde{u}_1\right) = 2c_0 + (6x - 2)c_1 - x^2.$$

If $x = \frac{1}{3}$ and $x = \frac{2}{3}$ are chosen as collocation points, then

$$r\left(\tilde{u}_1\left(\frac{1}{3}\right)\right) = r\left(\tilde{u}_1\left(\frac{2}{3}\right)\right) = 0,$$

which leads to the equations

$$\begin{bmatrix} 2 & 0 \\ 2 & 2 \end{bmatrix} \begin{bmatrix} c_0 \\ c_1 \end{bmatrix} = \begin{bmatrix} \frac{1}{9} \\ \frac{4}{9} \end{bmatrix},$$

giving $c_0 = \frac{1}{18}$, $c_1 = \frac{1}{6}$.

Thus the approximate solution is

$$\tilde{u}_1(x) = \frac{1}{18} x(1-x)(1+3x).$$

This example illustrates the conventional use of the collocation method. The idea may also be used with collocation at m points, where $m > n$, so that an overdetermined system of equations is obtained for the unknown parameters. These equations may then be solved by the method of least squares.

Example 2.4 Suppose, in Example 2.3, the same cubic approximation is used but the chosen collocation points are $x = \frac{1}{4}$, $x = \frac{1}{2}$ and $x = \frac{3}{4}$. Then, forcing the residual to vanish at the collocation points yields the system

$$
\begin{bmatrix} 2 & -\frac{1}{2} \\ 2 & 1 \\ 2 & \frac{5}{2} \end{bmatrix} \begin{bmatrix} c_0 \\ c_1 \end{bmatrix} = \begin{bmatrix} \frac{1}{16} \\ \frac{1}{4} \\ \frac{9}{16} \end{bmatrix}.
$$

This is an overdetermined set of algebraic equations of the form

$$(2.9) \qquad\qquad \mathbf{Ac} = \mathbf{h}.$$

The usual method of least squares yields the following square set of equations:

$$\mathbf{A}^T \mathbf{Ac} = \mathbf{A}^T \mathbf{h}.$$

Thus eqn (2.9) becomes

$$
\begin{bmatrix} 12 & 6 \\ 6 & \frac{5}{2} \end{bmatrix} \begin{bmatrix} c_0 \\ c_1 \end{bmatrix} = \begin{bmatrix} \frac{7}{4} \\ \frac{13}{8} \end{bmatrix},
$$

giving $c_0 = \frac{1}{16}$, $c_1 = \frac{1}{6}$.

Thus the approximate solution is

$$\tilde{u}_1(x) = \frac{1}{48} x(1 - x)(3 + 8x).$$

Example 2.5 The second approach is the method of least squares applied directly to the residual. Again the trial functions are chosen to satisfy the boundary conditions, and the residual is minimized in the sense that the parameters are chosen so that

$$I[\tilde{u}] = \iint_D r(\tilde{u})^2 \, dx \, dy$$

is a minimum.

Thus

$$\frac{\partial I}{\partial c_i} = 0, \quad i = 0, \ldots, n.$$

Now

$$I(c_0, \ldots, c_n) = \iint_D \left\{ \mathcal{L}\left(\sum c_j v_j\right) - f \right\}^2 dx \, dy.$$

If $\mathcal{L}$ is a linear operator, then

$$\frac{\partial I}{\partial c_i} = 2 \iint_D \left\{ \mathcal{L}\left(\sum c_j v_j\right) \mathcal{L}v_i - f \mathcal{L}v_i \right\} dx \, dy,$$

so that

(2.10) $$\sum_{j=0}^n c_j \iint_D (\mathcal{L}v_i \mathcal{L}v_j) \, dx \, dy = \iint_D f \mathcal{L}v_i \, dx \, dy, \quad i = 0, \ldots, n.$$

Consider the problem of Example 2.3 with trial function

$$\tilde{u}(x) = x(1 - x)(c_0 + c_1 x).$$

Here $v_0 = x(1 - x)$, $v_1 = x^2(1 - x)$; thus $\mathcal{L}v_0 = 2$, $\mathcal{L}v_1 = 6x - 2$.
    Therefore

$$\int_0^1 (\mathcal{L}v_0)^2 \, dx = 4, \quad \int_0^1 (\mathcal{L}v_1)^2 \, dx = 4,$$

$$\int_0^1 \mathcal{L}v_0 \mathcal{L}v_1 \, dx = \int_0^1 \mathcal{L}v_1 \mathcal{L}v_0 \, dx = 2,$$

$$\int_0^1 x^2 \mathcal{L}v_0 \, dx = \frac{2}{3}, \quad \int_0^1 x^2 \mathcal{L}v_1 \, dx = \frac{5}{6}.$$

Equation (2.10) then gives

$$\begin{bmatrix} 4 & 2 \\ 2 & 4 \end{bmatrix} \begin{bmatrix} c_0 \\ c_1 \end{bmatrix} = \begin{bmatrix} \frac{2}{3} \\ \frac{5}{6} \end{bmatrix},$$

giving $c_0 = \frac{1}{12}$, $c_1 = \frac{1}{6}$.
    The approximate solution is then

$$\tilde{u}_1(x) = \frac{1}{12} x(1 - x)(1 + 2x).$$

Example 2.6 The final method to be considered is the Galerkin method. In this method the integral of the residual, weighted by the basis functions, is set to zero, i.e.

$$\iint_D r(\tilde{u}) v_i \, dx \, dy = 0, \quad i = 0, \ldots, n.$$

This yields the following $n+1$ equations for the $n+1$ parameters c_i:

$$\iint_D (\mathcal{L}\tilde{u} - f)\, v_i \, dx\, dy = 0, \quad i = 0, \ldots, n,$$

i.e.

(2.11) $$\sum_{j=0}^n c_j \iint_D v_i \mathcal{L} v_j \, dx\, dy = \iint_D f v_i \, dx\, dy, \quad i = 0, \ldots, n.$$

In the special case $\mathcal{L} = -\nabla^2$, these equations are of the form

$$\mathbf{A}\mathbf{c} = \mathbf{h},$$

where

$$A_{ij} = -\iint_D v_i \, \nabla^2 v_j \, dx\, dy$$

and

$$h_i = \iint_D v_i f \, dx\, dy.$$

Again, consider the problem (2.8) with the basis functions v_0 and v_1 of Example 2.5. Then

$$\int_0^1 v_0 \mathcal{L} v_0 \, dx = \frac{1}{3}, \quad \int_0^1 v_1 \mathcal{L} v_1 \, dx = \frac{2}{15},$$

$$\int_0^1 v_0 \mathcal{L} v_1 \, dx = \frac{1}{6}, \quad \int_0^1 v_1 \mathcal{L} v_0 \, dx = \frac{1}{6},$$

$$\int_0^1 x^2 \mathcal{L} v_0 \, dx = \frac{1}{20}, \quad \int_0^1 x^2 \mathcal{L} v_1 \, dx = \frac{1}{30}.$$

Equation (2.11) thus gives

$$\begin{bmatrix} \frac{1}{3} & \frac{1}{6} \\ \frac{1}{6} & \frac{2}{15} \end{bmatrix} \begin{bmatrix} c_0 \\ c_1 \end{bmatrix} = \begin{bmatrix} \frac{1}{20} \\ \frac{1}{30} \end{bmatrix},$$

giving $c_0 = \frac{1}{15}$, $c_1 = \frac{1}{6}$.

Thus the approximate solution is

$$\tilde{u}_1(x) = \frac{1}{30} x(1-x)(2+5x).$$

For this particular problem, it is difficult to decide which is the 'best' method. It would appear that the Galerkin and the overdetermined collocation methods give the best distribution of error; but notice that there are regions in which the least squares method gives the best results, although, overall, it is probably the least accurate. It is interesting to note that Crandall (1956) came to similar conclusions for an initial-value problem involving a first-order equation; in the case presented there, the least squares method turns out to be the 'best' higher-order polynomial approximation, $\tilde{u}_n$. The right-hand side of problem (2.8) is quadratic and it is not difficult to see that the exact solution is quartic, so that approximations $\tilde{u}_n$ with $n \geq 4$ will necessarily recover the exact solution. We make a small change and consider the problem

$$-u'' = e^x, \qquad 0 < x < 1,$$
$$u(0) = u(1) = 0.$$

This problem has the exact solution

$$u_0(x) = 1 + (e-1)x - e^x,$$

and no polynomials will recover this exactly.

The approximations used in Examples 2.3–2.6 have all been sufficiently 'simple' to be amenable to hand calculation. In the next example we consider higher values of n which make hand calculation almost impossible. We have used a spreadsheet to develop the solutions. The interested reader could, of course, use any of the widely available computational packages such as MATLAB, Mathematica and Mathcad. These computer algebra packages are particularly helpful for obtaining the integrals in the least squares and the Galerkin methods.

Example 2.7 We shall implement spreadsheet solutions with $n = 4$, i.e.

(2.12) $$\tilde{u}_4 = x(1-x)(c_0 + c_1 x + c_2 x^2 + c_3 x^3 + c_4 x^4).$$

Collocation.

$$r(\tilde{u}_4) = 2c_0 + (6x - 2)c_1 + (12x^2 - 6x)c_2 + (20x^3 - 12x^2)c_3$$
$$+ (30x^4 - 20x^3)c_4 - e^x,$$

and we shall collocate at the five points $x = 0.1, 0.3, 0.5, 0.7, 0.9$. The spreadsheet implementation is shown in Fig. 2.2.

We see that

$$\tilde{u}_4 = x(1-x)(0.718285 + 0.218225x + 0.051914x^2 + 0.009267x^3 + 0.002305x^4).$$

Overdetermined collocation. In this case we shall collocate at the nine points $x = 0.1, 0.2, \ldots, 0.9$. The spreadsheet implementation is shown in Fig. 2.3.

We see that

$$\tilde{u}_4 = x(1-x)(0.718216 + 0.218224x + 0.051912x^2 + 0.009264x^3 + 0.002308x^4).$$

	A	B	C	D	E	F	G
1	Collocation						
2	A						h
3	2.000	-1.400	-0.480	-0.100	-0.017		1.105171
4	2.000	-0.200	-0.720	-0.540	-0.297		1.349859
5	2.000	1.000	0.000	-0.500	-0.625		1.648721
6	2.000	2.200	1.680	0.980	0.343		2.013753
7	2.000	3.400	4.320	4.860	5.103		2.459603
8	A_inverse						c
9	0.214844	0.060764	0.174479	0.026042	0.023872		0.71828521
10	-1.015625	1.701389	-1.302083	0.729167	-0.112847		0.21822505
11	2.213542	-6.250000	6.510417	-3.125000	0.651042		0.05191367
12	-2.256944	7.986111	-10.416667	5.902778	-1.215278		0.00926688
13	0.868056	-3.472222	5.208333	-3.472222	0.868056		0.00230520

Fig. 2.2 Spreadsheet for the collocation method.

	A	B	C	D	E	F	G
1	Overdetermined collocation						
2	A						h
3	2.000	-1.400	-0.480	-0.100	-0.017		1.105171
4	2.000	-0.800	-0.720	-0.320	-0.112		1.221403
5	2.000	-0.200	-0.720	-0.540	-0.297		1.349859
6	2.000	0.400	-0.480	-0.640	-0.512		1.491825
7	2.000	1.000	0.000	-0.500	-0.625		1.648721
8	2.000	1.600	0.720	0.000	-0.432		1.822119
9	2.000	2.200	1.680	0.980	0.343		2.013753
10	2.000	2.800	2.880	2.560	2.048		2.225541
11	2.000	3.400	4.320	4.860	5.103		2.459603
12	A_trans*A						d=A_trans*h
13	36.000000	18.000000	14.400000	12.600000	10.998000		30.676
14	18.000000	30.600000	28.800000	25.596000	22.491000		21.3909291
15	14.400000	28.800000	31.795200	30.988800	28.756800		18.6322072
16	12.600000	25.596000	30.988800	32.197200	31.217700		16.6151154
17	10.998000	22.491000	28.756800	31.217700	31.292997		14.6620662
18	B_inverse						c=B_inverse*d
19	0.061858	-0.284593	0.881882	-1.167300	0.536886		0.71828605
20	-0.284593	3.091179	-10.402840	13.878997	-6.407558		0.21822436
21	0.881882	-10.402840	38.045722	-53.216459	25.292994		0.05191197
22	-1.167300	13.878997	-53.216459	77.295390	-37.770871		0.00926446
23	0.536886	-6.407558	25.292994	-37.770871	18.885436		0.00230834

Fig. 2.3 Spreadsheet for the overdetermined collocation method.

Least squares. The basis functions are, from eqn (2.12),

$$v_i(x) = (1 - x)x^{i+1}, \quad i = 0, \ldots, 4,$$

so that

$$\mathcal{L}v_i = -(i + 1)ix^{i-1} + (i + 2)(i + 1)x^i, \quad i = 0, \ldots, 4,$$

and the integrals are given in Table 2.2. The spreadsheet implementation is shown in Fig. 2.4.

We see that

$$\tilde{u}_4 = x(1 - x)(0.718282 + 0.218256x + 0.051847x^2 + 0.009299x^3 + 0.002316x^4).$$

Table 2.2 The integrals $\int_0^1 \mathcal{L}v_i \mathcal{L}v_j \, dx$ and $\int_0^1 e^x \mathcal{L}v_i \, dx$ for the least squares method

i,j	0	1	2	3	4	
0	4	2	2	2	2	$2e - 2$
1	2	4	4	4	4	$8 - 2e$
2	2	4	$\dfrac{24}{5}$	$\dfrac{26}{5}$	$\dfrac{38}{7}$	$12e - 30$
3	2	4	$\dfrac{26}{5}$	$\dfrac{28}{35}$	$\dfrac{45}{7}$	$144 - 52e$
4	2	4	$\dfrac{38}{7}$	$\dfrac{45}{7}$	$\dfrac{50}{7}$	$310e - 840$

	A	B	C	D	E	F	G
1	Least squares						
2	A						h
3	4.000	2.000	2.000	2.000	2.000		3.436564
4	2.000	4.000	4.000	4.000	4.000		2.563436
5	2.000	4.000	4.800	5.200	5.429		2.619382
6	2.000	4.000	5.200	5.943	6.429		2.649345
7	2.000	4.000	5.429	6.429	7.143		2.667367
8	A_inverse						c
9	0.333	-0.167	0.000	0.000	0.000		0.71828183
10	-0.167	5.583	-19.250	24.500	-10.500		0.21825551
11	0.000	-19.250	89.250	-129.500	59.500		0.05184745
12	0.000	24.500	-129.500	203.000	-98.000		0.00929918
13	0.000	-10.500	59.500	-98.000	49.000		0.00231604

Fig. 2.4 Spreadsheet for the least squares method.

Galerkin. As in the least squares method, the basis functions are

$$v_i(x) = (1 - x)x^{i+1},$$

and the integrals are given in Table 2.3. The spreadsheet implementation is shown in Fig. 2.5.

We see that

$$\tilde{u}_4 = x(1 - x)(0.718284 + 0.218234x + 0.051895x^2 + 0.009272x^3 + 0.002312x^4).$$

So far, the problems considered have involved Dirichlet boundary conditions only and the trial functions have been assumed to satisfy them. Let us now see how to handle a Robin boundary condition.

Consider the equation

$$(2.13) \qquad\qquad -u'' = f, \qquad 0 < x < 1,$$

Table 2.3 The integrals $\int_0^1 v_i \mathcal{L} v_j \, dx$ and $\int_0^1 e^x v_i \, dx$ for the Galerkin method

i,j	0	1	2	3	4	
0	$\dfrac{1}{3}$	$\dfrac{1}{6}$	$\dfrac{1}{10}$	$\dfrac{1}{15}$	$\dfrac{1}{21}$	$3-e$
1	$\dfrac{1}{6}$	$\dfrac{2}{15}$	$\dfrac{1}{10}$	$\dfrac{8}{105}$	$\dfrac{5}{84}$	$3e-8$
2	$\dfrac{1}{10}$	$\dfrac{1}{10}$	$\dfrac{3}{35}$	$\dfrac{1}{14}$	$\dfrac{5}{84}$	$30-11e$
3	$\dfrac{1}{15}$	$\dfrac{8}{105}$	$\dfrac{1}{14}$	$\dfrac{4}{63}$	$\dfrac{1}{18}$	$53e-144$
4	$\dfrac{1}{21}$	$\dfrac{5}{84}$	$\dfrac{5}{84}$	$\dfrac{1}{18}$	$\dfrac{5}{99}$	$840-309e$

	A	B	C	D	E	F	G
1	Galerkin						
2	A						h
3	0.333	0.167	0.100	0.067	0.048		0.281718
4	0.167	0.133	0.100	0.076	0.060		0.154845
5	0.100	0.100	0.086	0.071	0.060		0.098900
6	0.067	0.076	0.071	0.063	0.056		0.068937
7	0.048	0.060	0.060	0.056	0.051		0.050915
8	A_inverse						c
9	35.000	-280.000	840.000	-1050.000	462.000		0.71828423
10	-280.000	3080.000	-10500.000	14070.000	-6468.000		0.21823433
11	840.000	-10500.000	38640.000	-54390.000	25872.000		0.05189461
12	-1050.000	14070.000	-54390.000	79380.000	-38808.000		0.00927222
13	462.000	-6468.000	25872.000	-38808.000	19404.000		0.00231200

Fig. 2.5 Spreadsheet for the Galerkin method.

with the boundary conditions

$$(2.14) \qquad u(0) = g,$$

$$(2.15) \qquad u'(1) + \sigma u(1) = h.$$

We choose the trial function $\tilde{u}$ to satisfy the Dirichlet condition (2.14), and the weighting function v to satisfy the homogeneous form of the Dirichlet boundary condition, i.e. to satisfy

$$(2.16) \qquad v(0) = 0.$$

The weighted residual formulation

$$(2.17) \qquad \int_0^1 r(u)v \, dx = 0$$

is true for all functions $v(x)$ if $u(x)$ is the solution of eqn (2.13). We now use integration by parts in eqn (2.17) with $r(u) = -u'' - f$:

$$[-u'v]_{x=0}^{x=1} + \int_0^1 (u'v' - fv)\, dx = 0.$$

Using the boundary condition (2.15), we obtain

(2.18) $$u'(0)v(0) + \int_0^1 (u'v' - fv)\, dx + [(\sigma u - h)\, v]_{x=1} = 0.$$

In eqn (2.18), we have an integral formulation of the boundary-value problem in which the order of the highest derivative occurring has been reduced. This formulation is often called a *weak form* of the problem. If we return to eqn (2.18) and use the fact that $v(0) = 0$, eqn (2.16), we obtain

(2.19) $$\int_0^1 (u'v' - fv)\, dx + [(\sigma u - h)\, v]_{x=1} = 0.$$

Here we see that the Robin condition is automatically satisfied in the weak form (2.19). Such a condition is called a *natural boundary condition*. The Dirichlet condition, which it was necessary to impose, is called an *essential boundary condition*.

We can use eqn (2.19) to develop a Galerkin approach to an approximate solution $\tilde{u}_n$, eqn (2.7), with basis functions $v_i(x)$:

(2.20) $$\int_0^1 (\tilde{u}'v_i' - fv_i)\, dx + [(\sigma \tilde{u} - h)\, v_i]_{x=1} = 0.$$

Example 2.8

$$-u'' = x, \qquad 0 < x < 1,$$
$$u(0) = 2, \qquad u'(1) = 3.$$

Choose a quadratic trial function which satisfies the essential boundary condition

$$\tilde{u}_2(x) = 2 + c_1 x + c_2 x^2.$$

Then eqn (2.20) yields

$$\int_0^1 \{(c_1 + 2c_2 x)\, 1 - xx\}\, dx + (-3)\, x\, |_{x=1} = 0$$

and

$$\int_0^1 \{(c_1 + 2c_2 x)\, 2x - xx^2\}\, dx + (-3)\, x^2\, |_{x=1} = 0.$$

Thus

$$\begin{bmatrix} 1 & 1 \\ 1 & \frac{4}{3} \end{bmatrix} \begin{bmatrix} c_1 \\ c_2 \end{bmatrix} = \begin{bmatrix} \frac{10}{3} \\ \frac{13}{4} \end{bmatrix},$$

giving $c_1 = \frac{43}{12}$, $c_2 = -\frac{1}{4}$ so that

$$\tilde{u}_2(x) = 2 + \frac{43}{12}x - \frac{1}{4}x^2.$$

Notice that $\tilde{u}_2'(1) = \frac{37}{12} \approx 3.08$, compared with the exact value $u'(1) = 3$.

This problem has an exact solution which is cubic in x, which would be recoverable exactly by $\tilde{u}_n(x)$ with $n \geq 3$.

If we change the right-hand side to e^x, then the exact solution is

$$u(x) = 3 + (3 + e)x - e^x.$$

A spreadsheet implementation yields

$$\tilde{u}_2(x) = 2 + 4.845155x - 0.845155x^2,$$

so that $\tilde{u}_2'(1) \approx 3.154845$, compared with $u'(1) = 3$.

Similarly, we can show that

$$\tilde{u}_5(x) = 2 + 4.718229x - 0.499224x^2 - 0.170193x^3 - 0.034916x^4 - 0.013896x^5,$$

so that $\tilde{u}_5'(1) \approx 3.000058$.

Finally, then, consider Poisson's equation

$$-\nabla^2 u = f \quad \text{in } D$$

subject to the boundary conditions

(2.21) $$u = g(s) \quad \text{on } C_1$$

and

(2.22) $$\frac{\partial u}{\partial n} + \sigma(s)u = h(s) \quad \text{on } C_2.$$

The weighted residual formulation is

$$-\iint_D \left(\nabla^2 \tilde{u} + f \right) v_i \, dx \, dy = 0, \qquad i = 0, \ldots, n,$$

which becomes, using the first form of Green's theorem,

$$\iint_D \operatorname{grad} \tilde{u} \cdot \operatorname{grad} v_i \, dx \, dy - \oint_C v_i \frac{\partial \tilde{u}}{\partial n} ds - \iint_D f v_i \, dx \, dy, \qquad i = 0, \ldots, n.$$

On C_2, the Robin boundary condition (2.22) holds, and on C_1 the trial function must satisfy the essential Dirichlet condition (2.21), while the basis functions satisfy the homogeneous form of this condition, i.e. on C_1, $v_i = 0$.

Thus the Galerkin equations become

(2.23)
$$\iint_D \left(\frac{\partial \tilde{u}}{\partial x} \frac{\partial v_i}{\partial x} + \frac{\partial \tilde{u}}{\partial y} \frac{\partial v_i}{\partial y} - f v_i \right) dx\, dy + \oint_{C_2} (\sigma \tilde{u} - h)\, v_i\, ds = 0, \quad i = 0, \ldots, n.$$

The procedure adopted to solve the boundary-value problem is very similar to the one-dimensional case and is illustrated in Exercise 2.15.

2.4 Extremum formulation: homogeneous boundary conditions

Although in this text we shall concentrate most of our attention on weighted residual methods for the development of finite element equations, there is much to be gained by setting the finite element method in a variational context. In this section we develop the idea in terms of a simple mechanical example.

Many problems of practical interest are modelled by equations such as eqn (2.1), and examples are given in Appendix A. These equations are often equivalent to the problem of the minimization of a functional, which itself may be interpreted in terms of the total energy of the system under consideration.

In any physical situation, an expression for the total energy could be obtained and then minimized to find the equilibrium solution. However, instead of finding the energy explicitly, it would be useful to be able to start with the governing partial differential equation and develop the corresponding functional. Generally, the functional may be obtained without directly determining an expression for the total energy of the system. Indeed, the procedure could then be considered as a mathematical technique independent of the physics of the problem under consideration. To develop the general ideas, the following specific problem is considered.

Example 2.9 It is required to find the equilibrium displacement of a membrane stretched across a frame, in the shape of a curve C, which is subjected to a pressure loading $p(x,y)$ per unit area. If the tension T in the membrane is assumed constant, then the transverse deflection w satisfies Poisson's equation

(2.24)
$$-\nabla^2 w = \frac{p}{T}.$$

Suppose that the membrane is given a small displacement Δw at the point (x,y). If D is the surface area of the membrane, then the total work done by the applied pressure force is

$$\Delta \iint_D pw \, dx \, dy = \iint_D p \, \Delta w \, dx \, dy$$

$$= \iint_D -T \, \nabla^2 w \, \Delta w \, dx \, dy$$

$$= \iint_D -T \, \{\operatorname{div}(\Delta w \operatorname{grad} w) - \operatorname{grad} \Delta w \cdot \operatorname{grad} w\} \, dx \, dy$$

$$= - \oint_C T \, \Delta w \frac{\partial w}{\partial n} ds + \iint_D \frac{T}{2} \Delta \mid \operatorname{grad} w \mid^2 dx \, dy.$$

The first integral is obtained using the divergence theorem, and the second using Exercise 2.9.

Thus

$$(2.25) \qquad \Delta \iint_D pw \, dx \, dy = -T \oint_C \Delta w \frac{\partial w}{\partial n} ds + \Delta \iint_D \frac{T}{2} \mid \operatorname{grad} w \mid^2 dx \, dy.$$

If the boundary conditions are of the homogeneous Dirichlet type (2.2) on a part C_1 of C, then w is fixed, and hence $\Delta w = 0$ on C_1. If the boundary conditions are of the homogeneous Neumann type (2.3) on C_2, then $\partial w / \partial n = 0$ on C_2. This represents the vanishing of the restraining force on C_2 and is often referred to as a *free boundary condition*. In either case, the boundary integral vanishes, and

$$\Delta \iint_D pw \, dx \, dy = \Delta \iint_D \frac{T}{2} \mid \operatorname{grad} w \mid^2 dx \, dy,$$

or

$$\Delta \iint_D \left\{ \mid \operatorname{grad} w \mid^2 - \frac{2p}{T} w \right\} dx \, dy = 0.$$

Thus if $I[w]$ is the functional given by

$$(2.26) \qquad I[w] = \iint_D \left\{ \left(\frac{\partial w}{\partial x}\right)^2 + \left(\frac{\partial w}{\partial y}\right)^2 - \frac{2p}{T} w \right\} dx \, dy,$$

then the solution of eqn (2.24), subject to the homogeneous boundary conditions, is such that $\Delta I = 0$.

To interpret I, the integrand may be seen to be proportional to

$$\frac{1}{2} T \left\{ \left(\frac{\partial w}{\partial x}\right)^2 + \left(\frac{\partial w}{\partial y}\right)^2 \right\} - pw.$$

The first term represents the potential energy per unit area stored in the membrane, and the second term is the potential energy per unit area of the applied pressure force. Thus $I[w]$ as given by eqn (2.26) is proportional to the total potential energy of the system. $\Delta I = 0$ for equilibrium is equivalent to saying that at equilibrium, the potential energy is stationary.

If a part C_3 of the boundary is elastically supported, then neither a Dirichlet nor a Neumann boundary condition is suitable. In this case the boundary condition is of the Robin type (2.4), and

$$\frac{\partial w}{\partial n} = -\sigma w \quad \text{on } C_3.$$

This leads, from eqn (2.25), to an extra term

$$\int_{C_3} \sigma w^2 \, ds$$

in $I\,[w]$. This term is proportional to the potential energy stored in the elastic support on C_3.

Thus the modified functional is

$$(2.27) \qquad I\,[w] = \iint_D \left\{ \left(\frac{\partial w}{\partial x}\right)^2 + \left(\frac{\partial w}{\partial y}\right)^2 - \frac{2p}{T} w \right\} dx\,dy + \int_{C_3} \sigma w^2 \, ds.$$

In this particular case it was possible to transform the integral from one involving second derivatives in the integrand to another containing only first derivatives. This was accomplished by way of Green's theorem and the use of the homogeneous boundary condition. It is not difficult to see that in eqn (2.27) the term $(\partial w/\partial x)^2 + (\partial w/\partial y)^2$ could be transformed back again to give

$$(2.28) \qquad I\,[w] = \iint_D \left\{ w\,(-\nabla^2 w) - \frac{2p}{T} w \right\} dx\,dy;$$

see Exercise 2.10.

The final form (2.28) of the functional for eqn (2.24) would suggest that, in general, for eqn (2.1), i.e. $\mathcal{L}u = f$ in some region D, with homogeneous boundary conditions, the functional should be

$$(2.29) \qquad I\,[u] = \iint_D u\mathcal{L}u \, dx\,dy - 2 \iint_D uf \, dx\,dy.$$

It will now be shown that if $\mathcal{L}$ is a self-adjoint, positive definite operator, then the unique solution of $\mathcal{L}u = f$, with homogeneous boundary conditions, occurs at a minimum value of $I\,[u]$ as given by eqn (2.29).

Suppose that u_0 is the exact solution; then

$$\mathcal{L}u_0 = f.$$

Thus

$$I\left[u\right] = \iint_D u\mathcal{L}u\,dx\,dy - 2\iint_D u\mathcal{L}u_0\,dx\,dy$$

$$= \iint_D u\mathcal{L}\left(u - u_0\right)dx\,dy - \iint_D u\mathcal{L}u_0\,dx\,dy$$

$$= \iint_D u\mathcal{L}\left(u - u_0\right)dx\,dy - \iint_D \left(u - u_0\right)\mathcal{L}u_0\,dx\,dy - \iint_D u_0\mathcal{L}u_0\,dx\,dy.$$

Since $\mathcal{L}$ is self-adjoint and the boundary conditions are homogeneous,

$$I\left[u\right] = \iint_D \left(u - u_0\right)\mathcal{L}u\,dx\,dy - \iint_D \left(u - u_0\right)\mathcal{L}u_0\,dx\,dy - \iint_D u_0\mathcal{L}u_0\,dx\,dy$$

$$= \iint_D \left(u - u_0\right)\mathcal{L}\left(u - u_0\right)dx\,dy - \iint_D u_0\mathcal{L}u_0\,dx\,dy.$$

Since $\mathcal{L}$ is positive definite and u_0 is non-trivial,

$$\iint_D u_0\mathcal{L}u_0\,dx\,dy > 0$$

and

$$\iint_D \left(u - u_0\right)\mathcal{L}\left(u - u_0\right)dx\,dy \geq 0,$$

equality occurring if and only if $u \equiv u_0$. Thus $I\left[u\right]$ takes its minimum value when $u = u_0$.

This result gives a method for finding an approximate solution to the equation $\mathcal{L}u = f$ with homogeneous boundary conditions. A systematic method of finding such an approximate solution is the Rayleigh–Ritz method, which, as will be seen in Section 2.7, seeks a stationary value of I by finding its derivatives with respect to a chosen set of parameters.

Suppose that u_0 is a function which yields a stationary value for $I\left[u\right]$. Consider variations around u_0 given by the so-called *trial function*

$$\tilde{u} = u_0 + \alpha v,$$

where v is an arbitrary function and α is a variable parameter. Then $I[u]$ is stationary when $\alpha = 0$, i.e.

(2.30)
$$\left.\frac{dI}{d\alpha}\right|_{\alpha=0} = 0.$$

Now,

$$I[\tilde{u}] = \iint_D \left(u_0 + \alpha v\right)\mathcal{L}\left(u_0 + \alpha v\right)dx\,dy - 2\iint_D \left(u_0 + \alpha v\right)f\,dx\,dy.$$

The Finite Element Method

Thus

$$\frac{dI}{d\alpha} = \iint_D \left\{ v\mathcal{L}\left(u_0 + \alpha v\right) + \left(u_0 + \alpha v\right)\mathcal{L}v \right\} dx \, dy - 2 \iint_D vf \, dx \, dy,$$

since $\mathcal{L}$ and $d/d\alpha$ commute, so that eqn (2.30) gives

$$\iint_D \left(v\mathcal{L}u_0 + u_0\mathcal{L}v\right) dx \, dy - 2 \iint_D vf \, dx \, dy.$$

Since $\mathcal{L}$ is self-adjoint and the boundary conditions are homogeneous, it then follows that

$$\iint_D v\left(\mathcal{L}u_0 - f\right) dx \, dy = 0.$$

Finally, since v is arbitrary, the integral vanishes if and only if $\mathcal{L}u_0 = f$, i.e. u_0 is the unique solution of eqn (2.1).

In practice, the choice of trial functions is restricted and it is usually impossible to choose a function u which locates the exact minimum; the best that can be done is to set up a sequence of approximations to it. An important question which then arises is 'does this sequence converge to the unique solution?' The answer for a self-adjoint, positive definite operator is yes, provided that the set of trial functions is complete, since a stationary point corresponds to a solution of the equation and this solution is unique. There can be one stationary point only for the functional, and this yields its minimum value. Thus the approximating sequence will provide a monotonically decreasing sequence of values for $I[u]$ bounded below by its minimum value $I[u_0]$.

It is important to remember that the results in this section relate to positive definite operators, such as those associated with steady-state problems which yield elliptic operators. However, for time-dependent problems, the associated operators are usually hyperbolic or parabolic and, as such, are not positive definite. Nevertheless, variational principles often do exist for such problems, and we shall consider them briefly in Section 2.9.

2.5 Non-homogeneous boundary conditions

In Section 2.4, the functional for $\mathcal{L}u = f$ was deduced assuming that the boundary conditions were homogeneous. It was due to this fact that $\mathcal{L}$ was seen to be linear, self-adjoint and positive definite. In general, of course, most problems involve non-homogeneous boundary conditions, and in this section the functional given by eqn (2.29) is extended to include such cases.

The boundary conditions to be considered are the non-homogeneous counterparts of eqns (2.2), (2.3) and (2.4), which are:

- Dirichlet boundary condition,

$$u = g(s);$$

- Neumann boundary condition,

$$\frac{\partial u}{\partial n} = j(s);$$

- Robin boundary condition,

$$\frac{\partial u}{\partial n} + \sigma(s)u = h(s).$$

The Neumann condition will be treated as a special case of the Robin condition with $\sigma(s) \equiv 0$. All three of these conditions are of the form

$$(2.31) \qquad\qquad \mathcal{B}u = b(s),$$

where $\mathcal{B}$ is a suitable linear differential operator. Thus the problem to be considered is that of finding the solution, u_0, of eqn (2.1), i.e. $\mathcal{L}u_0 = f$, subject to the boundary condition (2.31).

The procedure is to change the problem to one with homogeneous boundary conditions. Suppose that v is any function which satisfies the boundary condition (2.31), i.e. $\mathcal{B}v = b$.

Then, if

$$(2.32) \qquad\qquad w = u - v,$$

$\mathcal{B}w = \mathcal{B}u - \mathcal{B}v$, since $\mathcal{B}$ is linear. Thus

$$(2.33) \qquad\qquad \mathcal{B}w = 0,$$

provided the function u is chosen to satisfy the boundary conditions.

Let

$$w_0 = u_0 - v;$$

then

$$\mathcal{L}w_0 = \mathcal{L}u_0 - \mathcal{L}v, \text{ since } \mathcal{L} \text{ is linear,}$$
$$= f - \mathcal{L}v.$$

Now let

$$(2.34) \qquad\qquad F = f - \mathcal{L}v;$$

then

$$\mathcal{L}w_0 = F.$$

w_0 also satisfies the homogeneous boundary conditions (2.33), so that, using the results of Section 2.4, w_0 must be the unique function which minimizes the functional

$$I[w] = \int\!\!\int_D (w\mathcal{L}w - 2wF)\, dx\, dy.$$

But this functional can be rewritten in terms of u using eqns (2.32) and (2.34) as

$$I[u] = \int\!\!\int_D \{(u-v)\,\mathcal{L}\,(u-v) - 2\,(u-v)\,(f - \mathcal{L}v)\}\, dx\, dy$$

$$= \int\!\!\int_D (u\mathcal{L}u - 2uf + u\mathcal{L}v - v\mathcal{L}u)\, dx\, dy + \int\!\!\int_D (2vf - v\mathcal{L}v)\, dx\, dy.$$

Now the last integral on the right-hand side is a fixed quantity as far as the minimization is concerned, and as such cannot affect the function u_0 which gives the minimum value. Consequently, this term may be deleted from I, leaving the functional as

$$(2.35) \qquad I[u] = \int\!\!\int_D (u\mathcal{L}u - 2uf + u\mathcal{L}v - v\mathcal{L}u)\, dx\, dy,$$

which is minimized by u_0.

Example 2.10 Consider Poisson's equation

$$(2.36) \qquad\qquad -\nabla^2 u = f \qquad \text{in } D$$

subject to the non-homogeneous Dirichlet condition

$$(2.37) \qquad\qquad u = g(s) \qquad \text{on } C.$$

In this case the last two terms in eqn (2.35) may be integrated using the second form of Green's theorem to give

$$\int\!\!\int_D (v\,\nabla^2 u - u\,\nabla^2 v)\, dx\, dy = \oint_C \left(v\frac{\partial u}{\partial n} - u\frac{\partial v}{\partial n} \right) ds$$

$$= \oint_C \left(g\frac{\partial u}{\partial n} - g\frac{\partial v}{\partial n} \right) ds,$$

since both u and v satisfy the boundary condition.

Now the second term on the right-hand side is independent of u and thus cannot be varied in the minimization of the functional. This term may be deleted to give the functional as

$$I[u] = \iint_D \left(-u\,\nabla^2 u - 2uf \right) dx\,dy + \oint_C g\frac{\partial u}{\partial n}\,ds,$$

which, using the first form of Green's theorem, yields

$$I[u] = \iint_D |\text{grad } u|^2\,dx\,dy - \oint_C u\frac{\partial u}{\partial n}\,ds$$
$$- \iint_D 2uf\,dx\,dy + \oint_C g\frac{\partial u}{\partial n}\,ds.$$

But on C, $u = g$, so that the two boundary integrals cancel, leaving the functional as

$$(2.38) \qquad I[u] = \iint_D \left\{ \left(\frac{\partial u}{\partial x}\right)^2 + \left(\frac{\partial u}{\partial y}\right)^2 - 2uf \right\} dx\,dy.$$

This functional is minimized by the solution u_0 of eqn (2.36) with the boundary condition (2.37). It is interesting to note that this functional is identical with the functional for the problem with homogeneous boundary conditions, eqn (2.26). However, when using this functional to find approximate solutions to the boundary-value problem, it is necessary that all trial functions satisfy the essential non-homogeneous boundary condition.

Example 2.11 In this example we consider Poisson's eqn (2.36) in D, subject to the Robin boundary condition

$$(2.39) \qquad \frac{\partial u}{\partial n} + \sigma(s)u = h(s) \qquad \text{on } C.$$

As in Example 2.10, Green's theorem is used to give

$$\iint_D \left(v\,\nabla^2 u - u\,\nabla^2 v \right) dx\,dy = \oint_C \left(v\frac{\partial u}{\partial n} - u\frac{\partial v}{\partial n} \right) ds$$
$$= \oint_C \left(v(h - \sigma u) - u(h - \sigma v) \right) ds$$
$$= \oint_C (vh - uh)\,ds.$$

Once again, the first term on the right-hand side is independent of u and may be deleted from the functional, giving

$$I[u] = \iint_D \left\{ |\text{grad } u|^2 - 2uf \right\} dx\,dy - \oint_C \left(u\frac{\partial u}{\partial n} + uh \right) ds,$$

where the first term has been obtained, in the usual manner, from the first form of Green's theorem.

Finally, then, since u satisfies the boundary condition (2.39), it follows that

$$(2.40) \quad I[u] = \iint_D \left\{ \left(\frac{\partial u}{\partial x} \right)^2 + \left(\frac{\partial u}{\partial y} \right)^2 - 2uf \right\} dx\, dy + \oint_C \left(\sigma u^2 - 2uh \right) ds$$

is the functional which is minimized by the solution u_0 of Poisson's eqn (2.36) subject to the Robin boundary condition (2.39). The functional for the Neumann problem is found by setting $\sigma \equiv 0$, to give

$$I[u] = \iint_D \left\{ \left(\frac{\partial u}{\partial x} \right)^2 + \left(\frac{\partial u}{\partial y} \right)^2 - 2uf \right\} dx\, dy - \oint_C 2uh\, ds.$$

2.6   Partial differential equations: natural boundary conditions

In Exercise 2.2, it is shown that under certain circumstances the general linear second-order partial differential operator $\mathcal{L}$ may be given by the expression

$$\mathcal{L}u = -\text{div}\left(\kappa \, \text{grad}\, u \right) + \rho u.$$

The Robin boundary condition may also be generalized to give an expression of the form

$$(2.41) \qquad\qquad (\kappa \, \text{grad}\, u).\mathbf{n} + \sigma(s)u = h(s).$$

Suppose that u_0 is the solution of

$$\mathcal{L}u = f$$

subject to the Dirichlet boundary condition

$$u = g(s)$$

on some part C_1 of the boundary and the Robin condition (2.41) on the remainder C_2.

In the development which follows, generalizations of Green's theorem are needed. They are proved in Exercise 2.2 and stated here for convenience.

The first form is

$$\oint_C u(\kappa \, \text{grad}\, v) \cdot \mathbf{n}\, ds$$

$$(2.42) \qquad = \iint_D \text{grad}\, u \cdot \kappa \, \text{grad}\, v\, dx\, dy + \iint_D u\, \text{div}(\kappa \, \text{grad}\, v)\, dx\, dy.$$

The second form, for symmetric $\boldsymbol{\kappa}$ is

$$\oint_C \{u(\boldsymbol{\kappa}\operatorname{grad} v)\cdot\mathbf{n} - v(\boldsymbol{\kappa}\operatorname{grad} u)\cdot\mathbf{n}\}\,ds = \iint_D \{u\operatorname{div}(\boldsymbol{\kappa}\operatorname{grad} v)$$

(2.43)
$$-v\operatorname{div}(\boldsymbol{\kappa}\operatorname{grad} u)\}\,dx\,dy.$$

Now, in this case, the last two terms in the functional (2.35) are given by

$$I[u] = \iint_D \{v\operatorname{div}(\boldsymbol{\kappa}\operatorname{grad} u) - u\operatorname{div}(\boldsymbol{\kappa}\operatorname{grad} v)\}\,dx\,dy$$

using eqn (2.43)

$$I[u] = \oint_C \{u(\boldsymbol{\kappa}\operatorname{grad} v)\cdot\mathbf{n} - v(\boldsymbol{\kappa}\operatorname{grad} u)\cdot\mathbf{n}\}\,ds$$

$$= \oint_C \{v(\boldsymbol{\kappa}\operatorname{grad} u)\cdot\mathbf{n} - u(\boldsymbol{\kappa}\operatorname{grad} v)\cdot\mathbf{n}\}\,ds$$

$$= \int_{C_1} \{g(\boldsymbol{\kappa}\operatorname{grad} u)\cdot\mathbf{n} - g(\boldsymbol{\kappa}\operatorname{grad} v)\cdot\mathbf{n}\}\,ds + \int_{C_2} (vh - uh)\,ds$$

using the boundary conditions on C_1 and C_2.

Now the second and third terms are independent of the trial function u, and as such may be deleted from the functional to give

$$I[u] = \iint_D \{-u\operatorname{div}(\boldsymbol{\kappa}\operatorname{grad} u) + \rho u^2 - 2uf\}\,dx\,dy$$

$$+ \int_{C_1} g(\boldsymbol{\kappa}\operatorname{grad} u)\cdot\mathbf{n}\,ds - \int_{C_2} uh\,ds$$

$$= \iint_D \operatorname{grad} u\cdot(\boldsymbol{\kappa}\operatorname{grad} u)\,dx\,dy - \int_C u(\boldsymbol{\kappa}\operatorname{grad} u)\cdot\mathbf{n}\,ds$$

$$+ \iint_D (\rho u^2 - 2uf)\,dx\,dy + \int_{C_1} g(\boldsymbol{\kappa}\operatorname{grad} u)\cdot\mathbf{n}\,ds - \int_{C_2} uh\,ds.$$

Thus, using the boundary conditions on C_1 and C_2,

(2.44)
$$I[u] = \iint_D \{\operatorname{grad} u\cdot(\boldsymbol{\kappa}\operatorname{grad} u) + \rho u^2 - 2uf\}\,dx\,dy + \int_{C_2} (\sigma u^2 - 2uh)\,ds.$$

Again the derivation of the functional shows that all trial functions must satisfy the essential Dirichlet boundary condition.

Nothing has been said about the Neumann boundary condition. To illustrate the role of this condition, consider Poisson's equation, eqn (2.36),

$$-\nabla^2 u = f \qquad \text{in } D$$

subject to the Neumann boundary condition

$$\frac{\partial u}{\partial n} = h(s) \qquad \text{on } C.$$

Then, from the functional (2.40), with $\sigma \equiv 0$, the solution u_0 minimizes

$$I[u] = \iint_D \left\{ \left(\frac{\partial u}{\partial x}\right)^2 + \left(\frac{\partial u}{\partial y}\right)^2 - 2uf \right\} dx\, dy - \oint_C 2uh\, ds.$$

Using trial functions of the form

$$\tilde{u} = u_0 + \alpha v,$$

the functional becomes

$$I(\alpha) = \iint_D \left\{ \text{grad}(u_0 + \alpha v) \cdot \text{grad}(u_0 + \alpha v) - 2(u_0 + \alpha v)f \right\} dx\, dy$$
$$- \oint_C 2(u_0 + \alpha v)h\, ds,$$

so that

$$\frac{dI}{d\alpha} = \iint_D \left\{ \text{grad}\, v \cdot \text{grad}(u_0 + \alpha v) + \text{grad}(u_0 + \alpha v) \cdot \text{grad}\, v - 2vf \right\} dx\, dy$$
$$- \oint_C 2vh\, ds.$$

Since $dI/d\alpha = 0$ when $\alpha = 0$, it follows that

$$\iint_D \left\{ \text{grad}\, v \cdot \text{grad}\, u_0 - vf \right\} dx\, dy - \oint_C vh\, ds = 0.$$

Thus, using the first form of Green's theorem, we find

$$-\iint_D v\left(\nabla^2 u + f\right) dx\, dy + \oint_C v\left(\frac{\partial u_0}{\partial n} - h\right) ds = 0.$$

Now, since v is arbitrary, it follows that this equation can hold only if $-\nabla^2 u_0 = f$ and $\partial u_0/\partial n = h$. Thus the stationary point occurs at the solution of the differential equation and the Neumann boundary condition is satisfied naturally, i.e. it does not have to be enforced on the trial functions. This Neumann condition is a natural boundary condition.

In a similar manner (see Exercise 2.11), it may be shown that the Robin boundary condition (2.41) is a natural boundary condition for the functional (2.44), i.e. provided that the Dirichlet boundary condition is enforced where applicable, then at the stationary value of the functional (2.44), the equation $\mathcal{L}u = f$ is satisfied in D, and the Robin boundary condition (2.41) is satisfied naturally over that part of the boundary on which it applies. This result is extremely useful in that the boundary condition which is most difficult to enforce, namely the Robin condition, is satisfied naturally by choosing a suitable functional. This means that for any given degree of polynomial approximation, there are more variational parameters available for minimization than there would be if this condition were enforced (see Exercise 2.6).

In Section 2.3, we developed the concept of the essential and natural boundary conditions in a weighted-residual context. In this section we have seen how they arise in a variational context. The important point is that no matter how we develop methods for the approximate solution of boundary-value problems, our trial functions must satisfy the essential boundary conditions. A measure of the 'quality' of the approximate solution is how well the natural boundary conditions are satisfied.

2.7 The Rayleigh–Ritz method

The Rayleigh–Ritz method provides an algorithm for minimizing a given functional and requires the choice of a suitable complete set of linearly independent basis functions $v_i(x, y)$, $i = 0, 1, 2, \ldots$. The exact solution, u, is approximated by a sequence of trial functions

$$(2.45) \qquad \tilde{u}_n = \sum_{i=0}^{n} c_i v_i,$$

where the constants c_i are chosen to minimize $I[\tilde{u}_n]$ at each stage (*cf.* eqn (2.7)). A measure of the 'quality' of the solution is how well the natural conditions are satisfied.

If $\tilde{u}_n \to u$ as $n \to \infty$ in some sense (see Chapter 7), then the procedure is said to converge to the solution. At each stage, the problem is reduced to one of solving a system of linear algebraic equations.

Consider Poisson's equation $-\nabla^2 u = f$ with homogeneous Dirichlet or Neumann boundary conditions.

The functional is given by eqn (2.27) as

$$I[u] = \iint_D \left\{ \left(\frac{\partial u}{\partial x} \right)^2 + \left(\frac{\partial u}{\partial y} \right)^2 - 2uf \right\} \, dx \, dy.$$

Using the approximation (2.45), this functional may be written as

$$I(c_0, \ldots, c_n) = \iint_D \left\{ \left(\sum c_i \frac{\partial v_i}{\partial x} \right)^2 + \left(\sum c_i \frac{\partial v_i}{\partial y} \right)^2 - 2 \sum c_i v_i f \right\} dx\, dy$$

$$= c_i^2 \iint_D \left\{ \left(\frac{\partial v_i}{\partial x} \right)^2 + \left(\frac{\partial v_i}{\partial y} \right)^2 \right\} dx\, dy$$

$$+ 2 \sum_{j \neq i} c_i c_j \iint_D \left(\frac{\partial v_i}{\partial x} \frac{\partial v_j}{\partial x} + \frac{\partial v_i}{\partial y} \frac{\partial v_j}{\partial y} \right) dx\, dy$$

$$- 2 c_i \iint_D v_i f \, dx\, dy + \text{terms independent of } c_i.$$

Therefore

$$(2.46) \qquad \frac{\partial I}{\partial c_i} = 2 A_{ii} c_i + 2 \sum_{j \neq i} A_{ij} c_j - 2 h_i,$$

where

$$(2.47) \qquad A_{ij} = \iint_D \left(\frac{\partial v_i}{\partial x} \frac{\partial v_j}{\partial x} + \frac{\partial v_i}{\partial y} \frac{\partial v_j}{\partial y} \right) dx\, dy$$

and

$$(2.48) \qquad h_i = \iint_D v_i f \, dx\, dy.$$

Now the variational parameters c_i are to be chosen such that $I(c_0, \ldots, c_n)$ is a minimum. Thus

$$\frac{\partial I}{\partial c_i} = 0, \qquad i = 0, \ldots, n.$$

Equation (2.46) then gives

$$\sum_{j=0}^{n} A_{ij} c_j = h_i, \qquad i = 0, \ldots, n,$$

or

$$(2.49) \qquad \qquad \mathbf{Ac} = \mathbf{h},$$

where the elements of the matrices $\mathbf{A}$ and $\mathbf{h}$ are given by eqns (2.47) and (2.48) and $\mathbf{c} = [c_0, c_1, \ldots, c_n]^T$. Equation (2.49) is a system of linear algebraic equations for the unknown parameters which has a unique solution provided that $\mathbf{A}$ is non-singular.

At the end of this section, the general form of the Rayleigh–Ritz matrices will be presented and it will be shown that $\mathbf{A}$ is non-singular whenever the operator $\mathcal{L}$ is positive definite. First, however, the method will be illustrated by the following example.

Example 2.12 Consider the two-point boundary-value problem (2.8)

$$-u'' = x^2, \qquad 0 < x < 1,$$

$$u(0) = u(1) = 0.$$

Suppose we take a cubic approximation $\tilde{u}_2(x)$, remembering that the trial function must satisfy the essential boundary conditions. Thus

$$\tilde{u}_3 = x(1-x)(c_0 + c_1 x).$$

Using the one-dimensional forms of eqns (2.47) and (2.48),

$$A_{11} = \int_0^1 (1-2x)^2 \, dx = \frac{1}{3}, \quad A_{22} = \int_0^1 (2x-3x^2)^2 \, dx = \frac{2}{15},$$

$$A_{12} = A_{21} = \int_0^1 (1-2x)(2x-3x^2)^2 \, dx = \frac{1}{6},$$

$$h_1 = \int_0^1 (x-x^2)x^2 \ dx = \frac{1}{20}, \quad h_2 = \int_0^1 (x^2-x^3)x^2 \, dx = \frac{1}{30}.$$

Thus the Rayleigh–Ritz equations (2.49) are

$$\begin{bmatrix} \frac{1}{3} & \frac{1}{6} \\ \frac{1}{6} & \frac{2}{15} \end{bmatrix} \begin{bmatrix} c_0 \\ c_1 \end{bmatrix} = \begin{bmatrix} \frac{1}{20} \\ \frac{1}{30} \end{bmatrix}.$$

Solving yields $c_0 = \frac{1}{15}$, $c_1 = \frac{1}{6}$. Thus the Rayleigh–Ritz cubic approximation is given by

$$\tilde{u}_2(x) = \frac{1}{30} x(1-x)(2+5x).$$

The exact solution is $u_0 = \frac{1}{12}x(1-x^3)$, and a comparison between $\tilde{u}_2$ and u_0 is shown in Figure 2.6.

In this particular example, the Rayleigh–Ritz method gives a good approximation to the exact solution with only two parameters, c_0 and c_1. This is due to the simplicity of the problem and the fact that the exact solution would be recovered if a quartic trial function was used. This is a special case of the general result that if the exact solution is a linear combination of the basis functions used, then the approximating method will yield the exact solution (see Exercises 2.3 and 2.15). For problems in which the exact solution is a transcendental

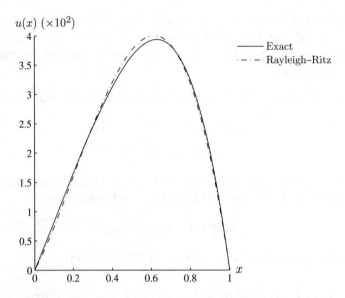

Fig. 2.6 Comparison of the Rayleigh–Ritz cubic approximate solution with the exact solution for the problem of Example 2.12.

function, a low-order polynomial approximation is not likely to yield a good approximation.

So far, only homogeneous boundary conditions have been considered, and the basis functions have been chosen to satisfy these conditions. However, if the boundary conditions are non-homogeneous, it is not just a matter of forcing the basis functions themselves to satisfy them, since a linear combination will not satisfy the given conditions. Thus the next stage is to develop the Rayleigh–Ritz method further to include non-homogeneous conditions.

Consider Poisson's equation $-\nabla^2 u = f$ with the non-homogeneous boundary conditions

$$u = g(s) \qquad \text{on } C_1$$

and

$$\frac{\partial u}{\partial n} + \sigma(s)u = h(s) \qquad \text{on } C_2.$$

The functional is given by eqn (2.40) as

$$I[u] = \iint_D \left\{ \left(\frac{\partial u}{\partial x}\right)^2 + \left(\frac{\partial u}{\partial y}\right)^2 - 2uf \right\} dx\, dy + \int_{C_2} (\sigma u^2 - 2uh)\, ds.$$

Choose a linearly independent set of basis functions v_i which satisfy the homogeneous Dirichlet condition $v_i = 0$ on C_1; then a sequence of trial functions which satisfy the non-homogeneous Dirichlet condition on C_1 is

$$(2.50) \qquad \tilde{u}_n = g + \sum_{i=1}^{n} c_i v_i.$$

Notice that this may be written as

$$\tilde{u}_n = \sum_{i=0}^{n} c_i v_i,$$

where $v_0 = g$ and $c_0 = 1$. Then

$$I(c_0, c_1, \ldots, c_n) = \iint_D \left\{ \left(\sum c_i \frac{\partial v_i}{\partial x} \right)^2 + \left(\sum c_i \frac{\partial v_i}{\partial y} \right)^2 - 2 \sum c_i v_i f \right\} dx\, dy$$
$$+ \int_{C_2} \left\{ \sigma \left(\sum c_i v_i \right)^2 - 2 \left(\sum c_i v_i \right) h \right\} ds.$$

All the parameters except c_0 are unknown, and may be used to minimize I in a similar manner to the homogeneous case to give

$$\frac{\partial I}{\partial c_i} = 2 A_{ii} c_i + 2 \sum_{j \neq 1} A_{ij} c_j - 2 h_i + 2 S_{ii} c_i + 2 \sum_{j \neq 1} S_{ij} c_j - 2 k_i, \quad i = 0, \ldots, n,$$

where A_{ij} and h_i are given by eqns (2.47) and (2.48), respectively. The coefficients S_{ij} and k_i are due to the non-homogeneous boundary terms, and are given by

$$(2.51) \qquad S_{ij} = \int_{C_2} \sigma v_i v_j \, ds$$

and

$$(2.52) \qquad k_i = \int_{C_2} v_i h \, ds.$$

Thus the set of equations $\partial I / \partial c_i = 0$, $i = 0, \ldots n$, is a rectangular set

$$\mathbf{Bc} = \mathbf{g},$$

where

$$B_{ij} = A_{ij} + S_{ij}$$

and

$$g_i = h_i + k_i.$$

$\mathbf{B}$ is an $n \times (n+1)$ matrix and $\mathbf{c} = [c_0 \; c_1 \; \cdots \; c_n]^T$. Consider the first equation of the set, *viz.*

$$B_{01}c_0 + B_{02}c_2 + \cdots + B_{0n}c_n = g_0.$$

Since $c_0 = 1$, this may be written as

$$B_{02}c_2 + \cdots + B_{0n}c_n = g_0 - B_{01}.$$

The other equations may be rewritten in a similar manner to yield the square set of equations

(2.53) $$\mathbf{B}'\mathbf{c}' = \mathbf{g}',$$

where

$$B'_{ij} = B_{ij}, \quad i, j = 1, \ldots, n,$$
$$c'_i = c_i, \quad i = 1, \ldots, n,$$
$$g'_i = g_i - B_{in}, \quad i = 1, \ldots, n.$$

This is exactly the procedure adopted in the finite element method in Chapter 3.

Example 2.13

$$-u'' = x, \quad 0 < x < 1,$$
$$u(0) = 2, \quad u'(1) = 3.$$

Take as trial function $\tilde{u}_2 = 2 + c_1 x + c_2 x^2$, which satisfies the essential boundary condition $\tilde{u}_2(0) = 2$. Then

$$A_{11} = \int_0^1 1 \, dx = 1, \quad A_{22} = \int_0^1 4x^2 \, dx = \frac{4}{3},$$

$$A_{12} = A_{21} = \int_0^1 1.2x \, dx = 1,$$

$$A_{13} = \int_0^1 x.0x \, dx = 0, \quad A_{23} = \int_0^1 x^2 0 \, dx = 0,$$

$$S_{ij} = 0, \quad i = 1, 2, \; j = 1, 2, 3, \quad \text{since here } \sigma = 0,$$

$$h_1 = \int_0^1 xx \, dx = \frac{1}{3}, \quad h_2 = \int_0^1 x^2 x \, dx = \frac{1}{4},$$

$$k_1 = [x3]_{x=1} = 3, \quad k_2 = [x^2 3]_{x=1} = 3.$$

Just as in Example 2.12, the one-dimensional forms of eqns (2.47), (2.48), (2.51) and (2.52) have been used.

Table 2.4 Comparison of the Rayleigh–Ritz quadratic approximation with the exact solution for Example 2.13

x	0	0.2	0.4	0.6	0.8	1
Rayleigh–Ritz	2	2.707	3.393	4.060	4.707	5.333
Exact solution	2	2.699	3.389	4.064	4.715	5.333

The Rayleigh–Ritz equations (2.53) are thus

$$\begin{bmatrix} 1 & 1 \\ 1 & \frac{4}{3} \end{bmatrix} \begin{bmatrix} c_1 \\ c_2 \end{bmatrix} = \begin{bmatrix} \frac{1}{3} + 3 - 0 \\ \frac{1}{4} + 3 - 0 \end{bmatrix},$$

which give $c_1 = \frac{43}{12}$, $c_2 = -\frac{1}{4}$. Thus the approximate solution is

$$\tilde{u}_2(x) = 2 + \frac{43}{12}x - \frac{1}{4}x^2.$$

This is compared with the exact solution

$$u_0(x) = 2 + \frac{7}{2}x - \frac{1}{5}x^3$$

in Table 2.4. The agreement is seen to be good and, in particular, $\tilde{u}_2'(1) = \frac{37}{12}$ may well be an acceptable approximation to the natural Neumann boundary condition $u'(1) = 3$.

So far, the Rayleigh–Ritz procedure has been followed in a purely formal manner. It will now be proved that if $\mathcal{L}$ is positive definite then the Rayleigh–Ritz method produces non-singular matrices and that the sequence of approximations converges to the exact solution. Now, suppose that u_0 is the solution of

(2.54) $$\mathcal{L}u = f \quad \text{in } D$$

subject to the boundary condition

(2.55) $$\mathcal{B}u = b(s) \quad \text{on } C.$$

Notice that the boundary condition is non-homogeneous; on some parts of the boundary an essential condition will hold, and on other parts a natural condition will hold. This will be dealt with as follows: the essential boundary condition will be enforced on the trial functions and the natural boundary conditions will be satisfied approximately by the choice of a suitable functional. The functional for eqn (2.54) is given by eqn (2.35),

$$I[u] = \iint_D (u\mathcal{L}u - 2uf + u\mathcal{L}w - w\mathcal{L}u) \, dx \, dy,$$

where w is any function which satisfies the boundary condition (2.55). Choose a complete set of linearly independent basis functions v_i and define the trial function $\tilde{u}_n$ in the usual way,

$$\tilde{u}_n = \sum_{i=0}^{n} c_i v_i.$$

Thus, in a similar manner to that leading to eqn (2.46),

$$I(c_0, c_1, \ldots, c_n) = c_i^2 \iint_D v_i \mathcal{L} v_i \, dx \, dy$$

$$+ \sum_{j \neq i} c_i c_j \iint (v_i \mathcal{L} v_j + v_j \mathcal{L} v_i) \, dx \, dy$$

$$+ c_i \iint_D (-2 v_i f + v_i \mathcal{L} w - w \mathcal{L} v_i) \, dx \, dy$$

$$+ \text{terms independent of } c_i, \qquad i = 0, \ldots n.$$

Thus

$$\frac{\partial I}{\partial c_i} = 2 c_i \iint_D v_i \mathcal{L} v_i \, dx \, dy + \sum_{j \neq i} c_j \iint_D (v_i \mathcal{L} v_j + v_j \mathcal{L} v_i) \, dx \, dy$$

$$+ \iint_D (-2 v_i f + v_i \mathcal{L} w - w \mathcal{L} v_i) \, dx \, dy, \qquad i = 0, \ldots, n.$$

The parameters c_i are chosen to make I stationary, so that $\partial I / \partial c_i = 0$. Hence the following set of equations is obtained:

$$(2.56) \qquad \sum_{j=0}^{n} A_{ij} c_j = h_j, \qquad i = 0, \ldots, n,$$

where

$$(2.57) \qquad A_{ij} = \frac{1}{2} \iint_D (v_i \mathcal{L} v_j + v_j \mathcal{L} v_i) \, dx \, dy$$

and

$$h_i = \iint_D \left\{ v_i f + \frac{1}{2} (w \mathcal{L} v_i - v_i \mathcal{L} w) \right\} dx \, dy.$$

In matrix form, eqns (2.56) may be written as

$$\mathbf{A} \mathbf{c} = \mathbf{h},$$

and the coefficients c_i may be found provided that $\mathbf{A}$ is non-singular.

If $\mathcal{L}$ is positive definite, then $\mathbf{A}$ is non-singular; the proof is as follows. Suppose that $\mathbf{A}$ is singular; then there exists a non-trivial solution $\mathbf{x}_0$ of

$$\mathbf{A}\mathbf{x} = \mathbf{0}.$$

Thus

(2.58) $$\mathbf{x}_0^T \mathbf{A}\mathbf{x}_0 = \mathbf{0}.$$

Now, if $\mathbf{v} = [v_0, \ldots, v_n]^T$, then using eqn (2.57),

$$\mathbf{A} = \frac{1}{2} \iint_D \{(\mathcal{L}\mathbf{v})\mathbf{v}^T + \mathbf{v}(\mathcal{L}\mathbf{v})^T\} \, dx \, dy,$$

so that eqn (2.58) becomes

$$\iint_D \mathcal{L}\left(\mathbf{x}_0^T \mathbf{v}\right) \mathbf{v}^T \mathbf{x}_0 \, dx \, dy = -\iint_D \mathbf{x}_0^T \mathbf{v} \left(\mathcal{L}\mathbf{v}\right)^T \mathbf{x}_0 \, dx \, dy$$

$$= -\iint_D \mathbf{x}_0^T \left(\mathcal{L}\mathbf{v}\right) \mathbf{v}^T \mathbf{x}_0 \, dx \, dy,$$

since the right-hand side is a scalar, so that transposition leaves it unchanged. Thus

$$\iint_D \mathcal{L}\left(\mathbf{x}_0^T \mathbf{v}\right) \mathbf{v}^T \mathbf{x}_0 \, dx \, dy = -\iint_D \mathcal{L}\left(\mathbf{x}_0^T \mathbf{v}\right) \mathbf{v}^T \mathbf{x}_0 \, dx \, dy.$$

It follows, then, that there exists a scalar function $V = \mathbf{x}_0^T \mathbf{v}$ such that

$$\iint_D V \mathcal{L} V \, dx \, dy = 0.$$

Now $\mathcal{L}$ is positive definite; thus $V \equiv 0$ and, since $\mathbf{x}_0$ is a non-trivial vector, the functions v_i must be linearly dependent. But the chosen set of functions v_i are linearly independent. The contradiction proves that $\mathbf{A}$ is non-singular and a unique solution exists whenever $\mathcal{L}$ is positive definite.

It may also be shown (Mikhlin 1964) that the Rayleigh–Ritz procedure yields a minimizing sequence and that it converges to the exact solution.

2.8   The 'elastic analogy' for Poisson's equation

In elasticity problems, the stresses may be derived from an elastic potential W by the equations (Sokolnikoff 1956)

$$\sigma_x = \frac{\partial W}{\partial \varepsilon_x}, \ \sigma_y = \frac{\partial W}{\partial \varepsilon_y}, \quad \sigma_z = \frac{\partial W}{\partial \varepsilon_z},$$

$$\sigma_{xy} \frac{\partial W}{\partial \varepsilon_{xy}}, \quad \sigma_{yz} = \frac{\partial W}{\partial \varepsilon_{yz}}, \ \sigma_{zx} = \frac{\partial W}{\partial \varepsilon_{zx}},$$

where W is given by

$$W = \frac{1}{2}\varepsilon^T\sigma.$$

W is sometimes called the strain energy density, since the strain energy is given by

$$\int_V W \, dV.$$

The strain and stress vectors are given respectively by

$$\varepsilon = \left[\varepsilon_x, \varepsilon_y, \varepsilon_z, \varepsilon_{xy}, \varepsilon_{yz}, \varepsilon_{zx}\right]^T,$$

$$\sigma = \left[\sigma_x, \sigma_y, \sigma_z, \sigma_{xy}, \sigma_{yz}, \sigma_{zx}\right]^T.$$

The relationship between the stress and strain takes the form

(2.59) $\sigma = \kappa\varepsilon,$

where

$$\kappa = \frac{E}{(1+v)(1-2v)}\begin{bmatrix} 1-v & v & v & 0 & 0 & 0 \\ v & 1-v & v & 0 & 0 & 0 \\ v & v & 1-v & 0 & 0 & 0 \\ 0 & 0 & 0 & \dfrac{1-2v}{2} & 0 & 0 \\ 0 & 0 & 0 & 0 & \dfrac{1-2v}{2} & 0 \\ 0 & 0 & 0 & 0 & 0 & \dfrac{1-2v}{2} \end{bmatrix};$$

E and v are the Young's modulus and Poisson's ratio, respectively. Now the strain–displacement, *compatibility*, relationship is

(2.60) $\varepsilon = \mathcal{D}^T\mathbf{u},$

where $\mathcal{D}$ is the matrix of differential operators given by

$$\boldsymbol{D} = \begin{bmatrix} \dfrac{\partial}{\partial x} & 0 & 0 & \dfrac{\partial}{\partial y} & 0 & \dfrac{\partial}{\partial z} \\[2mm] 0 & \dfrac{\partial}{\partial y} & 0 & \dfrac{\partial}{\partial x} & \dfrac{\partial}{\partial z} & 0 \\[2mm] 0 & 0 & \dfrac{\partial}{\partial z} & 0 & \dfrac{\partial}{\partial y} & \dfrac{\partial}{\partial x} \end{bmatrix}.$$

The stresses and body forces satisfy the following equilibrium equation (Przemie-niecki 1968):

$$(2.61) \qquad\qquad \boldsymbol{D}\boldsymbol{\sigma} = -\mathbf{w},$$

and $\mathbf{w}$ is the vector of body forces per unit volume, given by

$$\mathbf{w} = [w_x\ w_y\ w_z]^T.$$

By use of the stress–strain law (2.59) and the strain–displacement equation (2.60), eqn (2.61) may be used to relate the displacements to the body forces by

$$(2.62) \qquad\qquad \mathcal{L}\mathbf{u} = -\mathbf{w},$$

where the matrix differential operator $\mathcal{L}$ is given by

$$\mathcal{L}(\cdot) \equiv \boldsymbol{D}\left(\kappa\boldsymbol{D}^T(\cdot)\right).$$

Equation (2.62) is the vector equivalent of the scalar equation (2.1) and, to preserve the notation already adopted, it will be written as

$$\mathcal{L}\mathbf{u} = \mathbf{f}.$$

The corresponding vector form of the functional (2.29) is

$$(2.63) \qquad\qquad I[\mathbf{u}] = \int_V \mathbf{u}^T \mathcal{L}\mathbf{u}\, dV - 2\int_V \mathbf{u}^T \mathbf{f}\, dV.$$

There are two integral formulae in elasticity (Mikhlin 1964), known as Betti's formulae, which are very similar to the two forms of Green's theorem for scalar functions. The formula that is useful here is

$$(2.64) \qquad\qquad \int_V \mathbf{u}^T \mathcal{L}\mathbf{u}\, dV = 2\int_V W\, dV - \int_S \mathbf{u}^T \mathbf{t}\, dS,$$

where $\mathbf{t}$ is the stress vector defining the boundary tractions. In elasticity problems, the usual boundary conditions are prescribed boundary forces or displacements; thus the functional for homogeneous boundary conditions is, using eqn (2.63) and (2.64),

(2.65) $$I[\mathbf{u}] = \int_V \left(W - \mathbf{u}^T \mathbf{f}\right) dV.$$

For non-homogeneous boundary conditions, a similar procedure to that used in Section 2.5 yields the functional

$$I[\mathbf{u}] = \int_V \left(W - \mathbf{u}^T \mathbf{f}\right) dV - \int_S \mathbf{u}^T \mathbf{t} \, dS,$$

in which trial functions must satisfy the prescribed displacement condition

$$\mathbf{u} = \mathbf{g}(S)$$

on the boundary. Now consider the generalized Poisson equation

$$-\text{div}\left(\boldsymbol{\kappa} \, \text{grad} \, u\right) = f,$$

which holds in some volume V bounded by a closed surface S. Suppose that the following boundary conditions are prescribed: a Dirichlet condition $u = g(S)$ on a part S_1 of the boundary and a Neumann condition $\partial u / \partial n = h(S)$ on the remainder, S_2. The functional is given by eqn (2.44), extended to the three-dimensional case, as

(2.66) $$I[u] = \int_V \left\{\text{grad}\, u \cdot (\boldsymbol{\kappa}\, \text{grad}\, u) - 2uf\right\} dV - 2\int_{S_2} uh \, dS,$$

and all trial functions must satisfy the essential Dirichlet boundary condition.

By comparison of eqn (2.66) with eqn (2.65), there is a close analogy between the elasticity problem and problems modelled by Poisson's equation. A fixed displacement is equivalent to a Dirichlet boundary condition, which is an essential condition and must be enforced on the trial functions. A given boundary force is analogous to the natural Neumann boundary condition.

This analogy can in fact be set up directly in the modelling process, as shown by the following example.

Example 2.14 Suppose that $T(x, y)$ is the steady-state temperature in a material occupying a region D in the xy plane; the extension to three dimensions follows in a similar manner. Heat is generated in D by means of a source distribution $Q(x, y)$. An element of the material is shown in Fig. 2.7.

The total heat generated in the element is

$$Q(x, y) \, \Delta x \, \Delta y,$$

and the total amount of heat leaving the element is

$$\left(q_x + \frac{\partial q_x}{\partial x}\Delta x\right) \Delta y + \left(q_y + \frac{\partial q_y}{\partial y}\Delta y\right) \Delta x - q_x \, \Delta y - q_y \, \Delta x.$$

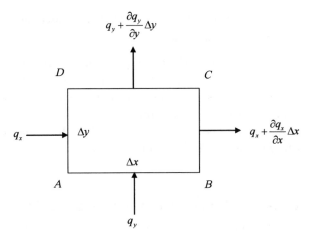

Fig. 2.7 An element $ABCD$ of a two-dimensional thermal conductor, showing the heat flux over its sides.

Since, in the steady state, the amount of heat generated in the element must be equal to that leaving it across its boundaries, it follows that

(2.67)
$$\frac{\partial q_x}{\partial x} + \frac{\partial q_y}{\partial y} = Q.$$

This equation is called the *continuity equation* and represents the conservation of heat in the material.

Now the heat flux is related to the temperature by the equations

(2.68)
$$q_x = -\kappa_{xx}\frac{\partial T}{\partial x} - \kappa_{xy}\frac{\partial T}{\partial y},$$
$$q_y = -\kappa_{yx}\frac{\partial T}{\partial x} - \kappa_{yy}\frac{\partial T}{\partial y},$$

where the coefficients are the elements of a tensor of the form (2.5). Thus eqn (2.67) becomes the generalized Poisson equation

$$-\mathrm{div}(\kappa \, \mathrm{grad} \, T) = Q.$$

In matrix form, eqn (2.68) may be written as

(2.69)
$$\sigma = \kappa\varepsilon,$$

where $\sigma = [q_x, q_y]^T$ and $\varepsilon = [-\partial T/\partial x, -\partial T/\partial y]^T$ are directly analogous to the stresses and strains in an elastic material; compare eqns (2.69) and (2.59).

Also, the continuity equation (2.67) may be written as

$$[\partial/\partial x \; \partial/\partial y] \begin{bmatrix} q_x \\ q_y \end{bmatrix} = [Q],$$

which, by comparison with eqn (2.61), shows that the source term Q is analogous to a body force in an elastic material.

Thus it has been shown that there is a direct analogy between field problems described by Poisson's equation and problems in elasticity. This is of historical interest because, in the 1960s, engineering computer codes had been developed for the solution of sophisticated structural problems. The elastic analogy allowed these codes to be used for the first application of the finite element method to field problems by Zienkiewicz and Cheung (1965).

2.9 Variational methods for time-dependent problems

Probably the most important time-dependent variational principle is that of Hamilton. It states that the motion of a system from time $t = 0$ to time $t = t_0$ is such that

$$I = \int_0^{t_0} L\, dt$$

is stationary.

L is the Lagrangian for the system and is related to the kinetic energy T and the potential energy V by

$$L = T - V.$$

Consider the wave equation

$$(2.70) \qquad \frac{\partial^2 u}{\partial x^2} = \frac{1}{c^2} \frac{\partial^2 u}{\partial t^2}, \qquad 0 < x < 1 \quad t > 0,$$

subject to the boundary conditions

$$(2.71) \qquad u(0, t) = 0,$$

$$(2.72) \qquad u(1, t) = 0$$

and the initial conditions

$$(2.73) \qquad u(x, 0) = f(x),$$

$$(2.74) \qquad \frac{\partial u}{\partial t}(x, 0) = g(x).$$

This problem models many situations, including the small transverse vibrations of a uniform elastic string stretched between the fixed points $x = 0$, $x = 1$ subject to an initial displacement distribution $f(x)$ and an initial velocity distribution $g(x)$. If the tension in the string is F and its mass per unit length is ρ, then

$$c^2 = F/\rho.$$

The kinetic energy of the string is

$$T = \frac{1}{2}\rho \int_0^1 \left(\frac{\partial u}{\partial t}\right)^2 dx$$

and the potential energy in the string is

$$V = \frac{1}{2}\rho \int_0^1 c^2 \left(\frac{\partial u}{\partial x}\right)^2 dx.$$

Thus the functional is given by

$$(2.75) \qquad I[u] = \int_0^{t_0} \int_0^1 \left\{ c^2 \left(\frac{\partial u}{\partial x}\right)^2 - \left(\frac{\partial u}{\partial t}\right)^2 \right\} dx\, dt.$$

The factor $\frac{1}{2}\rho$ has no influence on the function u_0 which makes $I[u]$ stationary, and has thus been removed from the functional.

The major drawback in the use of Hamilton's principle is that the operator

$$\mathcal{L} \equiv \frac{\partial^2}{\partial x^2} - \frac{1}{c^2}\frac{\partial^2}{\partial t^2}$$

is hyperbolic and is not positive definite. The solution u_0 of $\mathcal{L}u = 0$ does not yield an extremum for the functional (2.75); it gives only a stationary value. This means that numerical algorithms based on this functional cannot be guaranteed to converge as was the case for the positive definite operators associated with elliptic problems in Section 2.7. This is true in general for initial-value problems; there are no extremal variational principles; however, they may still be used to develop approximate methods.

Now, to find a stationary value for eqn (2.75), the trial functions must satisfy essential conditions on the boundary of the region in xt space. These conditions are of the form given by eqn (2.71), (2.72) and (2.73) together with

$$u(x, t_0) = h(x),$$

where t_0 is some arbitrary fixed time.

However, for properly posed hyperbolic problems, only initial values in time, such as eqn (2.74), are given; $h(x)$ is not known. Noble (1973) proposed that the functional be used assuming that $h(x)$ is known and then, at the end, the initial condition (2.74) be used to eliminate the unknown function. He also gave the following functional for the wave equation (2.70):

$$I[u] = \int_0^{t_0} \int_0^1 \left(c^2 \frac{\partial v}{\partial x}\frac{\partial u}{\partial x} - \frac{\partial v}{\partial t}\frac{\partial u}{\partial t} \right) dx\, dt - 2\int_0^1 g(x)u(x, t_0)\, dx,$$

where $v(x, t) = u(x, t_0 - t)$. All trial functions must satisfy the essential conditions (2.71), (2.72) and (2.73). The second initial condition (2.74) is taken into account in the functional itself, and it is not necessary to assume that $u(x, t_0)$ is known.

Finally, for the diffusion equation

$$\frac{\partial^2 u}{\partial x^2} = \frac{1}{\alpha} \frac{\partial u}{\partial t}, \qquad 0 < x < 1, \quad t > 0,$$

subject to the boundary conditions

$$u(0, t) = u(1, t) = 0$$

and the initial condition

$$u(x, 0) = f(x),$$

Noble (1973) gave the functional

$$(2.76) \qquad I[u] = \int_0^{t_0} \int_0^1 \left(\alpha \frac{\partial v}{\partial x} \frac{\partial u}{\partial x} + v \frac{\partial u}{\partial t} \right) dx \, dt - \int_0^1 f(x) u(x, t_0) \, dx,$$

where again $v(x, t) = u(x, t_0 - t)$ and the trial functions must satisfy the essential Dirichlet boundary conditions.

The application of the functionals presented in this section for the numerical solution of the wave and diffusion equations will be discussed in Section 5.4.

2.10 Exercises and solutions

Exercise 2.1 $\mathcal{L}$ is the Helmholtz operator $\nabla^2 + k^2$. Show that $\mathcal{L}$ is self-adjoint.

Exercise 2.2 Show that, in certain circumstances, the linear second-order partial differential operator $\mathcal{L}$ in two independent variables may be given by the expression

$$\mathcal{L}u = -\mathrm{div}(\boldsymbol{\kappa} \, \mathrm{grad} \, u) + \rho u,$$

where

$$\boldsymbol{\kappa} = \begin{bmatrix} \kappa_{xx} & \kappa_{xy} \\ \kappa_{yx} & \kappa_{yy} \end{bmatrix}.$$

Show that the first form of Green's theorem may be generalized to give eqn (2.42) and that if $\boldsymbol{\kappa}$ is symmetric, the second form is given by eqn (2.43).

Exercise 2.3 For the two-point boundary-value problem

$$-u'' = x^2, \qquad 0 < x < 1,$$
$$u(0) = u(1) = 0,$$

use the collocation method at the points $x = \frac{1}{4}, \frac{1}{2}, \frac{3}{4}$ with the trial function

$$\tilde{u}_2(x) = x(1 - x)(c_0 + c_1 x + c_2 x^2).$$

Compare with the exact solution given in Example 2.3 and comment on the result.

Exercise 2.4 Consider the two-point boundary-value problem

$$-u'' = \cosh x,$$
$$u(0) = u(1) = 0,$$

where we seek an approximate solution of the form

$$\tilde{u}_4 = x(1 - x)(c_0 + c_1 x + c_2 x^2 + c_3 x^3 + c_4 x^4),$$

with the exact solution $u = 1 + (\cosh 1 - 1)x - \cosh x$.
    Use the collocation method at the five points

$$x = 0.1, 0.3, 0.5, 0.7, 0.9$$

and overdetermined collocation at the nine points

$$x = 0.1, 0.2, \ldots, 0.9$$

to find $\tilde{u}_4$. Compare the results with the exact solution.

Exercise 2.5 For the two-point boundary-value problem

$$u'' = e^x, \qquad 0 < x < 1,$$
$$u(0) = u(1) = 0,$$

use the Rayleigh–Ritz method with the two trial functions

$$\tilde{u}_A = x(1 - x)c_0 + c_1 x \quad \text{and} \quad \tilde{u}_B = x(e^x - e).$$

Compare the results with the exact solution.

Exercise 2.6 Use the Rayleigh–Ritz method to solve the two-point boundary-value problem

$$-u'' = x^2, \qquad 0 < x < 1,$$
$$u(0) = u'(1) = 0,$$

using (i) a trial function $\tilde{u}_A(x) = c_0 x (1 - x/2)$ which satisfies both boundary conditions, and (ii) a trial function $\tilde{u}_B = c_0 x + c_1 x^2$ in which only the Dirichlet boundary condition is enforced, the Neumann boundary condition being a natural boundary condition.

Compare the results with the exact solution.

Exercise 2.7 Find approximate solutions of the form $x(1 - x)(c_0 + c_1 x)$ to the two-point boundary-value problem

$$-(1 + x)u'' - u' = x, \qquad 0 < x < 1,$$

$$u(0) = u(1) = 0,$$

using the collocation, overdetermined collocation, least-squares, Galerkin and Rayleigh–Ritz methods. Compare the results with the exact solution.

Exercise 2.8 Solve the two-point boundary-value problem

$$-u'' = x^2, \qquad 0 < x < 1,$$

$$u(0) = 1, \qquad u'(1) + 2u(1) = 1,$$

using both the Galerkin and the Rayleigh–Ritz methods with quadratic trial functions. Compare the results with the exact solution.

Exercise 2.9 Show that with the notation of Section 2.4,

$$\text{grad } \Delta w \cdot \text{grad } w = \frac{1}{2} \Delta \mid \text{grad } w \mid^2 .$$

Exercise 2.10 Equation (2.27) gives the functional

$$I[w] = \iint_D \left\{ \left(\frac{\partial w}{\partial x} \right)^2 + \left(\frac{\partial w}{\partial y} \right)^2 - \frac{2p}{T} w \right\} dx\, dy + \int_{C_3} \sigma w^2 \, ds,$$

where w satisfies the Robin boundary condition $\partial w / \partial n + \sigma(s)w = 0$ on C_3, and homogeneous Dirichlet and Neumann conditions on C_1 and C_2 respectively. Use the first form of Green's theorem to show that $I[w]$ may be given by eqn (2.28)

Exercise 2.11 Show that the boundary condition (2.41) is a natural boundary condition for the functional (2.44).

Exercise 2.12 Consider the functional (Hazel and Wexler 1972)

$$I[u] = \iint_D \{\mid \text{grad } u \mid^2 - 2uf\} \, dx\, dy + \int_{C_1} 2(g - u) \frac{\partial u}{\partial n} ds + \int_{C_2} (\sigma u^2 - 2hu) \, ds,$$

in which the Dirichlet boundary condition $u = g$ holds on C_1 and the Robin boundary condition $\partial u / \partial n + \sigma u = h$ holds on C_2. Show that both the Robin and the Dirichlet boundary conditions are natural conditions for this functional, but that the functional is not necessarily minimized by the solution u_0 of the corresponding boundary-value problem.

Exercise 2.13 Given the two-point boundary-value problem

$$-u'' = 2, \qquad 0 < x < 1,$$

$$u(0) = u'(1) = 0,$$

show that the exact solution is $u_0 = 2x - x^2$. Find $I[u_0]$, where $I[u]$ is given by eqn (2.29), and verify that $I[u_0] < I[\tilde{u}]$ for a choice of various trial functions $\tilde{u}$ which satisfy the boundary conditions.

Exercise 2.14 Consider Poisson's equation

$$-\nabla^2 u = f \qquad \text{in } D$$

with boundary conditions

$$u = g(s) \qquad \text{on } C_1$$

and

$$\frac{\partial u}{\partial n} + \sigma(s)u = h(s) \qquad \text{on } C_2.$$

Show that the Galerkin and Rayleigh–Ritz methods yield the same approximate solution.

Exercise 2.15 Consider Poisson's equation,

$$-\nabla^2 u = 2(x + y) - 4,$$

in the square whose vertices are at the points $(0,0)$, $(1,0)$, $(1,1)$, $(0,1)$. The boundary conditions are

$$u(0, y) = y^2, \qquad u(x, 0) = x^2,$$

$$\frac{\partial u}{\partial x}(1, y) = 2 - 2y - y^2, \qquad \frac{\partial u}{\partial y}(x, 1) = 2 - x - x^2.$$

Solve the boundary-value problem using trial functions of the form

$$\tilde{u}_A = x^2 + y^2 + c_0 xy,$$

$$\tilde{u}_B = x^2 + y^2 + c_0 xy + c_1 xy(x + y).$$

Exercise 2.16 For a beam on an elastic foundation, simply supported at its ends, show that the corresponding functional is given by

(2.77) $$I[u] = \int_0^1 \left\{ ku^2 + EI(u'')^2 + 2uf \right\} dx,$$

where EI is the flexural rigidity of the beam, k is the stiffness of the foundation and f is the loading per unit length.

The corresponding differential equation is

$$EIu^{iv} + ku = -f.$$

There are four boundary conditions: the fixed-end conditions $u(0) = u(l) = 0$, which are essential boundary conditions, and zero bending moment at the ends gives $u''(0) = u''(l) = 0$, which are natural boundary conditions.

Solution 2.1

$$\iint_D v\mathcal{L}u \, dx \, dy - \iint_D u\mathcal{L}v \, dx \, dy$$

$$= \iint_D v(\nabla^2 + k^2)u \, dx \, dy - \iint_D u(\nabla^2 + k^2)v \, dx \, dy$$

$$= \iint_D (v \, \nabla^2 u - u \, \nabla^2 v) \, dx \, dy$$

$$= \int_C \left(v\frac{\partial u}{\partial n} - u\frac{\partial v}{\partial n} \right) ds,$$

using the second form of Green's theorem. Thus $\mathcal{L}$ is self-adjoint.

Solution 2.2 The general linear second-order partial differential operator in two independent variables is

$$(2.78) \qquad \mathcal{L} \equiv a\frac{\partial^2}{\partial x^2} + b\frac{\partial^2}{\partial x \partial y} + c\frac{\partial^2}{\partial y^2} + d\frac{\partial}{\partial x} + e\frac{\partial}{\partial y} + f,$$

where the coefficients $a, \ldots, f$ are functions of x and y. Now,

$$-\mathrm{div}(\kappa \, \mathrm{grad} \, u) + \rho u$$

$$= -\mathrm{div}\left[\kappa_{xx}\frac{\partial u}{\partial x} + \kappa_{xy}\frac{\partial u}{\partial y} \quad \kappa_{yx}\frac{\partial u}{\partial x} + \kappa_{yy}\frac{\partial u}{\partial y} \right]^T + \rho u$$

$$= -\kappa_{xx}\frac{\partial^2 u}{\partial x^2} - (\kappa_{xy} + \kappa_{yx})\frac{\partial^2 u}{\partial x \partial y}$$

$$-\kappa_{yy}\frac{\partial^2 u}{\partial y^2} - \left(\frac{\partial \kappa_{xx}}{\partial x} + \frac{\partial \kappa_{yx}}{\partial y} \right)\frac{\partial u}{\partial x}$$

$$-\left(\frac{\partial \kappa_{xy}}{\partial x} + \frac{\partial \kappa_{yy}}{\partial y} \right)\frac{\partial u}{\partial y} + \rho u.$$

Thus, provided $\kappa_{xx}, \kappa_{xy}, \kappa_{yx}, \kappa_{yy}$ and ρ may be chosen such that

$$\kappa_{xx} = -a, \quad \kappa_{yy} = -c, \quad \rho = f, \quad \kappa_{xy} + \kappa_{yx} = -b,$$

$$\frac{\partial \kappa_{xx}}{\partial x} + \frac{\partial \kappa_{yx}}{\partial y} = -d \quad \text{and} \quad \frac{\partial \kappa_{xy}}{\partial x} + \frac{\partial \kappa_{yy}}{\partial y} = -e,$$

the general operator $\mathcal{L}$ in eqn (2.78) is given by the expression

$$\mathcal{L}u = -\text{div}(\boldsymbol{\kappa}\,\text{grad}\,u) + \rho u.$$

In some cases where the coefficients do not satisfy these conditions, an 'integrating factor', μ, may be found. This is such that when the original equation is multiplied by μ, the resulting coefficients do satisfy the required condition (Ames 1972).

To obtain the generalization of Green's theorem, consider the identity

$$\text{div}(u\boldsymbol{\kappa}\,\text{grad}\,v) = \text{grad}\,u \cdot (\boldsymbol{\kappa}\,\text{grad}\,v) + u\,\text{div}(\boldsymbol{\kappa}\,\text{grad}\,v).$$

Integrating over the region D and using the divergence theorem gives

$$\oint_C u(\boldsymbol{\kappa}\,\text{grad}\,v) \cdot \mathbf{n}\,ds = \iint_D \text{grad}\,u \cdot (\boldsymbol{\kappa}\,\text{grad}\,v)dx\,dy$$

$$+ \iint_D u\,\text{div}(\boldsymbol{\kappa}\,\text{grad}\,v)dx\,dy,$$

which is the generalization of the first form of Green's theorem, eqn (2.42). Interchange u and v and subtract to obtain

$$\oint_C \{u(\boldsymbol{\kappa}\,\text{grad}\,v) \cdot \mathbf{n} - v(\boldsymbol{\kappa}\,\text{grad}\,u) \cdot \mathbf{n}\}ds$$

$$= \iint_D \{u\,\text{div}(\boldsymbol{\kappa}\,\text{grad}\,v) - v\,\text{div}(\boldsymbol{\kappa}\,\text{grad}\,u)\}\,dx\,dy$$

$$+ \iint_D \{\text{grad}\,u \cdot (\boldsymbol{\kappa}\,\text{grad}\,v) - \text{grad}\,v \cdot (\boldsymbol{\kappa}\,\text{grad}\,u)\}\,dx\,dy.$$

Now, if $\boldsymbol{\kappa}$ is symmetric, then the second integral on the right-hand side is zero, so that the second form of Green's theorem is given by eqn (2.43).

Solution 2.3 The residual is

$$r(\tilde{u}_2) = 2c_0 + (6x - 2)c_1 + (12x^2 - 6x)c_2 - x^2,$$

so that, using the notation of eqn (2.9) and the results of Example 2.4,

$$A_{00} = A_{10} = A_{20} = 2, \qquad h_0 = \tfrac{1}{12},$$

$$A_{01} = -\frac{1}{2}, \quad A_{11} = 1, \quad A_{21} = \frac{5}{2}, \quad h_1 = \tfrac{1}{4},$$

$$A_{02} = -\frac{3}{4}, \quad A_{12} = 0, \quad A_{22} = \frac{9}{4}, \quad h_2 = \tfrac{9}{16}.$$

So, the collocation equations are

$$\begin{bmatrix} 2 & -\tfrac{1}{2} & -\tfrac{3}{4} \\ 2 & 1 & 0 \\ 2 & \tfrac{5}{2} & \tfrac{9}{4} \end{bmatrix} \begin{bmatrix} c_0 \\ c_1 \\ c_2 \end{bmatrix} = \begin{bmatrix} \tfrac{1}{16} \\ \tfrac{1}{4} \\ \tfrac{9}{16} \end{bmatrix},$$

which yield the values $c_0 = c_1 = c_2 = \tfrac{1}{12}$.
Thus the approximate solution is

$$\tilde{u}_2(x) = x(1-x)\frac{1}{12}(1 + x + x^2)$$

$$= \frac{1}{12}x(1 - x^3).$$

The exact solution is $u_0(x) = \tfrac{1}{12}x(1-x^3)$, so that the approximate solution is identical to the exact solution. This is to be expected, since the exact solution is contained amongst all possible functions of the form $x(1-x)(c_0 + c_1 x + c_2 x^2)$, from which the trial function is chosen.

Solution 2.4 The residual is

$$r(\tilde{u}_4) = 2c_0 + (6x - 2)c_1 + (12x^2 - 6x)c_2 + (20x^3 - 12x^2)c_3$$

$$+ (30x^4 - 20x^3)c_4 - \cosh x.$$

Collocating at $0.1, 0.3, 0.5, 0.7, 0.9$ and using overdetermined collocation at the points $0.1, 0.2, \ldots, 0.9$ yields the following values for the coefficients $c_0, \ldots, c_4$:

0.543082, 0.043065, 0.043164, 0.001233, 0.001577,
0.543082, 0.043062, 0.043165, 0.001223, 0.001579.

The corresponding approximate solutions are almost identical and compare very well with the exact solution, as shown in Fig. 2.8.

Solution 2.5 For the trial function $\tilde{u}_A$, the Rayleigh–Ritz coefficient matrix is given in Example 2.12:

$$\int_0^1 (x - x^2)e^x \, dx = 1 - e, \qquad \int_0^1 (x^2 - x^3)e^x \, dx = 3e - 8.$$

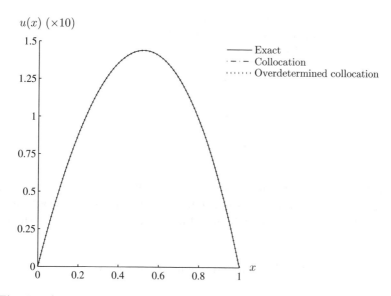

Fig. 2.8 A comparison of $\tilde{u}_4^c$ and $\tilde{u}_4^{oc}$ with the exact solution for Exercise 2.4.

Thus the Rayleigh–Ritz equations are

$$\begin{bmatrix} \frac{1}{3} & \frac{1}{6} \\ \frac{1}{6} & \frac{1}{3} \end{bmatrix} \begin{bmatrix} a \\ b \end{bmatrix} = \begin{bmatrix} 1 - e \\ 3e - 8 \end{bmatrix},$$

which yield $a = -15.283$, $b = 20.258$.

Thus $\tilde{u}_A(x) = x(1 - x)(-15.283 + 20.258x)$.

For the trial function $\tilde{u}_B(x) = ax(e^x - e)$,

$$A_{11} = \int_0^1 \left\{ e^{2x}(1 + 2x + x^2) - ee^x(1 + x) + e^2 \right\} dx = (5e^2 - 1)/4,$$

$$h_1 = \int_0^1 e^x x(e^x - e) dx = (e^2 + 1 - 4e)/4.$$

Thus

$$a = (e^2 + 1 - 4e)/(5e^2 - 1) = -0.0691,$$

so that

$$\tilde{u}_B(x) = 0.0691x(e - e^x).$$

The two solutions are compared in Table 2.5 with the exact solution

$$u_0(x) = 1 + (e - 1)x - e^x.$$

Table 2.5 Comparison of the two approximate solutions $\tilde{u}_A$ and $\tilde{u}_B$ with the exact solution u for Exercise 2.5

x	0	0.25	0.5	0.75	1
u_0	0	0.146	0.210	0.172	0
$\tilde{u}_A$	0	−1.916	−1.288	−0.168	0
$\tilde{u}_B$	0	0.025	0.037	0.031	0

Note that $\tilde{u}_A$ is extremely inaccurate. $\tilde{u}_B$, which incorporates the transcendental function e^x, is an improvement but is still completely unsatisfactory. In Exercise 3.5, the same problem is solved using the finite element method, and a comparison of the solutions is given in Fig. 3.36.

Solution 2.6 The Rayleigh–Ritz equations are as follows.

(i)

$$A_{00}c_0 = h_0,$$

where

$$A_{00} = \int_0^1 (1-x)^2 \, dx = \frac{1}{3}, \qquad h_0 = \int_0^1 x^2 \left(x - \frac{x^2}{2} \right) dx = \frac{3}{20};$$

thus

$$c_0 = \frac{9}{20} \quad \text{and} \quad \tilde{u}_A(x) = \frac{9}{20} x \left(1 - \frac{x}{2} \right).$$

(ii)

$$A_{00} = \int_0^1 1^2 \, dx = 1, \qquad A_{01} = A_{10} = \int_0^1 x^2 x^2 \, dx = \frac{1}{5},$$

$$A_{11} = \int_0^1 (2x)^2 \, dx = \frac{4}{3},$$

$$h_0 = \int_0^1 x^2 x \, dx = \frac{1}{4}, \qquad h_1 = \int_0^1 x^2 x^2 dx = \frac{1}{5}.$$

Thus the Rayleigh–Ritz equations are

$$\begin{bmatrix} 1 & 1 \\ 1 & \frac{4}{3} \end{bmatrix} \begin{bmatrix} c_0 \\ c_1 \end{bmatrix} = \begin{bmatrix} \frac{1}{4} \\ \frac{1}{5} \end{bmatrix},$$

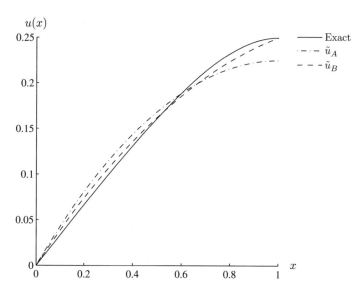

Fig. 2.9 Rayleigh–Ritz solutions for the two approximations in Exercise 2.6.

which yields $c_0 = \frac{2}{5}$, $c_1 = -\frac{3}{20}$, so that $\tilde{u}_B(x) = \frac{1}{20}x(8 - 3x)$. The exact solution is $u_0(x) = \frac{1}{20}x(4 - x^3)$, and the results are compared in Fig. 2.9.

From Fig. 2.9, it is seen that $\tilde{u}_B$ is the better approximation, which illustrates that it is better not to enforce natural boundary conditions but to retain as many variational parameters as possible in the trial function. $\tilde{u}'_B(1) = 0.1$, so that the natural boundary condition is reasonably well satisfield.

Solution 2.7 The differential equation may be written in self-adjoint form as

$$-\frac{d}{dx}\left((1+x)\frac{du}{dx}\right) = x.$$

Collocation at $x = \frac{1}{3}$, $x = \frac{2}{3}$. The residual is $c_0(1 + 4x) + c_1(-2 + 2x + 9x^2) - x$; thus the equations are

$$
\begin{bmatrix} \frac{7}{3} & -\frac{1}{3} \\ \frac{11}{3} & \frac{10}{3} \end{bmatrix}
\begin{bmatrix} c_0 \\ c_1 \end{bmatrix} =
\begin{bmatrix} \frac{1}{3} \\ \frac{2}{3} \end{bmatrix},
$$

which give $c_0 = \frac{4}{27}$, $c_1 = \frac{1}{27}$.

Thus the collocation approximation is

$$\tilde{u}_A(x) = x(1 - x)(0.148 + 0.037x).$$

Collocation at $x = \frac{1}{4}$, $x = \frac{1}{2}$, $x = \frac{3}{4}$ leads to the overdetermined set

$$
\begin{bmatrix} 2 & -\frac{5}{16} \\ 3 & \frac{5}{4} \\ 4 & \frac{73}{16} \end{bmatrix}
\begin{bmatrix} c_0 \\ c_1 \end{bmatrix} =
\begin{bmatrix} \frac{1}{4} \\ \frac{1}{2} \\ \frac{3}{4} \end{bmatrix}.
$$

The method of least squares is used to solve this set, giving

$$
\begin{bmatrix} 29 & 20.125 \\ 20.125 & 23.258 \end{bmatrix}
\begin{bmatrix} c_0 \\ c_1 \end{bmatrix} =
\begin{bmatrix} 5 \\ 3.813 \end{bmatrix},
$$

which yields $c_0 = 0.147$, $c_2 = 0.0369$.

Thus the overdetermined collocation approximation is

$$
\tilde{u}_B(x) = x(1 - x)(0.147 + 0.0369x).
$$

Least squares method.

$$
A_{00} = \int_0^1 (1 + 4x)^2 dx = \frac{31}{3},
$$

$$
A_{01} = A_{10} = \int_0^1 (1 + 4x)(-2 + 2x + 9x^2)dx = \frac{29}{3},
$$

$$
A_{11} = \int_0^1 (-2 + 2x + 9x^2)dx = \frac{218}{15},
$$

$$
h_0 = \int_0^1 x(1 + 4x)dx = \frac{11}{6},
$$

$$
h_1 = \int_0^1 x(-2 + 2x + 9x^2)dx = \frac{23}{12}.
$$

Thus

$$
\begin{bmatrix} \frac{31}{3} & \frac{29}{3} \\ \frac{29}{3} & \frac{218}{15} \end{bmatrix}
\begin{bmatrix} c_0 \\ c_1 \end{bmatrix} =
\begin{bmatrix} \frac{11}{6} \\ \frac{23}{12} \end{bmatrix},
$$

which gives $c_0 = 0.143$, $c_1 = 0.0367$.

Thus the least squares approximation is

$$
\tilde{u}_C(x) = x(1 - x)(0.143 + 0.0367x).
$$

Galerkin method.

$$A_{00} = -\int_0^1 (x - x^2)\frac{\partial}{\partial x}\{(1+x)(1-2x)\}\,dx = \frac{1}{2},$$

$$A_{01} = A_{10} = -\int_0^1 (x^2 - x^3)\frac{\partial}{\partial x}\{(1+x)(2-2x)\}\,dx = \frac{17}{60},$$

$$A_{11} = -\int_0^1 (x^2 - x^3)\frac{\partial}{\partial x}\{(1+x)(2x - 3x^2)\}\,dx = \frac{7}{30},$$

$$h_0 = \int_0^1 xx(1-x)\,dx = \frac{1}{12},$$

$$h_1 = \int_0^1 xx^2(1-x)\,dx = \frac{1}{20}.$$

Thus the Galerkin equations are

$$\begin{bmatrix} \frac{1}{2} & \frac{17}{60} \\ \frac{17}{60} & \frac{7}{30} \end{bmatrix}\begin{bmatrix} c_0 \\ c_1 \end{bmatrix} = \begin{bmatrix} \frac{1}{12} \\ \frac{1}{20} \end{bmatrix},$$

which give $c_0 = 0.145$, $c_1 = 0.0382$ and the Galerkin approximation is

$$\tilde{u}_D(x) = x(1-x)(0.145 + 0.382x).$$

Rayleigh–Ritz method.

$$A_{00} = \int_0^1 (1+x)(1-2x)^2\,dx = \frac{1}{2},$$

$$A_{01} = A_{10} = \int_0^1 (1+x)(1-2x)(2x - 3x^2)\,dx = \frac{17}{60},$$

$$A_{11} = \int_0^1 (1+x)(2x - 3x^2)^2\,dx = \frac{7}{30},$$

$$h_0 = \int_0^1 xx(1-x)\,dx = \frac{1}{12},$$

$$h_1 = \int_0^1 xx^2(1-x)\,dx = \frac{1}{20}.$$

Table 2.6 Comparison of the approximate solutions ($\times 10^2$) with the exact solution ($\times 10^2$) for Exercise 2.7

x	0	0.2	0.4	0.6	0.8	1
Galerkin } Rayleigh–Ritz }	0	2.443	3.847	4.031	2.809	0
Collocation	0	2.489	3.911	4.089	2.844	0
Overdetermined collocation	0	2.467	3.878	4.055	2.821	0
Least squares	0	2.407	3.786	3.962	2.759	0
Exact solution	0	2.424	3.864	4.048	2.800	0

Thus the Rayleigh–Ritz solution is identical with the Galerkin solution. The exact solution is

$$u_0(x) = -\frac{x^2}{4} + \frac{x}{2} - \frac{\ln(1+x)}{4\ln 2},$$

and a comparison of the results is given in Table 2.6.

Solution 2.8 A quadratic trial function which satisfies the essential boundary condition is

$$\tilde{u}_2(x) = 1 + c_1 x + c_2 x^2.$$

Galerkin method. Equation (2.20) gives

$$\int_0^1 \left\{ (c_1 + 2c_2 x)1 - x^2 x \right\} dx + \left[\left\{ 2(1 + c_1 x + c_2 x^2) - 1 \right\} x \right]_{x=1} = 0$$

and

$$\int_0^1 \left\{ (c_1 + 2c_2 x)2x - x^2 x^2 \right\} dx + \left[\left\{ 2(1 + c_1 x + c_2 x^2) - 1 \right\} x^2 \right]_{x=1} = 0.$$

Thus the Galerkin equations are

$$\begin{bmatrix} 3 & 3 \\ 3 & \frac{10}{3} \end{bmatrix} \begin{bmatrix} c_1 \\ c_2 \end{bmatrix} = \begin{bmatrix} -\frac{3}{4} \\ -\frac{4}{5} \end{bmatrix},$$

which give

$$c_1 = -\frac{1}{10}, \qquad c_2 = -\frac{3}{20}.$$

Rayleigh–Ritz method. With the notation of Section 2.7, the coefficients A_{ij} $(i = 0, 1; j = 0, 1, 2)$ are given in Example 2.13.

$$S_{11} = \left[2x^2\right]_{x=1} = 2, \quad S_{12} = S_{21} = \left[2x^3\right]_{x=1} = 2,$$

$$S_{13} = \left[2x1\right]_{x=1} = 2, \quad S_{22} = \left[2x^4\right]_{x=1} = 2,$$

$$S_{23} = \left[2x^21\right]_{x=1} = 2,$$

$$h_1 = \int_0^1 x^2 x \, dx = \frac{1}{4}, \quad h_2 = \int_0^1 x^2 x^2 \, dx = \tfrac{1}{5},$$

$$k_1 = [x1]_{x=1} = 1, \quad k_2 = \left[x^2 1\right]_{x=1} = 1.$$

Thus eqn (2.53) gives

$$\begin{bmatrix} 1 + 2\,1 + 2 \\ 1 + 2\,\frac{4}{3} + 2 \end{bmatrix} \begin{bmatrix} c_1 \\ c_2 \end{bmatrix} = \begin{bmatrix} \frac{1}{4} + 1 - (0 + 2) \\ \frac{1}{5} + 1 - (0 + 2) \end{bmatrix},$$

which is the same system as for the Galerkin method, i.e. in both cases the system of equations is the same, with the solution $c_1 = -\frac{1}{10}$, $c_2 = -\frac{3}{20}$. Hence the Galerkin and Rayleigh–Ritz quadratic approximations are given by

$$u_2(x) = 1 - \frac{1}{10}x - \frac{3}{20}x^2.$$

The exact solution is $u_0(x) = 1 - \frac{1}{6}x - \frac{1}{12}x^4$, and this is compared with the approximate solution in Table 2.7. Also,

$$\tilde{u}_2'(1) + 2\tilde{u}_2(1) = 1.1,$$

so that the natural Robin condition is satisfied reasonably well.

Table 2.7 Comparison of the approximate solutions with the exact solution for Exercise 2.8

x	0	0.2	0.4	0.6	0.8	1
Galerkin Rayleigh–Ritz	1	0.974	0.936	0.886	0.824	0.75
Exact solution	1	0.967	0.931	0.889	0.833	0.75

Solution 2.9

$$\text{grad } \Delta w = \left[\frac{\partial}{\partial x}(\Delta w) \quad \frac{\partial}{\partial y}(\Delta w) \right]^T$$

$$= \left[\Delta \left(\frac{\partial w}{\partial x} \right) \quad \Delta \left(\frac{\partial w}{\partial y} \right) \right]^T$$

$$= \Delta(\text{grad } w).$$

Thus

$$\text{grad } \Delta w \cdot \text{grad } w = \Delta(\text{grad } w) \cdot \text{grad } w$$

$$= \left[\frac{\partial}{\partial x} \left(\frac{\partial w}{\partial x} \right) \Delta x + \frac{\partial}{\partial y} \left(\frac{\partial w}{\partial x} \right) \Delta y \quad \frac{\partial}{\partial x} \left(\frac{\partial w}{\partial y} \right) \Delta x + \frac{\partial}{\partial y} \left(\frac{\partial w}{\partial y} \right) \Delta y \right] \begin{bmatrix} \dfrac{\partial w}{\partial x} \\ \dfrac{\partial w}{\partial y} \end{bmatrix}$$

$$= \left\{ \frac{\partial}{\partial x} \left(\frac{\partial w}{\partial x} \right) \frac{\partial w}{\partial x} + \frac{\partial}{\partial x} \left(\frac{\partial w}{\partial y} \right) \frac{\partial w}{\partial y} \right\} \Delta x$$

$$+ \left\{ \frac{\partial}{\partial y} \left(\frac{\partial w}{\partial x} \right) \frac{\partial w}{\partial x} + \frac{\partial}{\partial y} \left(\frac{\partial w}{\partial y} \right) \frac{\partial w}{\partial y} \right\} \Delta y$$

$$= \frac{1}{2} \frac{\partial}{\partial x} \mid \text{grad } w \mid^2 \Delta x + \frac{1}{2} \frac{\partial}{\partial y} \mid \text{grad } w \mid^2 \Delta y$$

$$= \frac{1}{2} \Delta \mid \text{grad } w \mid^2.$$

Solution 2.10

$$I[w] = \iint_D \left\{ \text{grad } w \cdot \text{grad } w - \frac{2p}{T} w \right\} dx \, dy + \int_{C_3} \sigma w^2 \, ds$$

$$= \iint_D (-w \, \nabla^2 w) \, dx \, dy + \oint_C w \frac{\partial w}{\partial n} ds$$

$$- \iint_D \frac{2p}{T} w \, dx \, dy + \int_{C_3} \sigma w^2 \, ds$$

$$= \iint_D \left\{ w(-\nabla^2 w) - \frac{2p}{T} w \right\} dx \, dy$$

$$+ \int_{C_3} w \left(\frac{\partial w}{\partial n} + \sigma w \right) ds,$$

since

$$\int_{C_1} w \frac{\partial w}{\partial n} ds = \int_{C_2} w \frac{\partial w}{\partial n} ds = 0.$$

Thus

$$I[w] = \iint_D \left\{ w(-\nabla^2 w) - \frac{2\rho}{T} w \right\} dx\, dy,$$

since

$$\frac{\partial w}{\partial n} + \sigma w = 0 \text{ on } C_3.$$

Solution 2.11 Suppose that u_0 is the solution of $\mathcal{L}u = f$, subject to the Robin boundary conditions (2.41) and the Dirichlet boundary condition $u = g(s)$ on C_1. Consider variations about u_0 given by

$$\tilde{u} = u_0 + \alpha v.$$

Then

$$\left. \frac{dI}{d\alpha} \right|_{\alpha=0} = \iint_D \left\{ 2\, \text{grad}\, v \cdot (\boldsymbol{\kappa}\, \text{grad}\, u_0) + 2\rho v u_0 - 2vf \right\} dx\, dy$$

$$+ \int_{C_2} (2\sigma v u_0 - 2hv)\, ds$$

$$= 0 \text{ at the stationary point of } I.$$

Thus, using the first form of Green's theorem, eqn (2.42),

$$\iint_D v \left\{ -\text{div}(\boldsymbol{\kappa}\, \text{grad}\, u_0) + \rho u_0 - f \right\} dx\, dy$$

$$+ \int_{C_1} v(\boldsymbol{\kappa}\, \text{grad}\, u_0) \cdot \mathbf{n}\, ds + \int_{C_2} v(\sigma u_0 - h)\, ds = 0,$$

i.e.

$$\iint_D v \left\{ -\text{div}(\boldsymbol{\kappa}\, \text{grad}\, u_0) + \rho u_0 - f \right\} dx\, dy$$

$$+ \int_{C_1} v(\boldsymbol{\kappa}\, \text{grad}\, u_0) \cdot \mathbf{n}\, ds + \int_{C_2} v(\boldsymbol{\kappa}\, \text{grad}\, u_0) \cdot \mathbf{n} + \sigma u_0 - h)\, ds = 0.$$

Each term vanishes separately, since u_0 satisfies $\mathcal{L}u = f$ and the boundary condition (2.41), and v must satisfy the homogeneous Dirichlet condition $v = 0$ on C_1, in order that the trial function $\tilde{u}$ satisfies the non-homogeneous condition on C_1. Thus the condition (2.41) is a natural boundary condition for the functional (2.44).

Solution 2.12 Consider variations about u_0 given by

$$\tilde{u} = u_0 + \alpha v.$$

Then

$$\frac{dI}{d\alpha} = \iint_D \{2 \operatorname{grad}(u_0 + \alpha v) \cdot \operatorname{grad} v - 2vf\} \, dx \, dy$$

$$+ \int_{C_1} 2 \left\{ (g - u_0)\frac{\partial v}{\partial n} - v\frac{\partial u_0}{\partial n} - 2\alpha v\frac{\partial v}{\partial n} \right\} ds$$

$$+ \int_{C_2} \{2\sigma v(u_0 + \alpha v) - 2hv\} \, ds.$$

Thus, using the first form of Green's theorem,

$$\frac{1}{2}\frac{dI}{d\alpha} = -\iint_D v\left\{\nabla^2(u_0 + \alpha v) + f\right\} dx \, dy$$

$$+ \oint_C \left\{ (g - u_0)\frac{\partial v}{\partial n} - \alpha v\frac{\partial v}{\partial n} \right\} ds$$

$$+ \int_{C_2} \left\{ v\left(\frac{\partial u_0}{\partial n} + \sigma u_0 - h\right) + \alpha\left(\sigma v^2 + v\frac{\partial v}{\partial n}\right) \right\} ds.$$

Now a stationary point for I is given by $(dI/d\alpha)\,|_{\alpha=0}$; thus

$$-\iint_D v\left(\nabla^2 u_0 + f\right) dx \, dy + \int_{C_1} (g - u_0)\frac{\partial v}{\partial n}\,ds + \int_{C_2} v\left(\frac{\partial u_0}{\partial n} + \sigma u_0 - h\right) ds = 0.$$

Since v, and hence $\partial v/\partial n$, is arbitrary, it follows that

$$-\nabla^2 u_0 = f \quad \text{in } D,$$

$$u_0 = g \quad \text{on } C_1$$

and

$$\frac{\partial u_0}{\partial n} + \sigma u_0 = h \quad \text{on } C_2,$$

i.e. both the Robin and the Dirichlet boundary conditions are natural boundary conditions for I.

$$\frac{d^2 I}{d\alpha^2}\bigg|_{\alpha=0} = \iint_D 2 \,|\operatorname{grad} v|^2 \, dx \, dy - \int_{C_1} 4v\frac{\partial v}{\partial n}\,ds + \int_{C_2} 2\sigma v^2 \, ds.$$

Now, in general, it is not true that $-\int_{C_1} 4v\frac{\partial v}{\partial n}\,ds \geq 0$, i.e., for some choice of v, $(d^2 I/d\alpha^2)\,|_{\alpha=0} < 0$, so that the stationary point is not necessarily a minimum.

Solution 2.13 The general solution is $u(x) = A + Bx - x^2 : u(0) = 0$ gives $A = 0$, and $u'(1) = 0$ gives $B = 2$, so that $u_0(x) = 2x - x^2$.

$$I[u_0] = \int_0^1 \{u_0 (-u_0'') - 2u_0 2\} \, dx$$

$$= \int_0^1 -2u_0 \, dx$$

$$= -\frac{4}{3}.$$

The simplest function which satisfies the boundary conditions is

$$\tilde{u}(x) \equiv 0, \qquad I[\tilde{u}_0] = 0.$$

A quadratic function which satisfies both boundary conditions is

$$\tilde{u}_2(x) = x - \frac{1}{2}x^2, \qquad I[\tilde{u}_2] = -1.$$

The function $\tilde{u}_s(x) = \sin(\pi x/2)$ satisfies both boundary conditions:

$$I[\tilde{u}_s] = \pi^3/8 - 8/\pi = 1.329.$$

Clearly,

$$I[u_0] < I[\tilde{u}_1] < I[\tilde{u}_2] < I[\tilde{u}_s].$$

Solution 2.14 Take trial functions as given by eqn (2.50),

$$\tilde{u}_n = g + \sum_{i=1}^n c_i v_i,$$

where $v_i = 0$ on C_1, so that $\tilde{u}_n$ satisfies the essential boundary condition.

In Section 2.7, the Rayleigh–Ritz method was shown to yield the algebraic equations

$$\mathbf{Bc} = \mathbf{g}$$

for the unknown coefficients c_i, where

$$B_{ij} = \iint_D \operatorname{grad} v_i \cdot \operatorname{grad} v_j \, dx \, dy + \int_{C_2} \sigma v_i v_j \, ds,$$

$$g_i = \iint_D f v_i \, dx \, dy + \int_{C_2} h v_i \, ds - \iint_D \operatorname{grad} v_i \cdot \operatorname{grad} g \, ds - \int_{C_2} \sigma v_i g \, ds.$$

The Galerkin method is expressed by eqn (2.23), which gives, on substitution of the trial function $\tilde{u}_n$,

$$\iint_D \left\{ \left(\frac{\partial g}{\partial x} + \sum_j c_j \frac{\partial v_j}{\partial x} \right) \frac{\partial v_i}{\partial x} + \left(\frac{\partial g}{\partial y} + \sum_j c_j \frac{\partial v_j}{\partial y} \right) \frac{\partial v_i}{\partial y} - f v_i \right\} dx\, dy$$

$$+ \int_{C_2} \left\{ \sigma \left(g + \sum_j c_j v_j \right) - h \right\} v_i\, ds = 0, \qquad i = 0, \ldots, n,$$

i.e.

$$\sum_{i=0}^n B_{ij} c_i = g_i, \qquad i = 0, \ldots, n,$$

where the coefficients B_{ij} and g_i are identical with those obtained from the Rayleigh–Ritz method, i.e. the two methods yield the same approximate solution.

Solution 2.15 *Rayleigh–Ritz method.*

$$A_{00} = \int_0^1 \int_0^1 (y^2 + x^2)\, dx\, dy = \frac{2}{3},$$

$$A_{01} = A_{10} = \int_0^1 \int_0^1 \{y(2xy + y^2) + x(x^2 + 2xy)\}\, dx\, dy = \frac{7}{6},$$

$$A_{11} = \int_0^1 \int_0^1 \{(2xy + y^2)^2 + (x^2 + 2xy)^2\}\, dx\, dy = \frac{103}{45},$$

$$A_{02} = \int_0^1 \int_0^1 (y2x + 2yx)\, dx\, dy = 1,$$

$$A_{12} = \int_0^1 \int_0^1 \{(2xy + y^2)2x + (x^2 + 2xy)2y\}\, dx\, dy = 2,$$

$$S_{ij} = 0 \quad (i = 0, 1, 2;\ j = 0, 1, 2, 3),$$

$$h_0 = \int_0^1 \int_0^1 \{2(x + y) - 4\} xy\, dx\, dy = -\frac{1}{3},$$

$$h_1 = \int_0^1 \int_0^1 \{2(x + y) - 4\} xy(x + y)\, dx\, dy = -\frac{7}{18},$$

$$k_0 = \int_0^1 1y(2 - 2y - y^2)\, dy + \int_0^1 1x(2 - 2x - x^2)(-dx) = \frac{1}{6},$$

$$k_1 = \int_0^1 1y(1 + y)(2 - 2y - y^2)\, dy + \int_0^1 1x(2 - 2x - x^2)(-dx) = \frac{1}{10}.$$

Thus, for the trial function $\tilde{u}_A$, we have

$$\frac{2}{3}c_0 = -\frac{1}{3} + \frac{1}{6} - 1,$$

i.e.

$$c_0 = -\frac{7}{4},$$

so that

$$\tilde{u}_A(x, y) = x^2 + y^2 - \frac{7}{4}xy.$$

For the trial function $\tilde{u}_B$, we have

$$\begin{bmatrix} \frac{2}{3} & \frac{7}{6} \\ \frac{7}{6} & \frac{103}{45} \end{bmatrix} \begin{bmatrix} c_1 \\ c_2 \end{bmatrix} = \begin{bmatrix} -\frac{1}{3} + \frac{1}{6} - 1 \\ -\frac{7}{18} + \frac{1}{10} - 2 \end{bmatrix},$$

which yields $c_0 = 0, c_1 = -1$. Thus

$$\tilde{u}_B(x, y) = x^2 + y^2 - xy(x + y).$$

Galerkin method.

$$\int_0^1 \int_0^1 \left\{ 2x + c_0 y + c_1(2xy + y^2) \right\} dx\, dy$$

$$+ \int_0^1 \int_0^1 \left\{ 2y + c_0 x + c_1(x^2 + 2xy) \right\} x\, dx\, dy$$

$$- \int_0^1 \int_0^1 \left\{ 2(x + y) - 4 \right\} xy\, dx\, dy$$

$$- \left\{ \int_0^1 (2 - 2y - y^2) 1y\, dy + \int_0^1 (2 - 2x - x^2) 1x(-dx) \right\} = 0,$$

$$\int_0^1 \int_0^1 \left\{ 2x + c_0 y + c_1(2xy + y^2) \right\} (2xy + y^2)\, dx\, dy$$

$$+ \int_0^1 \int_0^1 \left\{ 2y + c_0 x + c_1(x^2 + 2xy) \right\} (x^2 + 2xy)\, dx\, dy$$

$$- \int_0^1 \int_0^1 \left\{ 2(x + y) - 4 \right\} xy(x + y)\, dx\, dy$$

$$- \left\{ \int_0^1 (2 - 2y - y^2) 1y(1 + y)\, dy \right.$$

$$\left. + \int_0^1 (2 - 2x - x^2) 1x(1 + x)(-dx) \right\} = 0.$$

Thus

$$\begin{bmatrix} 1 \\ 2 \end{bmatrix} + \begin{bmatrix} \frac{2}{3} & \frac{7}{6} \\ \frac{7}{6} & \frac{103}{45} \end{bmatrix} \begin{bmatrix} c_0 \\ c_1 \end{bmatrix} - \begin{bmatrix} -\frac{1}{3} \\ -\frac{7}{18} \end{bmatrix} - \begin{bmatrix} \frac{1}{6} \\ \frac{1}{10} \end{bmatrix} = \begin{bmatrix} 0 \\ 0 \end{bmatrix},$$

i.e. the Rayleigh–Ritz equations are identical with the Galerkin equations and thus yield the same approximate solutions $\tilde{u}_A$ and $\tilde{u}_B$, as expected.

Notice that $\tilde{u}_B$ is in fact the exact solution, and the reason that it is recovered is that the exact solution is a linear combination of the chosen basis functions, just as in Exercise 2.3.

Solution 2.16 The required functional is given by eqn (2.65), where

$$\mathbf{u} = [u] \quad \text{and} \quad \mathbf{f} = [-f].$$

The strain energy density comprises two terms, one due to the elastic foundation and the other due to the bending of the beam. These give

$$W = \frac{1}{2}ku^2 + \frac{1}{2}EI(u'')^2.$$

Thus

$$I[u] = \frac{1}{2}\int_0^1 \left\{ ku^2 + EI(u'')^2 + 2uf \right\} dx.$$

The factor $\frac{1}{2}$ does not affect the function u_0 which minimizes I, and may thus be disregarded; I is then given by eqn (2.77).

3 The finite element method for elliptic problems

3.1 Difficulties associated with the application of weighted residual methods

Although the weighted residual methods introduced in Chapter 2 have been used with success in many areas of physics and engineering, there are certain difficulties which prevent them being more widely used for the solution of practical problems.

One obvious problem involves the choice of trial functions. It is clear that for an irregular-shaped boundary, such as in Fig. 3.1, it would in general be impossible to find one function, let alone a sequence of functions, which satisfies every essential boundary condition. Thus, immediately, the class of problems amenable to solution by this method is restricted to those problems with a 'simple' geometry.

Even if the geometry is suitable and a sequence of functions satisying essential boundary conditions is available, these functions are usually polynomials. It is not difficult to appreciate that, in general, very high-order polynomials would be required to approach the exact behaviour of the unknown over the whole region. A worse situation than this, however, concerns the case of discontinuous material properties.

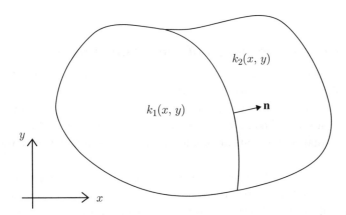

Fig. 3.1 A general curved region in two dimensions in which there is an interface between differing media.

Consider, for example, Poisson's equation

(3.1) $-\text{div}(k \text{ grad } u) = f(x, y),$

where $k(x, y)$ is a function of position which is discontinuous across some interface in the region of interest; see Fig. 3.1. We shall refer to equations such as this as Poisson's equation. Strictly speaking, Poisson's equation is $-\nabla^2 u = f$, i.e. $k = 1$. Such a situation arises in electromagnetic theory at the junction of two dielectrics with different permittivities. In these cases $\partial u/\partial n$, the normal derivative of u, would be discontinuous across the interface and polynomials, which are continuously differentiable, would not be suitable for accurate description of this situation.

In weighted residual methods, all parts of the region are treated with the same degree of importance, no undue attention being paid to areas which may in fact be of more interest than the rest.

Finally, weighted residual methods include coupling between points which are distant from one another, even though this coupling is weak. This yields dense matrices in the final analysis and is costly in terms of computer storage.

In this chapter, the ideas behind the weighted residual method are extended so that the above-mentioned difficulties may be overcome. In fact, we shall restrict ourselves to the Galerkin approach; the use of a variational approach will be discussed in Chapter 5.

Our examples in Chapter 2 suggest that the Galerkin approach is the most accurate in some sense. However, the importance of the method will be appreciated when we reach Chapters 5 and 7, where we shall see that it yields exactly the same equations as does the variational method and hence we can make specific statements about convergence and accuracy.

3.2   Piecewise application of the Galerkin method

We consider an approach in which the region of interest is subdivided into a finite set of *elements*, connected together at a set of points called the *nodes*. In each of these elements, the function behaviour is considered individually and then an overall set of equations is assembled from the individual components. These individual components are found by a piecewise application of the Galerkin method.

The distinction between element numbering and nodal numbering can sometimes lead to confusion, and there is no generally accepted notation. We shall adopt the following: subscripts will refer to nodal numbers and superscripts to element numbers; where it is important to distinguish between them, we shall write i for 'node i' and $[i]$ for 'element i'.

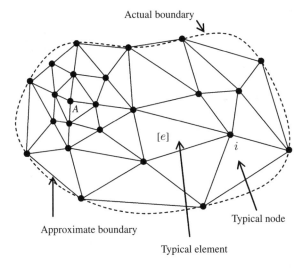

Fig. 3.2 Finite element idealization of a two-dimensional region using triangular elements, with a fine mesh-grading in the region of node A.

We can see that the difficulties encountered in the pure Galerkin approach are now overcome. Since the boundary geometry may be approximated as closely as required by a polygonal arc, or by polyhedra in three dimensions, even the most irregular boundaries are easily approximated; see Fig. 3.2. In Chapter 4, it will be shown how to use curved elements and get an even better boundary approximation. In the examples that follow, it will be seen that the enforcement of essential boundary conditions presents no problems. Because the trial functions are defined in a piecewise manner, discontinuities in normal derivatives over interfaces between different media are easily accounted for and relatively low-degree polynomials may be used to obtain suitable accuracy throughout the region; see Fig. 3.3. Also, grading of the finite element mesh may be used to concentrate on regions of specific interest; see Fig. 3.2 again. Finally, the piecewise nature of the application of the Galerkin method ensures that the influence of one element is limited to those elements actually connected to it, thus uncoupling remote regions. This yields sparse matrices when the overall system of equations is assembled, and hence the computational advantages of sparse matrices become available.

3.3 Terminology

The solution of boundary-value problems such as that given by eqns (2.1) and (2.39) frequently represents a quantity associated with a scalar field such as a potential. Consequently, we often refer to such problems as *field problems*.

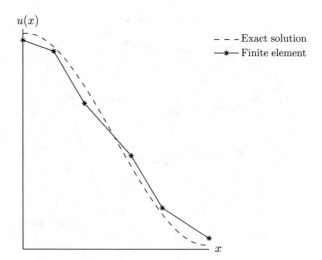

Fig. 3.3 Hypothetical finite element approximation to the solution of a two-point boundary-value problem using linear trial functions.

Because the finite element method was developed in its computational form by structural engineers (Argyris 1964, Zienkiewicz and Cheung 1965), the structural terminology has remained in the generalization to field problems. In Section 3.6 we shall develop equations of the form

$$KU = F,$$

where U is a vector of nodal variables, i.e. values of $u, \partial u/\partial x, \partial u/\partial y$ etc., evaluated at the nodes. The number of nodal variables associated with a particular node is often called the number of *degrees of freedom* at that node. u^e is the vector of element nodal variables. K and F are called the overall *stiffness matrix* and the overall *force vector*, respectively, and are assembled from element matrices k^e and f^e.

f^e is a column vector of known quantities obtained from the non-homogeneous terms in the boundary-value problem under consideration. F is the column vector of known quantities for the overall system. f^e is called the element force vector, and k^e is called the element stiffness matrix.

In some problems the field variables, q, are related to the potential function, u, by equations such as

$$q = -\operatorname{grad} u.$$

These relationships then yield, after the finite element analysis, equations of the form

$$q = SU.$$

S is called the overall *stress* matrix.

In recent years, some of the structural terminology has been lost and the matrices **K** and **F** are referred to according to the physical properties from which they are derived; for example, in thermal problems, **K** is sometimes called the conductivity matrix and **F** is referred to simply as the source term.

3.4 Finite element idealization

As in Chapter 2, two-dimensional problems will be considered for illustrative purposes; extensions of the method to three dimensions are, in principle, straightforward.

Consider the field problem

$$\mathcal{L}u = f \quad \text{in } D \tag{3.2}$$

with

$$\mathcal{B}u = g \quad \text{on } C. \tag{3.3}$$

$\mathcal{L}$ and $\mathcal{B}$ are differential operators. The finite element method seeks an approximation, $\tilde{u}(x, y)$, to the exact solution, $u(x, y)$, in a piecewise manner, the approximation being sought in each of a total of E elements. Thus, in the general element $[e]$, an approximation $\tilde{u}^e(x, y)$ is sought in such a manner that outside $[e]$,

$$\tilde{u}^e(x, y) = 0, \qquad e = 1, \ldots, E. \tag{3.4}$$

For example, in Fig. 3.4, $\tilde{u}^e(x_A, y_A)$ is in general non-zero, but $\tilde{u}^e(x_B, y_B) = 0$.

Using eqn (3.4), it follows that the approximate solution may be written as

$$\tilde{u}(x, y) = \sum_e \tilde{u}^e(x, y), \tag{3.5}$$

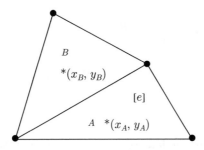

Fig. 3.4 A general element $[e]$ containing a point A (x_A, y_A). The point B (x_B, y_B) is not in $[e]$.

where the summation is taken over all the elements. The reason for writing the approximate solution in this way is not immediately apparent, but it makes setting up the overall system of equations a little easier to understand.

The first decision to be made is to choose a suitable subdivision of the region into finite elements which sufficiently approximates the boundary geometry. This is very much a matter of choice for the user. More will be said about this when we deal with individual elements in Sections 3.7 and 3.8 and with the isoparametric concept in Chapter 4. It also affects the accuracy of the solution, and this will be discussed in Chapter 6.

Once the discretization has been decided on, the next choice is that of the representation of the element approximation, $\tilde{u}^e(x, y)$, in terms of the weighted residual parameters. The most common form of approximation in each element is polynomial approximation. This is probably due to the fact that polynomials are relatively easily manipulated, both algebraically and computationally. Polynomials are also attractive from the point of view of the Weierstrass approximation theorem (Wade 1995), which states that any continuous function may be approximated arbitrarily closely by a suitable polynomial. The choice of polynomial is a matter for the user, but some guidelines are useful.

1. The number of terms in the polynomial must be equal to the total number of degrees of freedom associated with the element, otherwise the polynomial many not be unique. Thus, for a triangular element with three nodes, one degree of freedom at each node, a three-term polynomial is used, giving

$$\tilde{u}^e(x, y) = c_0 + c_1 x + c_2 y = [1 \ x \ y][c_0 \ c_1 \ c_2]^T.$$

   For an element with four nodes and two degrees of freedom at each node, an eight-term polynomial is used, giving

$$\tilde{u}^e(x, y) = c_0 + c_1 x + c_2 y + c_3 x^2 + c_4 xy + c_5 y^2 + c_6 x^2 y + c_7 xy^2$$
$$= [1 \ x \ y \ x^2 \ xy \ y^2 \ x^2 y \ xy^2][c_0, \ldots, c_7]^T.$$

   The coefficients c_i are called *generalized coordinates* and in each case the approximation is of the form

   (3.6) $\tilde{u}^e(x, y) = \mathbf{P}^e(x, y)\mathbf{\Delta}^e,$

   where $\mathbf{P}^e(x, y)$ is a row vector of linearly independent functions and $\mathbf{\Delta}^e$ is a column vector of constants.

2. There should be no preference for either the x or the y direction. This is often referred to by saying that the approximation must have *geometrical invariance*.

3. There are two other requirements, which will be dealt with in Chapter 7. The structural description of these requirements is that the element must be able to exactly reproduce rigid-body motions and constant-strain deformations. Mathematically, these requirements say that to ensure convergence of the method the unknown must be continuous and must be allowed to assume any arbitrary linear form.

Although it is not easy to give definite rules which are applicable in all cases, it is in general better not to retain high-order terms at the expense of lower ones. For this reason, complete polynomials are often favoured. Complete polynomials are those in which all possible terms up to any given degree are present. The necessary terms for all possible polynomials up to a complete quintic are shown in Table 3.1.

Thus a complete linear polynomial is of the form

$$c_0 + c_1 x + c_2 y$$

and requires an element with three degrees of freedom to uniquely define c_0, c_1, c_2. A complete cubic polynomial is of the form

$$c_0 + c_1 x + c_2 y + c_3 x^2 + c_4 xy + c_5 y^2 + c_6 x^3 + c_7 x^2 y + c_8 xy^2 + c_9 y^3,$$

and this requires an element with ten degrees of freedom to uniquely define $c_0, \ldots, c_9$.

Although the form given by eqn (3.6) is of the type used in the pure weighted residual approach of Chapter 2, it is in fact better to use the nodal degrees of freedom as parameters rather than the generalized coordinates. Suppose element $[e]$ has p nodes with one degree of freedom at each node; see Fig. 3.5. Using eqn (3.6), the approximation in element $[e]$ is given by

$$\tilde{u}^e(x, y) = \mathbf{P}^e(x, y)\mathbf{\Delta}^e.$$

Table 3.1 Complete polynomials up to order 5

					1					
				x		y				
			x^2		xy		y^2			
		x^2		x^2y		xy^2		y^3		
	x^4		x^3y		x^2y^2		xy^3		y^4	
x^5		x^4y		x^3y^2		x^2y^3		xy^4		y^5

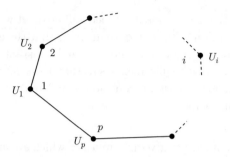

Fig. 3.5 Element $[e]$ with p nodes; U_i is the nodal variable associated with node i.

Therefore

$$U_i = \tilde{u}^e(x_i, y_i)$$
$$= \begin{bmatrix} 1 & x_i & y_i & x_i^2 & x_i y_i & y_i^2 & \cdots \end{bmatrix} \begin{bmatrix} c_0 \ldots c_p \end{bmatrix}^T, \quad i = 1, \ldots, p.$$

Thus the element vector $\mathbf{U}^e = \begin{bmatrix} U_1 \ldots U_p \end{bmatrix}^T$ may be written in the form

(3.7) $$\mathbf{U}^e = \mathbf{C}\boldsymbol{\Delta}^e,$$

where

$$\mathbf{C} = \begin{bmatrix} 1 & x_1 & y_1 & x_1^2 & \cdots \\ 1 & x_2 & y_2 & x_2^2 & \cdots \\ \vdots & \vdots & \vdots & \vdots & \\ 1 & x_p & y_p & x_p^2 & \cdots \end{bmatrix}.$$

Then

(3.8) $$\boldsymbol{\Delta}^e = \mathbf{C}^{-1}\mathbf{U}^e.$$

Thus, for this element, eqn (3.6) yields

$$\tilde{u}^e(x, y) = \mathbf{P}^e \mathbf{C}^{-1}\mathbf{U}^e,$$

i.e.

(3.9) $$\tilde{u}^e(x, y) = \mathbf{N}^e\mathbf{U}^e,$$

where $\mathbf{N}^e = \mathbf{P}^e\mathbf{C}^{-1}$ is called the *shape function matrix* for element $[e]$. It relates the unknown function $\tilde{u}^e(x, y)$ to the nodal variables given by $\mathbf{U}^e$. Outside element $[e]$, $\mathbf{N}^e \equiv \mathbf{0}$, so that eqn (3.4) is satisfied. There is a serious drawback to this approach, however, in that the matrix $\mathbf{C}$ has to be inverted, which may be computationally costly. Also, the matrix itself is often ill-conditioned and indeed may even be singular; see Exercise 3.1.

It is better to obtain eqn (3.9) directly by interpolation throughout the element. This is done by choosing a suitable set of interpolation polynomials $N_i^e(x, y)$, with the property that if (x_j, y_j) are the coordinates of node j, then

(3.10) $$N_i^e(x_j, y_j) = \delta_{ij}, \qquad i, j = 1, \ldots, p.$$

The shape function matrix is then given by

(3.11) $$\mathbf{N}^e(x, y) = \begin{bmatrix} N_1^e(x, y) & N_2^e(x, y) & \cdots & N_p^e(x, y) \end{bmatrix}.$$

The shape functions must be such that conditions 1, 2 and 3 are satisfied. Condition 3 leads to relationships between N_i^e and N_j^e; see Exercise 3.2.

It is often helpful in setting up the shape functions, as well as in the following analysis, to use a set of local coordinates, say (ξ, η), for each element rather than the global coordinates (x, y). This will be illustrated in the examples in Sections 3.5–3.8. Indeed, when we deal with isoparametric elements in Chapter 4, this procedure is essential.

At this stage, the discretization and the manner of element approximation have been decided. The next step in the procedure is to use the Galerkin method to set up the linear equations for the nodal variables U_i, $i = 1, \ldots, n$, and perhaps the derivatives $(\partial u/\partial x)_i$, $(\partial u/\partial y)_i$, etc. This particular step involves two distinct sets of operations; setting up the equation for the individual elements, followed by the assembly of the overall system of equations. It is best demonstrated by way of examples, and these will be illustrated in the following sections.

The solution of the linear equations yields the overall vector of unknowns

$$\mathbf{U} = [U_1 \ U_2 \ \ldots \ U_n]^T$$

as the approximate solution of eqn (3.1) at the nodal points. The approximate solution at non-nodal points is given by interpolation in each element as

$$\tilde{u}(x, y) = \sum_e \tilde{u}^e(x, y) = \sum_e \mathbf{N}^e(x, y)\mathbf{U}^e.$$

Finally, once the nodal values have been obtained, there may be field variables to detemine; for example, if $u(x, y)$ is the velocity potential for a fluid flow, then the velocity vector at any point (x, y) is given by

$$\mathbf{q}(x, y) = -\mathrm{grad}\, u.$$

In particular, in element $[e]$,

$$\mathbf{q}^e = \begin{bmatrix} -\partial/\partial x \\ -\partial/\partial y \end{bmatrix} \tilde{u}^e$$

$$= \begin{bmatrix} -\partial \mathbf{N}^e/\partial x \\ -\partial \mathbf{N}^e/\partial y \end{bmatrix} \mathbf{U}^e,$$

i.e.

$$\text{(3.12)} \qquad \mathbf{q}^e(x, y) = \mathbf{S}^e \mathbf{U}^e,$$

where

$$\text{(3.13)} \qquad \mathbf{S}^e = [-\partial \mathbf{N}^e / \partial x \quad -\partial \mathbf{N}^e / \partial y]^T$$

is the element *stress matrix*.

For problems involving material anisotropy described by a tensor $\boldsymbol{\kappa}$, it may be shown (see Exercise 3.8) that

$$\text{(3.14)} \qquad \mathbf{S}^e = \boldsymbol{\kappa} [-\partial \mathbf{N}^e / \partial x \quad -\partial \mathbf{N}^e / \partial y]^T.$$

So far, we have dealt with the approximation expressed by eqn (3.5) in an element-by-element manner, since the approach is to develop the region as a set of elements. It is, however, more helpful to consider the approximation (3.5) in a node-by-node manner. The shape functions $N_i^e(x, y)$ have the property that $N_i^e(x, y) = 0$ if $i \in [e]$, so that we can define a nodal function $w_i(x, y)$ which is local to node i, given by

$$\text{(3.15)} \qquad w_i(x, y) = \sum_{[e] \ni j} N_j^e(x, y).$$

So, instead of using an element-by-element approximation as in eqn (3.5), we use a node-by-node approximation

$$\text{(3.16)} \qquad \tilde{u}(x, y) = \sum_{j=1}^{n} w_j(x, y) U_j,$$

where n is the number of nodes in the finite element approximation. The set of nodal functions is thus a linearly independent set of basis functions for the approximation.

3.5 Illustrative problem involving one independent variable

The reason for the success of the finite element method is that it may be applied to problems of great complexity. However, to use such problems for illustrative purposes tends to obscure the underlying ideas. In this section, a simple problem involving an ordinary differential equation is solved. This is not to suggest that the finite element method is the best method for solving such problems; in fact, there are other numerical methods available which are superior. However, this particular problem involves only a small amount of algebra and the 'mechanics' of the method come through very well.

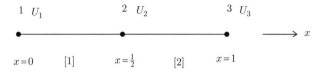

Fig. 3.6 Finite element idealization for the two-point boundary-value problem (3.17), showing the elements 1 and 2 and the global node numbering 1, 2, 3.

Example 3.1 Consider the two-point boundary-value problem

(3.17)
$$-u'' = 2, \qquad 0 < x < 1,$$
$$u(0) = u'(1) = 0.$$

Consider a two-element discretization of $[0, 1]$ with three nodes, as shown in Fig. 3.6, and suppose that there is one degree of freedom at each node.

   In each element there are just two nodal variables, and hence the interpolation polynomials for each element must be linear. Consider an element $[e]$ with midpoint x_m and length h, suppose that the nodes associated with element $[e]$ have local labels A, B, and that ξ is a local coordinate as shown in Fig. 3.7 and given by

(3.18)
$$\xi = \frac{2}{h}(x - x_m).$$

   The shape function matrix is

$$\mathbf{N}^e(\xi) = [N_A^e(\xi)\ N_B^e(\xi)],$$

where

$$N_A^e(-1) = N_B^e(1) = 1$$

and

$$N_A^e(1) = N_B^e(-1) = 0;$$

see Fig. 3.8.

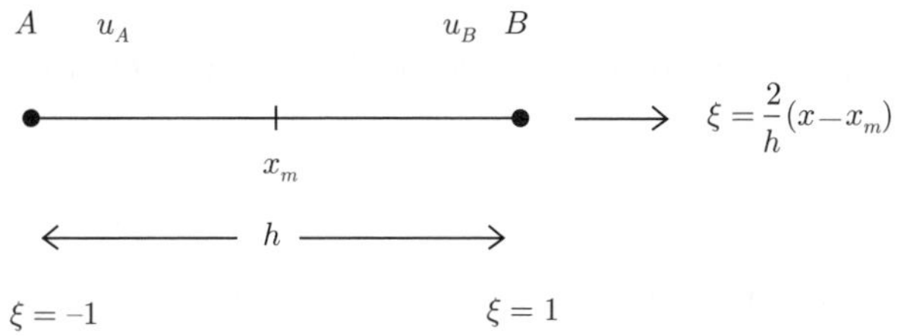

Fig. 3.7 Element $[e]$, showing the local node labels A,B and the local coordinate ξ.

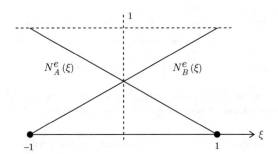

Fig. 3.8 Linear shape functions $N_A^e(\xi)$ and $N_B^e(\xi)$.

The shape functions are easily recognized as the Lagrange interpolation polynomials

(3.19) $$N_A^e(\xi) = \tfrac{1}{2}(1 - \xi), \quad N_B^e(\xi) = \tfrac{1}{2}(1 + \xi).$$

Since the interpolation is linear, this is often called a *linear element*. Then, in element $[e]$,

(3.20) $$\mathbf{U}^e = \mathbf{N}^e \mathbf{U}^e$$

with

(3.21) $$\mathbf{U}^e = [U_A \quad U_B]^T.$$

Therefore the overall finite element approximation as given by eqn (3.5) is

(3.22) $$\tilde{u} = \sum_{e=1}^{2} \tilde{u}^e.$$

In this case we have an essential Dirichlet condition at node 1 and a natural Neumann condition at node 3. Consequently, we must take $U_1 = u(0)$ and we have just two unknown nodal variables, U_2 and U_3.

The corresponding nodal functions are

$$w_1(x) = N_A^1(x), \qquad w_2(x) = N_B^1(x) + N_A^2(x), \qquad w_3(x) = N_B^2(x),$$

where the global coordinate x is related to the local coordinate ξ by eqn (3.18), and

(3.23) $$\tilde{u}(x) = w_1(x)U_1 + w_2(x)U_2 + w_3(x)U_3.$$

The Galerkin formulation is obtained by using a weighted residual approach, taking the basis functions w_2 and w_3 as the weighting functions; we need only those two functions associated with the unknowns U_2 and U_3:

$$\int_0^1 (-\tilde{u}'' - 2)w_i \, dx = 0, \quad i = 2, 3.$$

We integrate by parts, to reduce the order of derivative required:

$$[-\tilde{u}'w_i]_0^1 + \int_0^1 \tilde{u}'w_i' \, dx - \int_0^1 2w_i \, dx = 0, \quad i = 2, 3.$$

We now write down the two equations as follows:

$$i = 2: \quad \left[-\frac{d\tilde{u}}{dx}w_2 \right]_0^1 + \int_0^1 \frac{d\tilde{u}}{dx}\frac{dw_2}{dx} \, dx - \int_0^1 2w_2 \, dx = 0.$$

Now, $w_2(0) = 0$ since $N_B^1(0) = 0$ and $N_A^2(x) \equiv 0$ in element [1], and $w_2(1) = 0$ since $N_B^1(x) \equiv 0$ in element [2] and $N_A^2(1) = 0$. Hence the first term is zero and we have

$$\int_0^{1/2} \frac{d\tilde{u}}{dx}\frac{dw_2}{dx} \, dx + \int_{1/2}^1 \frac{d\tilde{u}}{dx}\frac{dw_2}{dx} \, dx - \int_0^1 2w_2 \, dx = 0,$$

i.e.

$$\int_0^{1/2} \left(\frac{dw_1}{dx}U_1 + \frac{dw_2}{dx}U_2 \right) \frac{dw_2}{dx} \, dx - \int_0^{1/2} 2w_2 \, dx$$

$$+ \int_{1/2}^1 \left(\frac{dw_2}{dx}U_2 + \frac{dw_3}{dx}U_3 \right) \frac{dw_2}{dx} \, dx - \int_{\frac{1}{2}}^1 2w_2 \, dx = 0;$$

$$i = 3: \quad \left[-\frac{d\tilde{u}}{dx}w_3 \right]_0^1 + \int_0^1 \frac{d\tilde{u}}{dx}\frac{dw_3}{dx} \, dx - \int_0^1 2w_3 \, dx = 0.$$

Now, $w_3(x) \equiv 0$ in element [1] and the Neumann condition $u'(0) = 0$ is a natural condition. Hence the first term is zero and we have

$$\int_{1/2}^1 \left(\frac{dw_2}{dx}U_2 + \frac{dw_3}{dx}U_3 \right) \frac{dw_3}{dx} \, dx - \int_{1/2}^1 2w_3 \, dx = 0.$$

It follows then that our two equations are of the form

(3.24)
$$k_{21}^1 U_1 + (k_{22}^1 + k_{11}^2)U_2 + k_{12}^2 U_3 = f_2^1 + f_1^2,$$
$$k_{21}^2 U_2 + k_{22}^2 U_3 = f_2^2,$$

where the element stiffness matrices and force vectors are given respectively by

(3.25)
$$k_{ij}^e = \int_{[e]} \frac{dN_i^e}{dx}\frac{dN_j^e}{dx} \, dx$$

$$\text{and} \quad f_i^e = \int_{[e]} 2N_i \, dx.$$

Now, $U_1 = u(0)$, a known value, so it follows that eqns (3.24) are sufficent to find U_2 and U_3, and our problem is solved. However, since the idea will be helpful in a more general setting, we develop the equation associated with the weighting function $w_1(x)$. This equation has the form

(3.26) $$k_{11}^1 U_1 + k_{12}^1 U_2 = f_1^1,$$

using the notation of eqn (3.25).

Consequently, the overall system of equations may be written as

(3.27) $$\mathbf{KU} = \mathbf{F},$$

where

$$\mathbf{K} = \begin{bmatrix} k_{11}^1 & k_{12}^1 & 0 \\ k_{21}^1 & k_{22}^1 + k_{11}^2 & k_{12}^2 \\ 0 & k_{21}^2 & k_{22}^2 \end{bmatrix} \quad \text{and} \quad \mathbf{F} = \begin{bmatrix} f_1^1 \\ f_2^1 + f_1^2 \\ f_2^2 \end{bmatrix}.$$

Before proceeding further, some remarks regarding the overall stiffness matrix may be made here, since they are generally applicable.

1. It is clearly symmetric, as indeed are the element stiffness matrices.
2. $K_{13} = K_{31} = 0$, showing that there is no coupling between nodes 1 and 3, i.e. the only coupling occurs between nodes associated with the same element. It is not difficult to see, then, that for a large number of elements $\mathbf{K}$ becomes sparse and banded.
3. $\mathbf{K}$ is singular. This, perhaps, is not so obvious. In structural terms, this simply says that $\mathbf{K}$ allows rigid-body motions, i.e. the structure is not fixed. This then suggests how the 'singularity' in $\mathbf{K}$ may be interpreted from the point of view of the solution of the boundary-value problem (3.17). Fixing a structure requires prescribing certain displacements, usually equal to zero, and this is equivalent to enforcing a Dirichlet boundary condition. By adding eqn (3.26) to the set of equations (3.24) we are, in essence, not enforcing the essential boundary condition. Before we see how to remove the singularity in $\mathbf{K}$, the element stiffness and forces will be evaluated.

Using the local coordinate ξ in the integrations in eqn (3.25) yields

$$k_{ij} = \int_{-1}^{1} \frac{2}{h} \frac{dN_i}{d\xi} \frac{2}{h} \frac{dN_j}{d\xi} \frac{h}{2} d\xi, \qquad i, j = 1, 2.$$

Whence, using eqn (3.19),

$$k_{11} = k_{22} = \frac{1}{h}, \qquad k_{12} = k_{21} = -\frac{1}{h}.$$

Also,

$$f_i = \int_{-1}^{1} 2N_i \frac{h}{2} d\xi, \qquad i = 1, 2,$$

whence

$$f_1 = f_2 = h.$$

A convenient way to write the element matrices is to label the rows and columns according to the nodal variable with which they are associated, as shown below:

$$\mathbf{k}^1 = \frac{1}{h} \begin{matrix} & 1 & 2 \\ \begin{bmatrix} 1 & -1 \\ -1 & 1 \end{bmatrix} & \begin{matrix} 1 \\ 2 \end{matrix} \end{matrix}, \qquad \mathbf{k}^2 = \frac{1}{h} \begin{matrix} & 2 & 3 \\ \begin{bmatrix} 1 & -1 \\ -1 & 1 \end{bmatrix} & \begin{matrix} 2 \\ 3 \end{matrix} \end{matrix},$$

$$\mathbf{f}^1 = h \begin{matrix} 1 & 2 \\ [\, 1 & 1 \,] \end{matrix}^T, \qquad \mathbf{f}^2 = h \begin{matrix} 2 & 3 \\ [\, 1 & 1 \,] \end{matrix}^T.$$

The overall stiffness and force matrices are thus

$$\mathbf{K} = \frac{1}{h} \begin{matrix} & 1 & 2 & 3 \\ \begin{bmatrix} 1 & -1 & 0 \\ -1 & 1+1 & -1 \\ 0 & -1 & 1 \end{bmatrix} & \begin{matrix} 1 \\ 2 \\ 3 \end{matrix} \end{matrix}$$

and

$$\mathbf{F} = h \begin{matrix} 1 & 2 & 3 \\ [\, 1 & 1+1 & 1 \,] \end{matrix}^T.$$

Notice that in $\mathbf{K}$ the contribution to the 2,2 position, and in $\mathbf{F}$ the contribution to the 2,1 position, is made up from two terms, one from each element. This is typical of the way that the overall matrices are assembled in general. If a node is associated with more than one element, then the contribution to the relevant parts of the overall matrix is merely a matter of addition of the corresponding terms. This will be discussed in Section 3.6. Notice also in this case that the element matrices have been defined in terms of h, the element length. Consequently, it is possible to solve this problem using elements of different lengths without the necessity of altering the analysis; see Exercise 3.3.

The equations (3.27) now become, with $h = \frac{1}{2}$,

$$(3.28) \qquad \frac{1}{12} \begin{bmatrix} 1 & -1 & 0 \\ -1 & 2 & -1 \\ 0 & -1 & 1 \end{bmatrix} \begin{bmatrix} U_1 \\ U_2 \\ U_3 \end{bmatrix} = \frac{1}{2} \begin{bmatrix} 1 \\ 2 \\ 1 \end{bmatrix}.$$

The next step in the procedure is to solve eqn (3.28) for the unknown nodal variables. As was seen earlier, the overall stiffness matrix is singular, so that a solution of the equations is not possible as they stand. However, the boundary conditions have still to be imposed, and we return to the original set of equations (3.24), which we obtain by removing row 1 from $\mathbf{K}$ and setting $U_1 = u(0)$ where appropriate to obtain

$$\text{(3.29)} \qquad \begin{aligned} 4U_2 - 2U_3 &= 1, \\ -2U_2 + 2U_3 &= \tfrac{1}{2}, \end{aligned}$$

which gives $U_2 = \tfrac{3}{4}$, $U_3 = 1$.

To express the solution $\tilde{u}(x)$, it is usually more convenient to return to the element-by-element form, eqns (3.5) and (3.20):

$$\tilde{u}^1(x) = \tfrac{1}{2}\left[1 - \xi \; 1 + \xi\right]\left[0 \; \tfrac{3}{4}\right]^T$$

$$= \frac{3x}{2}, \qquad \text{since } \xi = \frac{2}{\frac{1}{2}}\left(x - \tfrac{1}{4}\right) \text{ in element 1,}$$

and

$$\tilde{u}^2(x) = \tfrac{1}{2}\left[1 - \xi \; 1 + \xi\right]\left[\tfrac{3}{4} \; 1\right]^T$$

$$= \frac{x+1}{2}, \qquad \text{since } \xi = \frac{2}{\frac{1}{2}}\left(x - \tfrac{3}{4}\right) \text{ in element 2.}$$

Then eqn (3.22) yields the finite element solution as

$$\tilde{u}(x) = \begin{cases} 3x/2, & 0 \le x \le \tfrac{1}{2}, \\ (x+1)/2, & \tfrac{1}{2} \le x \le 1. \end{cases}$$

This is compared with a four-element solution and the exact solution in Fig. 3.9.

Now consider a four-element solution of eqn (3.17), with the discretization as shown in Fig. 3.10. In this case, with all elements of length $\tfrac{1}{4}$, the element stiffness matrices are

$$\mathbf{k}^1 = 4\begin{bmatrix} 1 & -1 \\ -1 & 1 \end{bmatrix}\begin{smallmatrix} 1 \\ 2 \end{smallmatrix}, \qquad \mathbf{k}^2 = 4\begin{bmatrix} 1 & -1 \\ -1 & 1 \end{bmatrix}\begin{smallmatrix} 2 \\ 3 \end{smallmatrix},$$

$$\mathbf{k}^3 = 4\begin{bmatrix} 1 & -1 \\ -1 & 1 \end{bmatrix}\begin{smallmatrix} 3 \\ 4 \end{smallmatrix}, \qquad \mathbf{k}^4 = 4\begin{bmatrix} 1 & -1 \\ -1 & 1 \end{bmatrix}\begin{smallmatrix} 4 \\ 4 \end{smallmatrix},$$

and the element force vectors are

$$\mathbf{f}^1 = \tfrac{1}{4}\begin{bmatrix} 1 \\ 1 \end{bmatrix}\begin{smallmatrix} 1 \\ 2 \end{smallmatrix}, \qquad \mathbf{f}^2 = \tfrac{1}{4}\begin{bmatrix} 1 \\ 1 \end{bmatrix}\begin{smallmatrix} 2 \\ 3 \end{smallmatrix}, \qquad \mathbf{f}^3 = \tfrac{1}{4}\begin{bmatrix} 1 \\ 1 \end{bmatrix}\begin{smallmatrix} 3 \\ 4 \end{smallmatrix}, \qquad \mathbf{f}^4 = \tfrac{1}{4}\begin{bmatrix} 1 \\ 1 \end{bmatrix}\begin{smallmatrix} 4 \\ 5 \end{smallmatrix},$$

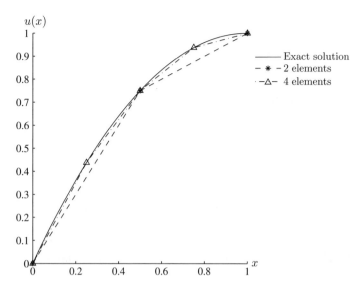

Fig. 3.9 Comparison of the two- and four-element solutions of eqn (3.17) with the exact solution $u_0(x) = 2x - x^2$.

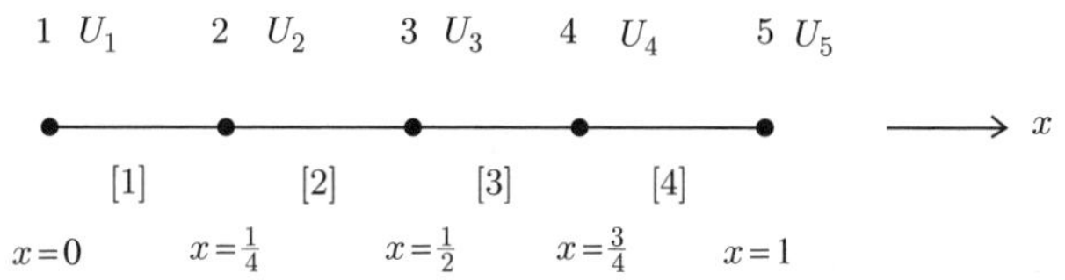

1 U_1 2 U_2 3 U_3 4 U_4 5 U_5

[1]   [2]   [3]   [4]

$x = 0$ $x = \frac{1}{4}$ $x = \frac{1}{2}$ $x = \frac{3}{4}$ $x = 1$

Fig. 3.10 Four-element idealization for the two-point boundary-value problem (3.17).

whence the overall system is

$$
4 \begin{bmatrix} 1 & -1 & 0 & 0 & 0 \\ & 2 & -1 & 0 & 0 \\ & & 2 & -1 & 0 \\ & & & 2 & -1 \\ & \text{symmetric} & & & 1 \end{bmatrix} \begin{bmatrix} U_1 \\ U_2 \\ U_3 \\ U_4 \\ U_5 \end{bmatrix} = \frac{1}{4} \begin{bmatrix} 1 \\ 2 \\ 2 \\ 2 \\ 1 \end{bmatrix}.
$$

Enforcing the homogeneous Dirichlet boundary condition $U_1 = 0$ then yields the set of equations

$$
\begin{aligned}
8U_2 - 4U_3 &= \tfrac{1}{2}, \\
-4U_2 + 8U_3 - 4U_4 &= \tfrac{1}{2}, \\
-4U_3 + 8U_4 - 4U_5 &= \tfrac{1}{2}, \\
-4U_4 + 4U_5 &= \tfrac{1}{4}.
\end{aligned}
$$

The solution is

$$U_2 = \tfrac{7}{16}, \quad U_3 = \tfrac{3}{4}, \quad U_4 = \tfrac{15}{16}, \quad U_5 = 1.$$

Then, as in the two-element case, the approximate solution throughout the interval $[0, 1]$ is found from eqns (3.20) and (3.22). It is easily seen to be given by

$$\mathbf{U}(x) = \begin{cases} \tfrac{7}{4}x, & 0 \leq x \leq \tfrac{1}{4}, \\ \tfrac{1}{8}(10x + 1), & \tfrac{1}{4} \leq x \leq \tfrac{1}{2}, \\ \tfrac{1}{8}(6x + 3), & \tfrac{1}{2} \leq x \leq \tfrac{3}{4}, \\ \tfrac{1}{4}(x + 3), & \tfrac{3}{4} \leq x \leq 1. \end{cases}$$

This solution is compared with the two-element solution and the exact solution in Fig. 3.9. There is also a comparison with three other finite element solutions in Table 3.2.

It should be noted here that the right-hand side in eqn (3.17) is a constant, so that the element force vector need only be found for element $[e]$. In general, it would be necessary to obtain the element force vectors separately for each element.

The exact solution of eqn (3.17) is

$$u_0(x) = 2x - x^2.$$

From Fig. 3.9 it is easy to see that the refined mesh, with four elements, gives a better approximation than does the original coarse two-element mesh. Indeed, this is the basis of the method; the more elements, the better the solution. In practice, a suitable number of elements is chosen to give satisfactory accuracy. Notice in this case that the convergence to the exact solution is monotonic; this is due to the fact that the refined mesh contains the original mesh as a subset. This is discussed in Chapter 7. Notice also that the mesh with graded elements

Table 3.2 A comparison of the four finite element solutions to the problem in Example 3.1 and Exercise 3.3

x	0.1	0.2	0.3	0.4	0.5	0.6	0.7	0.8	0.9	1.0
2 elements	0.15	0.3	0.45	0.6	0.75	0.8	0.85	0.9	0.95	1
4 elements	0.175	0.35	0.5	0.625	0.75	0.825	0.9	0.95	0.975	1
5 elements	0.18	0.36	0.5	0.64	0.74	0.84	0.91	0.96	0.99	1
10 elements	0.19	0.36	0.51	0.64	0.75	0.84	0.91	0.96	0.99	1
4 graded elements	0.188	0.356	0.5	0.625	0.75	0.8	0.85	0.9	0.95	1
Exact solution	0.19	0.36	0.51	0.64	0.75	0.84	0.91	0.96	0.99	1

gives better results than the equivalent ungraded mesh in the region in which u_0 is changing most rapidly.

There is one other point that may be made by reference to this example, which relates to the case of a non-homogeneous boundary condition. For a non-homogeneous Dirichlet condition, for example

$$u(0) = 1,$$

the overall set of equations is treated slightly differently. Consider the two-element approximation equations (3.28) as set up earlier. For the homogeneous Dirichlet condition $U_1 = 0$, the row and column corresponding to U_1 in the stiffness matrix were deleted, leaving eqn (3.29); in the non-homogeneous case this is not done. The first equation in eqn (3.28) is replaced by $U_1 = 1$, leaving a reduced set of two equations.

For a non-homogeneous Neumann boundary condition, for example

$$u'(1) = 1,$$

the term $[-(d\tilde{u}/dx)w_3]_0^1$ is now replaced by $-u'(1)w_3(1)$, since the Neumann condition is a natural boundary condition and $w_3(x) \equiv 0$ in element 1. This gives rise to a term $[0 \; -1]^T$ on the left-hand side of the reduced set of equations as follows:

$$\begin{bmatrix} 0 \\ -1 \end{bmatrix} + 2 \begin{bmatrix} -1 & 2 & -1 \\ 0 & -1 & 1 \end{bmatrix} \begin{bmatrix} 1 \\ U_2 \\ U_3 \end{bmatrix} = \tfrac{1}{2} \begin{bmatrix} 2 \\ 1 \end{bmatrix},$$

which yields

$$\begin{bmatrix} 4 & -2 \\ -2 & 2 \end{bmatrix} \begin{bmatrix} U_2 \\ U_3 \end{bmatrix} = \begin{bmatrix} 1 \\ \tfrac{1}{2} \end{bmatrix} - 2 \begin{bmatrix} -1 \\ 0 \end{bmatrix} - \begin{bmatrix} 0 \\ -1 \end{bmatrix},$$

whence $U_2 = \tfrac{9}{4}$, $U_3 = 3$; see the similar procedure adopted in the classical Rayleigh–Ritz method in Section 2.6. The resulting finite element solution is then

$$\tilde{u}(x) = \begin{cases} 1 + \tfrac{5}{2}x, & 0 \le x \le \tfrac{1}{2}, \\ \tfrac{3}{2} + \tfrac{3}{2}x, & \tfrac{1}{2} \le x \le 1. \end{cases}$$

We note that $\tilde{u}'(1) = \tfrac{3}{2}$, a relatively poor approximation to $u'(1) = 1$, but we have used only two elements.

This simple problem was deliberately chosen and worked through in detail to illustrate how the method works, step-by-step. Since the procedure for more general problems is identical with this one, the basic steps will be listed here and the generalization to problems involving partial differential equations will be presented in the next section.

1. Subdivide the region of interest into a finite number of subregions, the finite elements. In Example 3.1, the elements were simple subintervals of the interval $[0, 1]$. In two- and three-dimensional problems, there is a variety of elements to choose from, for example triangles and rectangles in two dimensions, and tetrahedra and rectangular bricks in three dimensions.

2. Choose nodal variables and shape functions so that the function behaviour throughout each element may be obtained. In Example 3.1, the nodal variables were simply the function values U_i, and the shape functions were chosen to give a linear variation for $\tilde{u}^e$. It is not necessary that only function values be determined at the nodes; derivatives may also be taken as nodal variables and suitable shape functions chosen; see Section 4.4.

3. Obtain the element stiffness and force matrices, and the stress matrices if field variables are required. In Chapter 5 we shall show how variational methods may be used as an alternative to the Galerkin method in certain cases.

4. Assemble the overall system of equations from the individual element matrices.

5. Enforce the essential boundary conditions. In Example 3.1, there were two boundary conditions: an essential homogeneous Dirichlet condition, which was enforced at this stage, and a natural homogeneous Neumann condition, which was handled automatically.

6. Solve the overall system of equations. In Example 3.1, four elements only were used and the resulting system of equations was easily solved by hand; for more equations, efficient computational methods must be used.

7. Compute further results. In many practical examples, field variables must be found from the function $\tilde{u}^e$. These are obtained using the overall stress matrix.

Example 3.2

$$-u'' = e^x,$$

$$u(0) = 1, \qquad u'(1) = 2.$$

We illustrate the use of a spreadsheet to solve this problem with five equal elements. Firstly, with the usual notation, we have element stiffness matrices

$$\mathbf{k}^e = \frac{1}{h} \begin{bmatrix} 1 & -1 \\ -1 & 1 \end{bmatrix}.$$

The element force vectors are given by (see Exercise 3.5)

$$e^{x_m} \begin{bmatrix} (1 + 2/h)\sinh(h/2) - \cosh(h/2) \\ (1 - 2/h)\sinh(h/2) + \cosh(h/2) \end{bmatrix} = e^{x_m} \begin{bmatrix} H_1 \\ H_2 \end{bmatrix}.$$

	A	B	C	D	E	F	G
1	K						F
2	10	-5	0	0	0		5.245096
3	-5	10	-5	0	0		0.299361
4	0	-5	10	-5	0		0.36564
5	0	0	-5	10	-5		0.446594
6	0	0	0	-5	5		2.254577
7	K_inverse						U
8	0.2	0.2	0.2	0.2	0.2		1.722254
9	0.2	0.4	0.4	0.4	0.4		2.395488
10	0.2	0.4	0.6	0.6	0.6		3.00885
11	0.2	0.4	0.6	0.8	0.8		3.549085
12	0.2	0.4	0.6	0.8	1		4

Fig. 3.11 Spreadsheet solution for the problem of Example 3.2.

Table 3.3 Comparison of the finite element solution with the exact solution for the problem of Example 3.2

x	0.2	0.4	0.6	0.8	1
5 elements	1.7223	2.3955	3.0089	3.5491	4
Exact solution	1.7223	2.3955	3.0089	3.5491	4

By analogy with Example 3.1, the overall system of equations is

$$
\begin{bmatrix} 0 \\ 0 \\ 0 \\ 0 \\ -2 \end{bmatrix} + \frac{1}{5} \begin{bmatrix} -1 & 2 & -1 & 0 & 0 & 0 \\ & -1 & 2 & -1 & 0 & 0 \\ & & -1 & 2 & -1 & 0 \\ & & & -1 & 2 & -1 \\ \text{symmetric} & & & & -1 & 1 \end{bmatrix} \begin{bmatrix} 1 \\ U_2 \\ U_3 \\ U_4 \\ U_5 \\ U_6 \end{bmatrix} = \begin{bmatrix} e^{x_1} H_2 + e^{x_2} H_1 \\ e^{x_2} H_2 + e^{x_3} H_1 \\ e^{x_3} H_2 + e^{x_4} H_1 \\ e^{x_4} H_2 + e^{x_5} H_1 \\ e^{x_5} H_2 \end{bmatrix},
$$

i.e.

$$
\mathbf{KU} = \mathbf{F}.
$$

The spreadsheet solution is shown in Fig. 3.11, and a comparison of the finite element solution with the exact solution, $u_0(x) = 2 + (2 + e)x - e^x$, is shown in Table 3.3, where we note that at the nodes the exact solution is recovered; we also notice this in Table 3.2. This phenomenon, called *superconvergence*, will be discussed in Section 7.1.

3.6 Finite element equations for Poisson's equation

In this section, the ideas outlined in Section 3.4 and illustrated with a simple problem in Section 3.5 are used to set up the element matrices for the solution of Poisson's equation in a general two-dimensional region D, subject to non-homogeneous Dirichlet or Robin conditions on the boundary curve C. It will be assumed that there is only degree of freedom at each node, namely U_i. The theory follows an identical argument when derivatives are also taken as nodal variables; the only change in practice is that different shape functions must be used. This will be discussed in Chapter 4.

Consider the following problem:

$$(3.30) \qquad\qquad -\mathrm{div}(k\,\mathrm{grad}\,u) = f(x,y) \quad \text{in } D$$

with the Dirichlet boundary condition

$$(3.31) \qquad\qquad u = g(s) \quad \text{on } C_1$$

and the Robin boundary condition

$$(3.32) \qquad\qquad k(s)\frac{\partial u}{\partial n} + \sigma(s)u = h(s) \quad \text{on } C_2.$$

A Neumann condition is obtained from the Robin condition as a special case when $\sigma \equiv 0$. In this case $k(x,y)$ is a scalar function of position, so that the problem is isotropic. In principle, it is not difficult to take anisotropy into account, in which case the material properties are defined by a tensor represented by κ (see eqn (2.5) and Exercise 3.7). The two-dimensional nature of the problem means that the finite element process is a little more complicated than that for one-dimensional problems. We shall illustrate the process by considering triangular elements with three nodes. It is very easy to generate elements of any shape with any number of nodes. A typical region is shown in Fig. 3.12.

The nodal numbering shown in Fig. 3.12 represents a *global* numbering system. It is convenient to set up a local numbering system as we did for the one-dimensional case in Section 3.5. The local system is shown in Fig. 3.13.

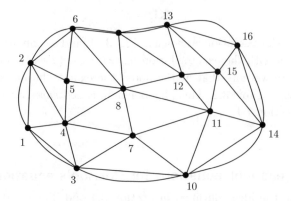

Fig. 3.12 Discretization of the two-dimensional region D using triangular elements.

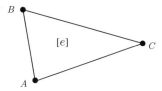

Fig. 3.13 Local node numbers A, B, C for the element $[e]$.

The element interpolation is given by

(3.33) $$\tilde{u}^e = N_A^e U_A + N_B^e U_B + N_C^e U_C,$$

where N_A^e, N_B^e and N_C^e are suitable interpolation polynomials, the shape functions.

Just as for one-dimensional elements, the shape functions have the property

$$N_i^e = \begin{cases} 0, & i \notin [e], \\ N_i^e(x, y), & i \in [e]. \end{cases}$$

The overall approximation in an element-by-element sense is

(3.34)
$$\tilde{u} = \sum_{e=1}^{E} \tilde{u}^e$$
$$= \sum_{e=1}^{E} (N_A^e U_A + N_B^e U_B + N_C^e U_C),$$

where E is the total number of elements.

In order to write our approximation in a node-by-node sense, we proceed as follows.

Suppose that node j is contained in the elements $[p]$, $[q]$, $[r]$ and $[s]$ as shown in Fig. 3.14. If we expand the summation on the right-hand side of eqn (3.34) and concentrate on node j, we have

$$\tilde{u} = \ldots + N_B^q U_B^q + \ldots + N_C^r U_C^r + \ldots N_A^s U_A^s + \ldots N_A^p U_A^p + \ldots$$
$$= \ldots (N_B^q + N_C^r + N_A^s + N_A^p) U_j + \ldots,$$

i.e.

(3.35) $$\tilde{u} = \sum_{j=1}^{n} w_j(x, y) U_j,$$

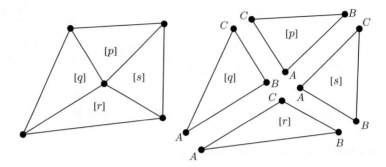

Fig. 3.14 Global node j in elements $[p]$, $[q]$, $[r]$ and $[s]$ with local node numbering.

where n is the total number of nodes and

$$(3.36) \qquad w_j(x, y) = N_B^q + N_C^r + N_A^s + N_A^p = \sum_{e \ni j} N_j^e.$$

The set $\{w_j(x, y)\}$ is a set of linearly independent nodal functions (*cf.* eqns (3.15) and (3.16)).

Suppose that there are N nodes in $C_2 \cup D$ and M nodes on C_1 so that $n = N + M$, and we choose the global numbering such that $j = 1, \ldots, N$ for the nodes in $C_2 \cup D$ and $j = N + 1, \ldots, n$ for nodes on C_1. Remember that the Dirichlet condition (3.31) is an essential condition, so we set $U_j = h(s_j)$ for $j = N + 1, \ldots, n$. Correspondingly, we do not set up the weighted residual equation associated with w_j; this is equivalent to setting $w_j = 0$ for $j = N + 1, \ldots, n$.

We now use the Galerkin method for eqn (3.30) with the weighting function given by eqn (3.36):

$$\iint_D \left(-\operatorname{div}(k \operatorname{grad} \tilde{u}) - f \right) w_i(x, y) \, dx \, dy = 0, \quad i = 1, 2, \ldots, n.$$

Using the first form of Green's theorem, this becomes

$$\iint_D k \operatorname{grad} w_i \cdot \operatorname{grad} \tilde{u} \, dx \, dy - \oint_C k \frac{\partial \tilde{u}}{\partial n} w_i \, ds - \iint_D f w_i \, dx \, dy = 0, \quad i = 1, 2, \ldots, n,$$

i.e.

$$(3.37) \quad \iint_D k \operatorname{grad} w_i \cdot \operatorname{grad} \tilde{u} \, dx \, dy - \int_{C_2} k \frac{\partial \tilde{u}}{\partial n} w_i \, ds - \iint_D f w_i \, dx \, dy = 0, \quad i = 1, 2, \ldots, N,$$

since we have $w_i = 0$ for $i = N + 1, \ldots, n$. Now, on C_2, we have the natural Neumann boundary condition, eqn (3.32), which we approximate by

$$k \frac{\partial \tilde{u}}{\partial n} = h - \sigma \tilde{u},$$

so that eqn (3.37) becomes

$$\iint_D k \operatorname{grad} w_i \cdot \operatorname{grad} \tilde{u} \, dx \, dy + \int_{C_2} \sigma \tilde{u} w_i \, ds = \iint_D f w_i \, dx \, dy + \int_{C_2} g w_i \, ds.$$

Finally, we use the nodal approximation (3.35) to obtain

$$\iint_D k \sum_{j=1}^{N+M} \operatorname{grad} w_i \cdot \operatorname{grad} w_j U_j \, dx \, dy + \int_{C_2} \sigma \sum_{j=1}^{N+M} w_i w_j U_j \, ds$$

$$= \iint_D f w_i \, dx \, dy + \int_{C_2} g w_i \, ds, \quad i = 1, 2, \ldots, N,$$

which we may write

$$\sum_{j=1}^{N+M} K_{ij} U_j = F_i, \quad i = 1, 2, \ldots, N,$$

or, in matrix form,

(3.38) $$\mathbf{K} \mathbf{U} = \mathbf{F},$$

where $\mathbf{K}$ is an $N \times (N + M)$ matrix, which may be developed from the element stiffness matrices as follows:

(3.39)
$$K_{ij} = \iint_D k \operatorname{grad} w_i \cdot \operatorname{grad} w_j \, dx \, dy + \int_{C_2} \sigma w_i w_j \, ds$$
$$= \sum_{[e] \ni j} \iint_{[e]} k \operatorname{grad} N_i^e \cdot \operatorname{grad} N_j^e \, dx \, dy + \sum_{[e] \in C_2} \int_{C_2^e} \sigma N_i^e N_j^e \, ds,$$

where C_2^e is that part of the approximation to C_2 which lies in element $[e]$, if appropriate; see Fig. 3.15.

Hence we may write

$$K_{ij} = \sum_e k_{ij}^e + \sum_{e \in C_2} \bar{k}_{ij}^e,$$

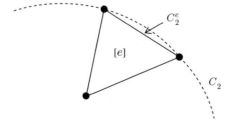

Fig. 3.15 Approximate boundary C_2^e.

where

(3.40)
$$k_{ij}^e = \int\int_{[e]} k \left(\frac{\partial N_i^e}{\partial x} \frac{\partial N_j^e}{\partial x} + \frac{\partial N_i^e}{\partial y} \frac{\partial N_j^e}{\partial y} \right) dx\, dy,$$

(3.41)
$$\bar{k}_{ij}^e = \int_{C_2^e} \sigma N_i^e N_j^e \, ds.$$

Also,

$$F_i = \sum_e f_i^e + \sum_{e \in C_2} \bar{f}_i^e,$$

where

(3.42)
$$f_i^e = \int\int_{[e]} f N_i^e \, dx\, dy,$$

(3.43)
$$\bar{f}_i^e = \int\int_{C_2^e} h N_i^e \, ds.$$

We could have proceeded in just the same way as for the one-dimensional element and included the equations for the weighting function associated with the Dirichlet boundary condition nodes to obtain an $(N + M) \times (N + M)$ square matrix, $\mathbf{K}_O$, which, just as before, would be singular. However, at this stage we shall not do this; instead, we shall work directly with the non-square matrix $\mathbf{K}$ in eqn (3.38). We shall say more about the matrix $\mathbf{K}_O$ at the end of this section.

We know that at nodes $i = N + 1, \ldots, n$ the values of U_i are prescribed, hence we can partition eqn (3.38) as follows:

$$[\mathbf{K}_1 \ \mathbf{K}_2] \begin{bmatrix} \mathbf{U}_1 \\ \mathbf{U}_2 \end{bmatrix} = \mathbf{F},$$

where $\mathbf{U}_1$ is a vector of the known values of u on C, i.e. we may write

$$\mathbf{K}_2 \mathbf{U}_2 = \mathbf{F} - \mathbf{K}_1 \mathbf{U}_1.$$

The $N \times N$ matrix $\mathbf{K}_2$ is called the *reduced* overall stiffness matrix, and it is this with which we shall work.

This overall system of equations we shall write in the form

(3.44)
$$\mathbf{KU} = \mathbf{F},$$

where $\mathbf{K}$ is the reduced stiffness matrix, $\mathbf{U}$ contains only the unknown nodal values and $\mathbf{F}$ is the modified force vector.

Finally, we need to address the existence or otherwise of solutions to eqn (3.44).

Example 3.3 Consider the differential operator given by

$$\mathcal{L}u = -\text{div}(k \text{ grad } u);$$

then

$$\iint_D u\mathcal{L}u \, dx \, dy = \iint_D \text{grad } u \cdot k \text{ grad } u \, dx \, dy - \oint_C uk\frac{\partial u}{\partial n} ds$$

$$= \iint_D k \mid \text{grad } u \mid^2 dx \, dy - \oint_C uk\frac{\partial u}{\partial n} ds.$$

For a homogeneous Dirichlet boundary condition, the boundary integral vanishes and hence $\mathcal{L}$ is positive definite provided that $k > 0$. For a homogeneous Robin boundary condition,

$$k\frac{\partial u}{\partial n} + \sigma u = 0.$$

It is also necessary that $\sigma > 0$ in order that $\mathcal{L}$ is positive definite (*cf.* Example 2.2).

Suppose that $v = \sum_{j=1}^{N} w_j v_j$, where v_j is arbitrary and w_j is the nodal function associated with node j; then

$$\iint_D v\mathcal{L}v \, dx \, dy = \iint_D k \text{ grad } v \cdot \text{grad } v \, dx \, dy - \int_{C_1} vk\frac{\partial v}{\partial n} ds + \int_{C_2} v\sigma v \, ds$$

$$= \iint_D k \sum \text{grad } w_i v_i. \sum \text{grad } w_j v_j \, dx \, dy + \int_{C_2} \sigma \sum w_i v_i \sum w_j v_j \, ds$$

$$= \mathbf{v}^T \left(\iint_D \begin{bmatrix} \partial w_1/\partial x & \partial w_1/\partial y \\ \partial w_2/\partial x & \partial w_2/\partial y \\ \vdots & \vdots \end{bmatrix} \begin{bmatrix} \partial w_1/\partial x & \partial w_2/\partial x & \cdots \\ \partial w_1/\partial y & \partial w_2/\partial y & \cdots \end{bmatrix} dx \, dy \right) \mathbf{v}$$

$$+ \mathbf{v}^T \left(\int_{C_2} \sigma \begin{bmatrix} w_1 \\ w_2 \\ \vdots \end{bmatrix} \begin{bmatrix} w_1 & w_2 & \cdots \end{bmatrix} ds \right) \mathbf{v},$$

i.e.

$$\iint_D v\mathcal{L}v \, dx \, dy = \mathbf{v}^T \mathbf{K} \mathbf{v},$$

where $\mathbf{K}$ is the reduced overall stiffness matrix.

Now we know that, provided $k > 0$ and $\sigma > 0$, $\mathcal{L}$ is positive definite. It follows then that, since $\mathbf{v}$ is arbitrary, under these conditions $\mathbf{K}$ is positive definite. Hence eqn (3.44) has a unique solution.

From a computational point of view, the explicit construction of the overall stiffness matrix from the element matrices is not immediately clear. This can be seen from eqn (3.36), in which the nodal functions w_j are developed in terms of the element shape functions. The shape functions are numbered locally, whereas the nodal functions are numbered globally. We shall now consider how these systems are related.

Suppose that $\mathbf{N}^e$ is the matrix of element shape functions so that

$$\tilde{u}^e = \begin{bmatrix} N_A^e & N_B^e & N_C^e \end{bmatrix} \begin{bmatrix} U_A \\ U_B \\ U_C \end{bmatrix}$$

$$= \mathbf{N}^e \mathbf{U}^e.$$

We shall write

$$\boldsymbol{\alpha}^e = \begin{bmatrix} \partial/\partial x \\ \partial/\partial y \end{bmatrix} \mathbf{N}^e.$$

The element matrices given by eqns (3.40)–(3.43) may be written in matrix form as

$$(3.45) \qquad \mathbf{k}^e = \iint_{[e]} k\boldsymbol{\alpha}^{e^T} \boldsymbol{\alpha}^e \, dx \, dy,$$

$$(3.46) \qquad \bar{\mathbf{k}}^e = \int_{C_2^e} \sigma \mathbf{N}^{e^T} \mathbf{N}^e \, ds,$$

$$(3.47) \qquad \mathbf{f}^e = \iint_{[e]} f \mathbf{N}^{e^T} \, dx \, dy,$$

$$(3.48) \qquad \bar{\mathbf{f}}^e = \int_{C_2^e} h \mathbf{N}^{e^T} \, ds.$$

Consider the element $[e]$ with m nodes $p, q, \ldots, i, \ldots, s$ as shown in Fig. 3.16.

Let us consider the setting up of the $N \times N$ matrix $\mathbf{K}_O$, which is obtained by using all n weighting functions, in which we take all w_j $(j = 1, \ldots, n)$ to be non-zero and given by eqn (3.36). We consider the contribution to $\mathbf{k}^e$ from the terms in eqn (3.39) and we include boundary contributions where appropriate;

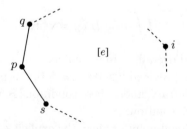

Fig. 3.16 Element $[e]$ with m nodes $p, q, \ldots, i, \ldots, s$.

$$\sum_{[e]\ni j}\iint_{[e]} k \text{ grad } N_i^e \cdot \text{grad } N_j^e \, dx \, dy + \sum_{[e]\in C_2}\int_{C_2} \sigma N_i^e N_j^e \, ds.$$

The only non-zero terms are those associated with nodes $p, q, \ldots, i, \ldots, s$, so the contribution to row i of $\mathbf{K}_O$ is

$$\begin{array}{ccccccc} 1 & 2 & p & q & i & s & n \end{array}$$
$$[0 \; 0 \ldots 0 \; k_{ip}^e \; 0 \ldots 0 \; k_{iq}^e \; 0 \ldots 0 \; k_{ii}^e \; 0 \ldots 0 \; k_{is}^e \; 0 \ldots 0] \; ,$$

and we have the contribution to all rows of $\mathbf{K}_O$ as

$$\begin{bmatrix} & 0 & & \\ & \vdots & & \\ k_{pp}^e + & \cdots & +k_{ps}^e & \\ & \vdots & & \\ k_{qp}^e + & \cdots & +k_{qs}^e & \\ & \vdots & & \\ k_{sp}^e + & \cdots & k_{ss}^e & \\ & \vdots & & \\ & 0 & & \end{bmatrix} \begin{matrix} 1 \\ \vdots \\ p \\ \vdots \\ q \\ \vdots \\ s \\ \vdots \\ n \end{matrix}$$

Now the element vector of nodal values

$$\mathbf{U}^e = [U_p \; U_q \; \cdots \; U_i \; \cdots \; U_s]^T$$

is related to the global vector $\mathbf{U} = [U_1 \; U_2 \; \cdots \; U_n]^T$ by

(3.49) $$\mathbf{U}^e = \mathbf{a}^e \mathbf{U},$$

where $\mathbf{a}$ is an $n \times n$ Boolean matrix given by

$$\mathbf{a}^e = \begin{bmatrix} 1 & 2 & p & q & s & n \\ 0 & 0 & \cdots & 1 & \cdots & 0 & \cdots & 0 & \cdots & 0 \\ 0 & 0 & \cdots & 0 & \cdots & 1 & \cdots & 0 & \cdots & 0 \\ \vdots & \vdots & & \vdots & & \vdots & & \vdots & & \vdots \\ 0 & 0 & \cdots & 0 & \cdots & 0 & \cdots & 1 & \cdots & 0 \end{bmatrix} \begin{matrix} p \\ q \\ \vdots \\ s \end{matrix} \; ,$$

i.e., in Fig. 3.17,

$$\mathbf{U} = [U_1 \ldots U_7]^T,$$

$$\mathbf{U}^3 = [U_3 \; U_5 \; U_6 \; U_7]^T$$

$$= \mathbf{a}^3 \mathbf{U},$$

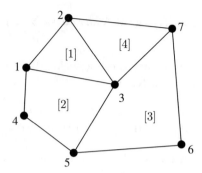

Fig. 3.17 A system of four elements and seven nodes.

where

$$
\mathbf{a}^3 =
\begin{array}{c}
\begin{array}{ccccccc} 1 & 2 & 3 & 4 & 5 & 6 & 7 \end{array} \\
\left[
\begin{array}{ccccccc}
 & & 1 & & & & \\
 & & & & 1 & & \\
 & & & & & 1 & \\
 & & & & & & 1
\end{array}
\right]
\begin{array}{c} 3 \\ 5 \\ 6 \\ 7 \end{array}
\end{array}.
$$

Consequently, we have the following structure for $\mathbf{K}_O$:

$$
\begin{array}{c}
\begin{array}{ccccc} 1 & p & q & s & n \end{array} \\
\left[
\begin{array}{ccccc}
\vdots & \vdots & \vdots & \vdots & \vdots \\
k^e_{pp} & k^e_{pq} & k^e_{ps} & & \\
k^e_{qp} & k^e_{qq} & k^e_{qs} & & \\
k^e_{sp} & k^e_{sq} & k^e_{ss} & & \\
\vdots & \vdots & \vdots & \vdots & \vdots
\end{array}
\right]
\begin{array}{c} \vdots \\ p \\ q \\ s \\ n \end{array}
\end{array},
$$

which may be written

$$
\mathbf{a}^{e^T} \mathbf{k}^e \mathbf{a}^e.
$$

Pre-multiplication by $\mathbf{a}^{e^T}$ affects only the rows of $\mathbf{k}^e$, and post-multiplication by $\mathbf{a}^e$ affects only the columns of $\mathbf{k}^e$. The net effect is to expand $\mathbf{k}^e$ from an $m \times m$ matrix to an $n \times n$ matrix.

Finally, then, we can write

(3.50)
$$
\mathbf{K}_O = \sum_e \mathbf{a}^{e^T} \mathbf{k}^e \mathbf{a}^e.
$$

Similarly, we may obtain

(3.51)
$$
\mathbf{F} = \sum_e \mathbf{a}^{e^T} \mathbf{f}^e.
$$

Equations (3.50) and (3.51) are convenient expressions for the computational development of $\mathbf{K}_O$ and $\mathbf{F}$. From a practical point of view, the system of equations is always produced in terms of the reduced stiffness matrix $\mathbf{K}$ (see eqn (3.44)).

The solution of eqn (3.44) then yields the unknown nodal values. It is not the intention in this text to discuss the methods of solution of the resulting algebraic equations, since this is adequately covered elsewhere (see e.g. Zienkiewicz and Taylor (2000a,b) and Smith and Griffiths (2004) and the references given therein), although we shall make some comments in Chapter 8. It is, however, interesting to note the structure of the equations, since their special form means that particular computational procedures may be used.

1. $\mathbf{K}$ is symmetric.

2. $\mathbf{K}$ is also sparse, since the i, j location contains a non-zero element only when nodes i and j are in the same element. Thus, for a system with a large number of elements, most of the K_{ij} will be zero.

3. $\mathbf{K}$ is positive definite.

4. Finally, suppose that overall the elements, the largest difference in node numbers in any element is $d - 1$; then $\mathbf{K}$ will have a semi-bandwidth d, and for even moderately sized problems the stiffness matrix will have all its non-zero elements banded around the leading diagonal; see Fig. 3.18. The bandwidth depends very much on the numbering system employed for the nodes and may be minimized by a careful choice of node numbering (see Exercise 3.12).

These four properties allow very large systems to be handled for any given amount of computer storage. They are also influential in the reduction of computation time, see Chapter 9.

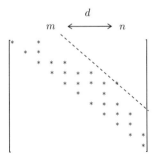

Fig. 3.18 Banded structure of a typical stiffness matrix. If nodes m and n are in the same element and max $| n - m | + 1 = d$, then the semi-bandwidth is d. Only the upper triangle is shown, since the matrix is symmetric.

3.7   A rectangular element for Poisson's equation

The simplest rectangular element is one with just four nodes, one at each corner. Choose local coordinates (ξ, η) as shown in Fig. 3.19. The element superscript e will be dropped in this section. Since there are four nodes with one degree of freedom at each node, the function variation throughout the element is of the following bilinear form:

$$(3.52) \qquad\qquad u(x, y) = c_0 + c_1 x + c_2 y + c_3 xy.$$

Writing this in terms of interpolation polynomials instead of generalized coordinates yields

$$(3.53) \qquad\qquad u(x, y) = \mathbf{NU}$$

$$= \tfrac{1}{4}[(1-\xi)(1-\eta) \ (1+\xi)(1-\eta) \ (1+\xi)(1+\eta) \ (1-\xi)(1+\eta)] \begin{bmatrix} U_1 \\ U_2 \\ U_3 \\ U_4 \end{bmatrix}.$$

Now,

$$\frac{\partial}{\partial x} \equiv \frac{2}{a}\frac{\partial}{\partial \xi} \quad \text{and} \quad \frac{\partial}{\partial y} \equiv \frac{2}{b}\frac{\partial}{\partial \eta}.$$

Thus

$$\alpha = \frac{1}{2}\begin{bmatrix} -\tfrac{1}{a}(1-\eta) & \tfrac{1}{a}(1-\eta) & \tfrac{1}{a}(1+\eta) & -\tfrac{1}{a}(1+\eta) \\ -\tfrac{1}{b}(1-\xi) & -\tfrac{1}{b}(1+\xi) & \tfrac{1}{b}(1+\xi) & \tfrac{1}{b}(1-\xi) \end{bmatrix}.$$

Now,

$$\mathbf{k} = \iint_{[e]} k\boldsymbol{\alpha}^T \boldsymbol{\alpha}\, dx\, dy, \qquad \text{using eqn (3.45).}$$

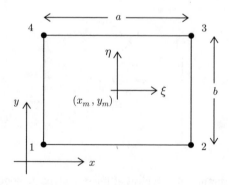

Fig. 3.19 The four-node rectangle. (x_m, y_m) are the coordinates of the centroid of the rectangle, $\xi = (2/a)(x - x_m)$ and $\eta = (2/b)(y - y_m)$.

For the special case $k = $ constant,

$$\mathbf{k} = k \int_{-1}^{1}\int_{-1}^{1} \boldsymbol{\alpha}^T \boldsymbol{\alpha}\frac{a}{2}d\xi\frac{b}{2}d\eta$$

(3.54)

$$= \frac{k}{6}\begin{bmatrix} 2(r+1/r) & r-2/r & -r-1/r & 1/r-2r \\ & 2(r+1/r) & 1/r-2r & -r-1/r \\ & & 2(r+1/r) & r-2/r \\ \text{symmetric} & & & 2(r+1/r) \end{bmatrix}\begin{matrix} 1 \\ 2 \\ 3 \\ 4 \end{matrix},$$

where $r = a/b$ is the aspect ratio of the element. In performing the integrations above, the algebra is simplified by noticing that

$$\int_{-1}^{1} \xi^{2n+1}d\xi = 0 \quad \text{and} \quad \int_{-1}^{1}\xi^{2n}d\xi = \frac{2}{2n+1}.$$

To evaluate the element force vector, consider the special case $f = $ constant. Using eqn (3.47),

(3.55)

$$\mathbf{f} = f \int_{-1}^{1}\int_{-1}^{1} f\mathbf{N}^T\frac{a}{2}d\xi\frac{b}{2}d\eta = \frac{ab}{2}f\begin{bmatrix} 1 & 2 & 3 & 4 \\ 1 & 1 & 1 & 1 \end{bmatrix}^T.$$

In this case $f(x, y)$ is a constant, so that the element force vector is the same for all elements. In general, this will not be the case, since $f(x, y)$ will be different in each element and so will yield a different force vector.

If the element is a boundary element and a non-homogeneous Robin boundary condition holds there, then additions are needed to the stiffness and force matrices as given by eqns (3.46) and (3.48). Suppose for example that nodes 3 and 4 are situated on a boundary along which

$$\partial u/\partial n + \sigma u = h,$$

with σ and h constants. On side 3,4, the arc length s is such that

$$ds = -dx = -\frac{1}{2}d\xi,$$

since the boundary integrals are traversed in a counterclockwise direction. Thus

$$\bar{\mathbf{k}} = \int_{-1}^{1}\sigma\frac{1}{16}\begin{bmatrix} 0 \\ 0 \\ 2(1+\xi) \\ 2(1-\xi) \end{bmatrix}[0 \; 0 \; 2(1+\xi) \; 2(1-\xi)]\left(-\frac{a}{2}\right)d\xi,$$

using eqn (3.46) and remembering that on side 3,4 $\eta = 1$ and that $\xi = -1$ at node 3 and $\xi = 1$ at node 4. Therefore

$$(3.56) \qquad \bar{k} = \frac{a}{6}\sigma \begin{array}{c} \\ \begin{array}{cccc} 1 & 2 & 3 & 4 \end{array} \\ \left[\begin{array}{cccc} 0 & 0 & 0 & 0 \\ 0 & 0 & 0 & 0 \\ 0 & 0 & 2 & 1 \\ 0 & 0 & 1 & 2 \end{array} \right] \begin{array}{c} 1 \\ 2 \\ 3 \\ 4 \end{array} \end{array}.$$

Also, using eqn (3.48),

$$\bar{f} = \int_{-1}^{1} h\frac{1}{4} [0 \quad 0 \quad 2(1+\xi) \quad 2(1-\xi)]^T \left(-\frac{a}{2}\right) d\xi$$

$$(3.57) \qquad = ah \begin{array}{c} \begin{array}{cccc} 1 & 2 & 3 & 4 \end{array} \\ [0 \quad 0 \quad 1 \quad 1]^T. \end{array}$$

Notice that the 'boundary' matrices contain non-zero terms only for those nodes which are themselves boundary nodes. This is, of course, as would be expected.

These results will now be used to obtain a one-element solution of the following boundary-value problem.

Example 3.4 Suppose that u satisfies Laplace's equation

$$\nabla^2 u = 0$$

in a region D which is a square with vertices at $(0,0)$, $(1,0)$, $(1,1)$, $(0,1)$. Suppose also that the boundary conditions are

$$\frac{\partial u}{\partial x} = 0 \quad \text{on} \quad x = 0 \quad \text{and} \quad x = 1,$$

$$\frac{\partial u}{\partial y} + u = 2 \quad \text{on} \quad y = 1,$$

$$u = 1 \quad \text{on} \quad y = 0.$$

We shall consider a one-element solution.

Using eqns (3.54), (3.56), and (3.57), the element matrices are

$$k = \frac{1}{6} \begin{array}{c} \\ \begin{array}{cccc} 1 & 2 & 3 & 4 \end{array} \\ \left[\begin{array}{cccc} 4 & -1 & -2 & -1 \\ & 4 & -1 & -2 \\ & & 4 & -1 \\ \text{sym} & & & 4 \end{array} \right] \begin{array}{c} 1 \\ 2 \\ 3 \\ 4 \end{array} \end{array},$$

$$\bar{k} = \frac{1}{6} \begin{array}{cccc} & \scriptstyle 1 & \scriptstyle 2 & \scriptstyle 3 & \scriptstyle 4 \\ \left[\begin{array}{cccc} 0 & 0 & 0 & 0 \\ & 0 & 0 & 0 \\ & & 2 & 1 \\ \mathrm{sym} & & & 2 \end{array}\right] & \begin{array}{c} \scriptstyle 1 \\ \scriptstyle 2 \\ \scriptstyle 3 \\ \scriptstyle 4 \end{array} \end{array},$$

$$\bar{f} = \begin{array}{cccc} \scriptstyle 1 & \scriptstyle 2 & \scriptstyle 3 & \scriptstyle 4 \\ [\,0 & 0 & 1 & 1\,]^T. \end{array}$$

In this case $f(x, y) \equiv 0$, so that $\mathbf{f} = \mathbf{0}$. Hence

$$K = \frac{1}{6} \begin{array}{cccc} & \scriptstyle 1 & \scriptstyle 2 & \scriptstyle 3 & \scriptstyle 4 \\ \left[\begin{array}{cccc} 4 & -1 & -2 & -1 \\ & 4 & -1 & -2 \\ & & 6 & 0 \\ \mathrm{sym} & & & 6 \end{array}\right] & \begin{array}{c} \scriptstyle 1 \\ \scriptstyle 2 \\ \scriptstyle 3 \\ \scriptstyle 4 \end{array} \end{array}$$

and

$$\mathbf{F} = \begin{array}{cccc} \scriptstyle 1 & \scriptstyle 2 & \scriptstyle 3 & \scriptstyle 4 \\ [\,0 & 0 & 1 & 1\,]^T. \end{array}$$

Thus the overall system of equations, after enforcing the non-homogeneous Dirichlet boundary conditions $U_1 = U_2 = 1$, is

$$\frac{1}{6} \begin{bmatrix} 6 & 0 \\ 0 & 6 \end{bmatrix} \begin{bmatrix} U_3 \\ U_4 \end{bmatrix} = \begin{bmatrix} 1 \\ 1 \end{bmatrix} - \frac{1}{6} \begin{bmatrix} -2 & -1 \\ -1 & -2 \end{bmatrix} \begin{bmatrix} 1 \\ 1 \end{bmatrix};$$

the solution then yields

$$U_3 = U_4 = \tfrac{3}{2}.$$

Interpolation through the element gives

$$\tilde{u}^e(x, y) = \frac{1}{4} \left[(1 - \xi)(1 - \eta)\ (1 + \xi)(1 - \eta)\ (1 + \xi)(1 + \eta)\ (1 - \xi)(1 + \eta) \right]$$

$$\times \begin{bmatrix} 1 \\ 1 \\ \frac{3}{2} \\ \frac{3}{2} \end{bmatrix}$$

$$= \frac{5}{4} + \frac{\eta}{4},$$

i.e. $u^e(x, y) = 1 + y/2$. The exact solution is $u(x, y) = 1 + y/2$. This has been recovered by the finite element solution because it is contained among all the possible forms (3.52).

A disadvantage of this element is that the orientation with respect to the coordinate axes determines whether the solution is continuous across interelement boundaries or not. Consider two adjacent elements whose sides are not parallel to the coordinate axes; see Fig. 3.20. Along the common side AB, we have $y = mx + c$, say, so that the variation of $\tilde{u}^e$ along this side is of the form

$$\tilde{u}^e = \alpha_0 + \alpha_1 x + \alpha_2 x^2,$$

and this must be determined by the nodal values along this side. Now, there are only two nodes on AB, so that the quadratic variation along this side is not unique, i.e., in general, $\tilde{u}^e$ is not continuous across AB except, of course, at the nodes. These elements are said to be non-conforming elements, or incompatible elements; see Fig. 3.21.

For rectangular elements whose sides are parallel to the coordinate axes, $\tilde{u}^e$ is continuous across interelement boundaries, and such elements are said to be conforming elements, or compatible elements. It should be noted here that the rectangular element of Fig. 3.20 can be a conforming element provided that, in each element, $\tilde{u}^e$ is expressed in terms of the local coordinates (ξ, η). In this case $\eta = $ constant along AB, so that along this line $\tilde{u}^e$ is a linear function of ξ and thus is uniquely determined by its values at the nodes A and B.

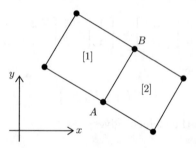

Fig. 3.20 Two adjacent rectangular elements whose sides are not parallel to the coordinate axes.

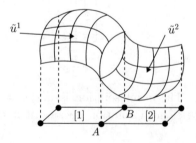

Fig. 3.21 Non-conformity between $\tilde{u}^1$ and $\tilde{u}^2$ along the interelement boundary AB for two rectangular elements whose sides are not parallel to the coordinate axes.

3.8 A triangular element for Poisson's equation

The rectangular element developed in Section 3.7 was shown to be a non-conforming element in certain circumstances, and this is a disadvantage if interelement continuity is required. A second disadvantage, which is probably more important, however, is that rectangular elements are applicable only to problems whose geometry is sufficiently regular; see Fig. 3.22(a). For irregular boundaries, rectangular elements are not appropriate, since it is difficult to approximate the boundary geometry with such elements; see Fig. 3.22(b).

A more versatile element, as far as boundary geometry approximation is concerned, is the triangle, since any curve can be approximated arbitrarily closely by a polygonal arc and the area enclosed by a polygon can be exactly covered by triangles; see Fig. 3.23. This, of course, is also true for rectangles but a larger number will be required to achieve a given accuracy.

It is possible to set up the element matrices using the global coordinates (x, y); however, the algebra is simplified by using a set of triangular coordinates

(a)   (b)

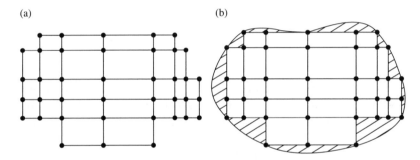

Fig. 3.22 (a) A typical geometry suitable for approximation with rectangular elements. (b) An irregular geometry, for which rectangles are not suitable by virtue of the very poor boundary geometry approximation. The region between the exact boundary and the approximate boundary is the shaded area.

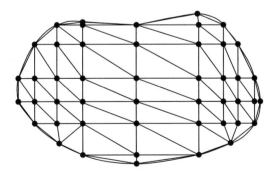

Fig. 3.23 The region of Fig. 3.22(b) approximated with triangular elements using the same number of nodes, showing the much improved boundary approximation.

The Finite Element Method

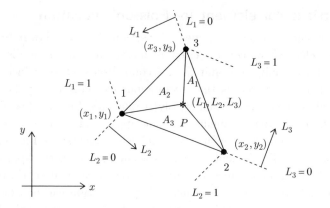

Fig. 3.24 Area coordinates for a triangular element.

(L_1, L_2, L_3) as shown in Fig. 3.24. These coordinates are often called area coordinates.

In Fig. 3.24,

$$(3.58) \qquad L_1 = \frac{A_1}{A}, \qquad L_2 = \frac{A_2}{A}, \qquad L_3 = \frac{A_3}{A},$$

where A is the area of the triangle and A_1, A_2, A_3 are the areas shown there. The position of P may thus be given by the coordinates (L_1, L_2, L_3). It follows that the three coordinates are not independent, since they satisfy the equation

$$(3.59) \qquad L_1 + L_2 + L_3 = 1.$$

The relationship between the global coordinates (x, y) and the local triangular coordinates (L_1, L_2, L_3) is given by

$$(3.60) \qquad x = L_1 x_1 + L_2 x_2 + L_3 x_3,$$

$$(3.61) \qquad y = L_1 y_1 + L_2 y_2 + L_3 y_3.$$

Equations (3.60) and (3.61) may be solved to obtain L_i in terms of x and y as

$$(3.62) \qquad L_i = \frac{(a_i + b_i x + c_i y)}{2A},$$

where the area A is given by

$$(3.63) \qquad A = \frac{1}{2} \begin{vmatrix} 1 & x_1 & y_1 \\ 1 & x_2 & y_2 \\ 1 & x_3 & y_3 \end{vmatrix}.$$

The constants a_i, b_i and c_i are given in terms of the nodal coordinates by

$$a_1 = x_2 y_3 - x_3 y_2,$$

(3.64) $$b_1 = y_2 - y_3,$$

$$c_1 = x_3 - x_2,$$

the others being obtained by cyclic permutation. From eqn (3.62), the following relationships between derivatives may be obtained:

(3.65) $$\frac{\partial L_i}{\partial x} = \frac{b_i}{2A},$$

(3.66) $$\frac{\partial L_i}{\partial y} = \frac{c_i}{2A}.$$

Finally, a result concerning an integral involving the area coordinates is required; the proof is given in Appendix C.

(3.67) $$\iint_A L_1^m L_2^n L_3^p \, dx \, dy = \frac{2Am! \, n! \, p!}{(m+n+p+2)!}.$$

Since the element has three nodes with one degree of freedom at each node, the function variation throughout the element is linear, i.e. it is of the form

(3.68) $$\tilde{u}^e(x,y) = a_0 + a_1 x + a_2 y.$$

Using nodal values and interpolating through the element in the usual manner gives

$$\tilde{u}^e = \mathbf{N}^e \mathbf{U}^e$$

$$= [L_1 \ L_2 \ L_3] \, [U_1 \ U_2 \ U_3]^T.$$

The shape functions are easily found, since the value of L_i at node j is δ_{ij} and each L_i varies linearly with x and y through the element.

Using eqn (3.40), the element stiffness matrix is given by

$$k_{ij} = \iint_A k \left(\frac{\partial L_i}{\partial x} \frac{\partial L_j}{\partial x} + \frac{\partial L_i}{\partial y} \frac{\partial L_j}{\partial y} \right) dx \, dy$$

$$= \iint_A k \left(\frac{b_i b_j}{4A^2} + \frac{c_i c_j}{4A^2} \right) dx \, dy.$$

In the special case $k = \text{constant}$,

(3.69) $$k_{ij} = \frac{k}{4A} (b_i b_j + c_i c_j).$$

The element force vector is given by

$$(3.70) \qquad\qquad f_i = \iint_A L_i \, f(x, y) \, dx \, dy.$$

In the special case $f = $ constant,

$$(3.71) \qquad\qquad f_i = \frac{fA}{3}$$

or, by using eqns (3.60) and (3.61), $f(x, y)$ can be written in terms of L_1, L_2, L_3 and hence f_i may be obtained.

Sometimes it is useful to interpolate $f(x, y)$ through the element from its nodal values; see Exercise 3.11. For example, if $f(x, y) = \sin(x + y)$, it is difficult to evaluate the integrals for f_i.

For boundary elements where the Robin boundary condition (3.32) holds, contributions to the element stiffness and force matrices are required in the following form (see eqns (3.46) and (3.48)):

$$\bar{\mathbf{k}} = \int_{C_2^e} \sigma(s) \mathbf{N}^{e^T} \mathbf{N}^e \, ds,$$

$$\bar{\mathbf{f}} = \int_{C_2^e} h(s) \mathbf{N}^{e^T} \, ds.$$

Suppose for example that side 3,1 is a boundary side; see Fig. 3.25. On side 3,1, $L_2 = 0$ and $s = \left(b_2^2 + c_2^2\right)^{\frac{1}{2}} L_1$, so that

$$ds = \left(b_2^2 + c_2^2\right)^{1/2} dL_1.$$

Then, since $L_3 = 1 - L_1$,

$$(3.72) \qquad \bar{\mathbf{k}} = \int_0^1 \sigma \begin{bmatrix} L_1^2 & 0 & L_1 - L_1^2 \\ 0 & 0 & 0 \\ L_1 - L_1^2 & 0 & (1 - L_1)^2 \end{bmatrix} \left(b_2^2 + c_2^2\right)^{1/2} dL_1$$

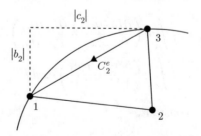

Fig. 3.25 Element with side 3,1 approximating the boundary.

and

(3.73)
$$\bar{\mathbf{f}} = \int_0^1 h\,[L_1 \quad 0 \quad 1 - L_1]^T\,(b_2^2 + c_2^2)^{1/2}\,dL_1.$$

Similar results are obtained by cyclic permutation when sides 1,2 and 2,3 are boundary sides.

These results will now be used to obtain a two-element solution of Example 3.4, which was solved using a single rectangular element.

Example 3.5 In this problem, a solution to Laplace's equation is sought in the square with vertices at $(0,0)$, $(1,0)$, $(1,1)$, $(0,1)$, subject to the boundary conditions

$$\frac{\partial u}{\partial x} = 0 \quad \text{on} \quad x = 0 \quad \text{and} \quad x = 1,$$

$$\frac{\partial u}{\partial y} + \sigma = 2 \quad \text{on} \quad y = 1,$$

$$u = 1 \quad \text{on} \quad y = 0.$$

Suppose that element 1 has nodes 1, 2, 4 and element 2 has nodes 4, 2, 3 as shown in Fig. 3.26. Since $f(x,y) = 0$,

$$\mathbf{f}^1 = \mathbf{f}^2 \equiv \mathbf{0}.$$

For element 1,

$$a_1 = 1, \quad a_2 = 0, \ a_3 = 0,$$
$$b_1 = -1, \ b_2 = 1, \ b_3 = 0,$$
$$c_1 = -1, \ c_2 = 0, \ c_3 = 1,$$

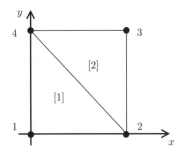

Fig. 3.26 Unit square divided into two triangular elements.

so that

$$
\mathbf{k}^1 = \frac{1}{2}
\begin{bmatrix}
\overset{1}{2} & \overset{2}{-1} & \overset{4}{-1} \\
 & 1 & 0 \\
\text{sym} & & 1
\end{bmatrix}
\begin{matrix} 1 \\ 2 \\ 4 \end{matrix} .
$$

There is no contribution from boundary terms in element 1, since one boundary has a homogeneous Neumann condition and the other requires a non-homogeneous Dirichlet condition.

For element 2,

$$
\begin{aligned}
a_1 &= 1, \quad a_2 = 1, \quad a_3 = -1, \\
b_1 &= -1, \, b_2 = 0, \quad b_3 = 1, \\
c_1 &= 0, \quad c_2 = -1, \, c_3 = 1,
\end{aligned}
$$

so that

$$
\mathbf{k}^1 = \frac{1}{2}
\begin{bmatrix}
\overset{4}{1} & \overset{2}{0} & \overset{3}{-1} \\
 & 1 & -1 \\
\text{sym} & & 2
\end{bmatrix}
\begin{matrix} 4 \\ 2 \\ 3 \end{matrix} .
$$

Notice that this result could easily have been obtained directly from $\mathbf{k}^1$ by symmetry considerations.

Element 2 has a boundary side 3,4, along which there is the Robin condition

$$
\frac{\partial u}{\partial n} + u = 2.
$$

Thus there is a contribution to both the stiffness and the force matrices given by

$$
\bar{\mathbf{k}}^2 = \int_0^1
\begin{bmatrix}
L_1^2 & 0 & L_1 - L_1^2 \\
0 & 0 & 0 \\
\text{sym} & & (1 - L_1)^2
\end{bmatrix}
1 \, dL_1
$$

$$
= \frac{1}{6}
\begin{bmatrix}
\overset{4}{2} & \overset{2}{0} & \overset{3}{1} \\
 & 0 & 0 \\
\text{sym} & & 2
\end{bmatrix}
\begin{matrix} 4 \\ 2 \\ 3 \end{matrix}
$$

and

$$
\bar{\mathbf{f}}^2 = \int_0^1 2 \left[L - 1 \quad 0 \quad -L_1 \right]^T 1 \, dL_1
$$

$$
= \left[\overset{4}{1} \quad \overset{2}{0} \quad \overset{3}{1} \right]^T .
$$

Assembling the overall matrices yields

$$
\mathbf{K} = \frac{1}{6}
\begin{bmatrix}
\overset{1}{6} & \overset{2}{-3} & \overset{3}{0} & \overset{4}{-3} \\[2pt]
 & 6 & -3 & 0 \\[2pt]
 & & 8 & -2 \\[2pt]
\text{sym} & & & 8
\end{bmatrix}
\begin{matrix} 1 \\ 2 \\ 3 \\ 4 \end{matrix}
$$

and

$$
\mathbf{F} = \begin{bmatrix} \overset{1}{0} & \overset{2}{0} & \overset{3}{1} & \overset{4}{1} \end{bmatrix}^T .
$$

Thus the overall system of equations becomes, after enforcing the essential Dirichlet boundary condition $U_1 = U_2 = 1$,

$$
\tfrac{8}{6}U_3 - \tfrac{2}{5}U_4 = 1 - \left(-\tfrac{3}{6}\right)1,
$$

$$
-\tfrac{2}{6}U_3 + \tfrac{8}{6}U_4 = 1 - \left(-\tfrac{3}{6}\right)1,
$$

which yields $U_3 = U_4 = \tfrac{3}{2}$, as before.

The solution at non-nodal points is given by

$$
U^1(x,y) = [L_1 \ L_2 \ L_3] \begin{bmatrix} 1 & 1 & \tfrac{3}{2} \end{bmatrix}^T
$$

$$
= L_1 + L_2 + \frac{3L_3}{2}
$$

$$
= 1 + \frac{L_3}{2}, \quad \text{using eqn (3.59)},
$$

$$
= 1 + \frac{y}{2}, \quad \text{using eqn (3.62)},
$$

and

$$
U^2(x,y) = [L_1 \ L_2 \ L_3] \begin{bmatrix} \tfrac{3}{2} & 1 & \tfrac{3}{2} \end{bmatrix}^T
$$

$$
= \frac{3L_1}{2} + L_2 + \frac{3L_3}{2}
$$

$$
= 1 + \frac{(L_1 + L_2)}{2}, \quad \text{using eqn (3.59)},
$$

$$
= 1 + \frac{y}{2}, \quad \text{using eqn (3.62)}.
$$

Thus this two-element solution again yields the exact solution, which is to be expected since the exact solution is linear.

In Section 3.7 it was seen that rectangular elements are, in general, non-conforming elements. This triangle, however, is always a conforming element.

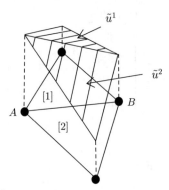

Fig. 3.27 Two adjacent triangular elements with a common side AB showing continuity of u^e across AB.

Consider two adjacent elements as shown in Fig. 3.27. Suppose that the equation of side AB is

$$y = mx + c;$$

then, using eqn (3.68), $\tilde{u}^e$ is given along AB in the form

$$\tilde{u}^e = \alpha_0 + \alpha_1 x,$$

i.e. $\tilde{u}^e$ is linear and hence α_0 and α_1 can be obtained uniquely in terms of the nodal values at A and B. Hence $\tilde{u}^e$ is continuous across the common side, so that the triangular element is a conforming element.

In this chapter, the finite element method has been developed in terms of a piecewise application of the Galerkin weighted residual method, and the general procedure was summarized at the end of Section 3.5. An alternative method of setting up the finite element equations is discussed in Chapter 5.

Two elements have been developed in detail for Poisson's equation. In each of these elements the function variation was linear; such elements are called first-order elements. In a similar manner, higher-order elements, i.e. those with function variation which is quadratic, cubic etc., may be developed using suitable shape functions. These will be discussed in the next chapter.

3.9   Exercises and solutions

Exercise 3.1 A triangular element has nodes at the points $(0, 0)$, $(1, 0)$ and $(0, 1)$. There are three degrees of freedom at each node, these being $u, \partial u/\partial x$ and $\partial u/\partial y$, giving a total of nine degrees of freedom for the element. If u is given throughout the element in terms of the generalized coordinates by

$$u(x, y) = c_0 + c_1 x + c_2 y + c_3 x^2 + c_4 xy + c_5 y^2 + c_6 x^3 + c_7(x^2 y + xy^2) + c_8 y^3,$$

obtain the matrix $\mathbf{C}$, as given by eqn (3.7), and show that it is singular.

Exercise 3.2 The shape functions for a finite element are $N_i^e(x, y)$, $i = 1, \ldots, m$. Show that with these interpolation functions, if the element is capable of recovering an arbitrary linear form for the unknown, then

$$\sum_{i \in e} N_i^e = 1, \quad \sum_{i \in e} N_i^e x_i = x, \quad \sum_{i \in e} N_i^e y_i = y.$$

Exercise 3.3 Consider the two-point boundary-value problem of Example 3.1,

$$-u'' = 2, \quad 0 < x < 1, \quad u(0) = u'(1) = 0.$$

Find finite element solutions using (i) four elements with nodes at $x = 0$, $\frac{1}{8}, \frac{1}{4}, \frac{1}{2}, 1$; (ii) five identical elements; (iii) ten identical elements.

Exercise 3.4 Given the two-point boundary-value problem

$$-u'' = 2, \quad 0 < x < 1, \quad u(0) = 1, \quad u'(1) = 0,$$

find the solution at $x = \frac{1}{4}$ using (i) two elements; (ii) three elements.

Exercise 3.5 Consider the two-point boundary-value problem

$$-u'' = e^x, \quad 0 < x < 1, \quad u(0) = u(1) = 0.$$

Find the solution using (i) four linear elements; (ii) ten linear elements.

Exercise 3.6 Repeat Exercise 3.5, but instead of integrating $\mathbf{N}^{e^T} e^x$ exactly throughout each element, replace e^x by a linear interpolation between its nodal values.

Exercise 3.7 Obtain the element stiffness matrix for the generalized Poisson equation

$$-\mathrm{div}(\kappa \ \mathrm{grad} \ u) = f.$$

Show that for a Robin boundary condition given by $\kappa \ \mathrm{grad} \ u.\mathbf{n} + \sigma u = h$ on C_2, the matrices $\bar{\mathbf{k}}^e$ and $\bar{\mathbf{f}}^e$ are still given by eqns (3.46) and (3.48).

Exercise 3.8 For problems involving material anisotropy, the field variable $\mathbf{q}$ is related to the 'potential' function u by

$$\mathbf{q} = -\kappa \ \mathrm{grad} \ u.$$

Obtain an expression for the element stress matrix.

Exercise 3.9 The one-dimensional form of the Poisson equation is

$$-\frac{d}{dx}\left(k(x)\frac{du}{dx}\right) = f(x).$$

This equation is to be solved using linear elements. In one of these elements, the material property changes discontinuously from k_1 to k_2; see Fig. 3.28. Obtain the stiffness matrix for this element.

Exercise 3.10 For the problem of Exercise 3.3(i), set up the overall stiffness and nodal force matrices using suitable Boolean selection matrices.

Exercise 3.11 Poisson's equation, $-\nabla^2 u = f$, is to be solved using linear triangular elements. By interpolating $f(x, y)$ throughout the element in terms of its nodal values, find the element nodal force vector.

Exercise 3.12 A region is divided into 12 elements with the node numbering as shown in Fig. 3.29. What is the value d of the semi-bandwidth of the overall stiffness matrix?

Renumber the nodes so as to reduce the value of d. What is the minimum possible value of d?

Exercise 3.13 Consider the problem

$$-\nabla^2 u = 2(x + y) - 4$$

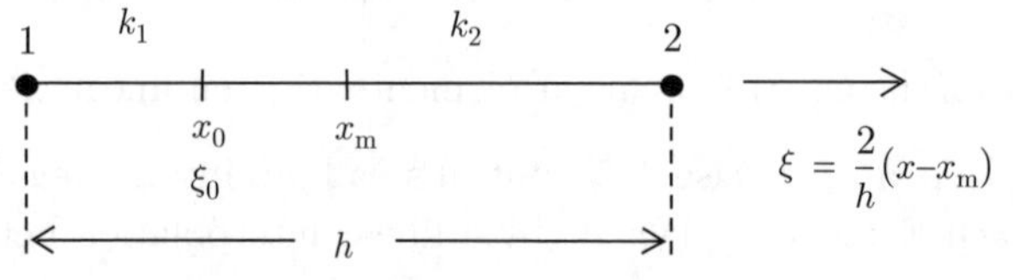

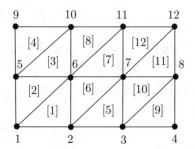

Fig. 3.28 Linear element in which $k(x)$ changes discontinuously at x_0; k_1 and k_2 are constants.

Fig. 3.29 A finite element grid consisting of 12 elements and 12 nodes.

in the square whose vertices are at $(0,0)$, $(1,0)$, $(1,1)$, $(0,1)$. The boundary conditions are

$$u(0,y) = y^2, \quad u(x,0) = x^2, \quad u(1,y) = 1 - y, \quad u(x,1) = 1 - x.$$

Find the finite element solution using four bilinear rectangular elements.

Exercise 3.14 The linear triangle of Section 3.8 is to be used to solve Poisson's equation within a region bounded by a closed curve C. The boundary condition on a part C_2 of the boundary is of the Robin type $\partial u/\partial n + \sigma u = h$. Suppose that element $[e]$ has side 3,1 as part of the polygonal approximation to C (see Fig. 3.25), and σ and h are interpolated along this side in terms of their nodal values. Obtain expressions for $\bar{\mathbf{k}}$ and $\bar{\mathbf{f}}$.

Exercise 3.15 Repeat Exercise 3.13 using four triangular elements.

Exercise 3.16 In this problem, u satisfies Laplace's equation in the region in the first quadrant bounded by the parabola $y = 1 - x^2$ and the coordinate axes. The boundary conditions are as follows.

On the parabola, $u = -1 + 3x - x^2$; on the x-axis, $u = 3x - 2$; and on the y-axis, $\partial u/\partial n + 2u = 2y - 7$. Find the solution using the four triangular elements shown in Fig. 3.30.

Exercise 3.17 Find the solution to Poisson's equation $-\nabla^2 u = 2$ in the region of Exercise 3.16 subject to the boundary conditions $u = -1 + 3x - 2x^2$ on the parabola, $u = -2 + 3x - x^2$ on the x-axis and $\partial u/\partial n + 2u = 2y - 7$ on the y-axis.

Exercise 3.18 Poisson's equation $-\nabla^2 u = 2$ is to be solved using triangular elements. Two different meshes are to be used, parts of which are shown in Fig. 3.31. Obtain row 5 of the overall equilibrium equation in each case. This form of Poisson's equaion is often called the *torsion equation.*

Exercise 3.19 A three-dimensional problem possesses axial symmetry. Obtain the element stiffness matrix for the triangular element shown in Fig. 3.32 when the governing equation is Poisson's equation,

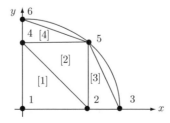

Fig. 3.30 Finite element idealization for the problem of Exercise 3.16.

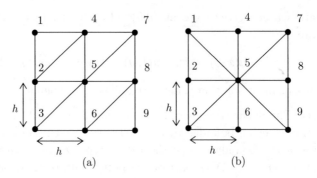

Fig. 3.31 Finite element idealizations of a square region using triangular elements.

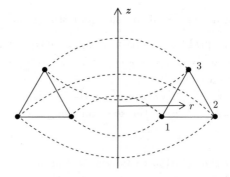

Fig. 3.32 The triangular element for the solution of Poisson's equation with axial symmetry.

(3.74)
$$-\frac{\partial^2 u}{\partial r^2} - \frac{1}{r}\frac{\partial u}{\partial r} - \frac{\partial^2 u}{\partial z^2} = f(r,z),$$

where r and z are the usual cylindrical polar coordinates.

Exercise 3.20 For the axisymmetric problem of Exercise 3.19, obtain the element force vector.

Exercise 3.21 The three-dimensional problem of Exercise 3.19 is subject to a Robin boundary condition

$$\frac{\partial u}{\partial n} + \sigma(S)u = h(S)$$

on a part S_2 of the boundary. If the element $[e]$ has side 3,1 as a boundary side, obtain the contributions to the element stiffness and force matrices from the boundary condition on S_2.

Exercise 3.22 Poisson's equation in three dimensions is to be solved using rectangular brick elements as shown in Fig. 3.33. Using the variables $\xi_0 = \xi\xi_i$,

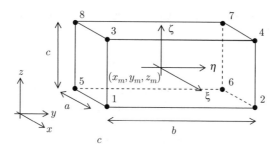

Fig. 3.33 The rectangular brick element; local coordinates are given by $\xi = (2/a)(x - x_m), \eta = (2/b)(y - y_m), \zeta = (2/c)(z - z_m)$.

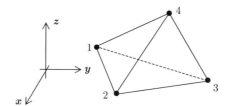

Fig. 3.34 Tetrahedral element with four nodes.

$\eta_0 = \eta\eta_i$ and $\zeta_0 = \zeta\zeta_i$, where (ξ_i, η_i, ζ_i) are the coordinates of node i (Zienkiewicz *et al.* 2005), obtain an expression for the element stiffness matrix.

Exercise 3.23 The three-dimensional equivalent of the plane triangle is the tetrahedral element as shown in Fig. 3.34. Analogous to the area coordinates given by eqns (3.58) and (3.59) are the volume coordinates (L_1, L_2, L_3, L_4). The relationship between the global coordinates and volume coordinates is given by

$$x = x_1 L_1 + x_2 L_2 + x_3 L_3 + x_4 L_4,$$

with similar formulae for y and z; also, $L_1 + L_2 + L_3 + L_4 = 1$.

Given the integration formula

$$\iiint_V L_1^m L_2^n L_3^p L_4^q \, dx \, dy \, dz = \frac{6V m! n! p! q!}{(m+n+p+q+3)!},$$

obtain the element stiffness matrix for Poisson's equation.

Exercise 3.24 Consider eqn (3.2) in the special case $f \equiv \lambda u$. If the boundary conditions are homogeneous, then the problem is an eigenvalue problem

$$\mathcal{L}u = \lambda u \quad \text{in } D, \quad \mathcal{B}u = 0 \quad \text{on } C.$$

Show that the finite element method leads to a generalized matrix eigenvalue problem of the form

$$\mathbf{KU} = \lambda \mathbf{MU},$$

where $\mathbf{K}$ is the usual overall stiffness matrix. $\mathbf{M}$ is often called the overall mass matrix and is obtained from the element mass matrices, which are given by

$$\mathbf{m}^e = \iint_{[e]} \mathbf{N}^{e^T} \mathbf{N}^e \, dx \, dy.$$

Use this result to find an approximation to the lowest eigenvalue for the problem

$$-u'' = \lambda u, \quad 0 < x < 1, \quad u(0) = u(1) = 0,$$

(i) using two elements; (ii) using three elements; (iii) using ten elements.

Solution 3.1

$$u = \begin{bmatrix} 1 & x & y & x^2 & xy & y^2 & x^3 & (x^2 y + y^2 x) & y^3 \end{bmatrix} \begin{bmatrix} c_0 & \cdots & c_8 \end{bmatrix}^T.$$

At node 1, $(0, 0)$,

$$u = c_0, \quad \frac{\partial u}{\partial x} = c_4, \quad \frac{\partial u}{\partial y} = c_2.$$

At node 2, $(1, 0)$,

$$u = c_0 + c_1 + c_3 + c_6, \quad \frac{\partial u}{\partial x} = c_1 + 2c_3 + 3c_6, \quad \frac{\partial u}{\partial y} = c_2 + c_4 + c_7.$$

At node 3, $(0, 1)$,

$$u = c_0 + c_2 + c_5 + c_8, \quad \frac{\partial u}{\partial x} = c_1 + c_4 + c_7, \quad \frac{\partial u}{\partial y} = c_2 + 2c_5 + 3c_8.$$

Thus the non-zero terms in $\mathbf{C}$ are

$$
\begin{array}{c}
\begin{array}{ccccccccc} 1 & 2 & 3 & 4 & 5 & 6 & 7 & 8 & 9 \end{array} \\
\begin{bmatrix}
1 & & & & & & & & \\
 & 1 & & & & & & & \\
 & & 1 & & & & & & \\
1 & 1 & & 1 & & 1 & & & \\
 & 1 & & 2 & & 3 & & & \\
 & & 1 & & 1 & & 1 & & \\
1 & & 1 & & & & 1 & & 1 \\
 & 1 & & & 1 & & & 1 & \\
 & & 1 & & 2 & & & & 3
\end{bmatrix}.
\end{array}
$$

Since columns 5 and 8 are equal, it follows that $\mathbf{C}$ is singular.

Solution 3.2 In terms of the nodal values, the unknown is given in element $[e]$ by

$$\tilde{u}^e = \sum_{i \in e} N_i^e U_i.$$

If the representation is capable of recovering an arbitrary linear form for $\tilde{u}^e$, then it must certainly be able to recover the three forms $\tilde{u}^e \equiv 1$, $\tilde{u}^e \equiv x$ and $\tilde{u}^e \equiv y$. Thus

$$\sum_{i \in e} N_i^e = 1, \quad \sum_{i \in e} N_i^e x_i = x, \quad \sum_{i \in e} N_i^e y_i = y.$$

Solution 3.3 (i) Using the results of Example 3.1,

$$k^1 = \frac{1}{1/8} \begin{bmatrix} 1 & -1 \\ -1 & 1 \end{bmatrix} \begin{matrix} 1 \\ 2 \end{matrix}, \qquad k^2 = \frac{1}{1/8} \begin{bmatrix} 1 & -1 \\ -1 & 1 \end{bmatrix} \begin{matrix} 2 \\ 3 \end{matrix},$$

$$k^3 = \frac{1}{1/4} \begin{bmatrix} 1 & -1 \\ -1 & 1 \end{bmatrix} \begin{matrix} 3 \\ 4 \end{matrix}, \qquad k^4 = \frac{1}{1/2} \begin{bmatrix} 1 & -1 \\ -1 & 1 \end{bmatrix} \begin{matrix} 4 \\ 5 \end{matrix},$$

$$f^1 = \frac{1}{8} \begin{bmatrix} 1 \\ 1 \end{bmatrix} \begin{matrix} 1 \\ 2 \end{matrix}, \quad f^2 = \frac{1}{8} \begin{bmatrix} 1 \\ 1 \end{bmatrix} \begin{matrix} 2 \\ 3 \end{matrix}, \quad f^3 = \frac{1}{4} \begin{bmatrix} 1 \\ 1 \end{bmatrix} \begin{matrix} 3 \\ 4 \end{matrix}, \quad f^4 = \frac{1}{2} \begin{bmatrix} 1 \\ 1 \end{bmatrix} \begin{matrix} 4 \\ 5 \end{matrix}.$$

Thus the overall matrices are, after suppressing the fixed freedom given by $U_1 = 0$,

$$K = \begin{bmatrix} 16 & -8 & & \\ & 12 & -4 & \\ & & 6 & -2 \\ \text{sym} & & & 2 \end{bmatrix} \begin{matrix} 2 \\ 3 \\ 4 \\ 5 \end{matrix}, \qquad F = \begin{bmatrix} \frac{1}{6} \\ \frac{3}{8} \\ \frac{3}{4} \\ \frac{1}{2} \end{bmatrix} \begin{matrix} 2 \\ 3 \\ 4 \\ 5 \end{matrix}.$$

Thus the overall system of equations $KU = F$ yields the solution

$$U_2 = \tfrac{15}{64}, \quad U_3 = \tfrac{7}{16}, \quad U_4 = \tfrac{3}{4}, \quad U_5 = 1.$$

(ii) In this case all element matrices are identical and are given by

$$k^e = \frac{1}{1/5} \begin{bmatrix} 1 & -1 \\ -1 & 1 \end{bmatrix}, \qquad f^e = \frac{1}{5} \begin{bmatrix} 1 \\ 1 \end{bmatrix}.$$

Thus the overall matrices are, after suppressing the fixed freedom given by $U_1 = 0$,

$$
\mathbf{K} = \begin{bmatrix}
10 & -5 & & & \\
& 10 & -5 & & \\
& & 10 & -5 & \\
& & & 10 & -5 \\
& & & & 5
\end{bmatrix}
\begin{matrix} 2 \\ 3 \\ 4 \\ 5 \\ 6 \end{matrix},
\qquad
\mathbf{F} = \frac{1}{5} \begin{bmatrix} 2 \\ 2 \\ 2 \\ 2 \\ 1 \end{bmatrix}
\begin{matrix} 2 \\ 3 \\ 4 \\ 5 \\ 6 \end{matrix}.
$$

(column headers for $\mathbf{K}$: 2 3 4 5 6)

Thus the overall system of equations yields the solution

$$
U_2 = \tfrac{9}{25}, \quad U_3 = \tfrac{16}{25}, \quad U_4 = \tfrac{21}{25}, \quad U_5 = \tfrac{24}{25}, \quad U_6 = 1.
$$

A comparison of these results with the two- and four-element solutions (Example 3.1), together with the exact solution, is given in Table 3.2.

    (iii) The spreadsheet is shown in Fig. 3.35 and a comparison with other solutions is shown in Table 3.2.

Solution 3.4 (i) Take the three nodes at $x = 0, \tfrac{1}{4}, 1$; then the element stiffness and force matrices are, using the results of Example 3.1,

$$
\mathbf{k}^1 = 4 \begin{bmatrix} 1 & -1 \\ -1 & 1 \end{bmatrix} \begin{matrix} 1 \\ 2 \end{matrix},
\qquad
\mathbf{k}^2 = \frac{4}{3} \begin{bmatrix} 1 & -1 \\ -1 & 1 \end{bmatrix} \begin{matrix} 2 \\ 3 \end{matrix},
$$

(column headers: $\mathbf{k}^1$: 1 2; $\mathbf{k}^2$: 2 3)

$$
\mathbf{f}^1 = \frac{1}{4} \begin{bmatrix} 1 \\ 1 \end{bmatrix} \begin{matrix} 1 \\ 2 \end{matrix},
\qquad
\mathbf{f}^2 = \frac{3}{4} \begin{bmatrix} 1 \\ 1 \end{bmatrix} \begin{matrix} 2 \\ 3 \end{matrix}.
$$

The overall matrices are thus given by

$$
\mathbf{K} = 4 \begin{bmatrix}
1 & -1 & 0 \\
& \frac{4}{3} & -\frac{1}{3} \\
\text{sym} & & \frac{1}{3}
\end{bmatrix}
\begin{matrix} 1 \\ 2 \\ 3 \end{matrix},
\qquad
\mathbf{F} = \frac{1}{4} \begin{bmatrix} 1 \\ 4 \\ 3 \end{bmatrix}
\begin{matrix} 1 \\ 2 \\ 3 \end{matrix}.
$$

(column headers for $\mathbf{K}$: 1 2 3)

Enforcing the non-homogeneous Dirichlet boundary condition $U_1 = 1$ gives the overall system of equations

$$
\tfrac{16}{3}U_2 - \tfrac{4}{3}U_3 = 1 - (-4),
$$

$$
-\tfrac{4}{3}U_2 + \tfrac{4}{3}U_3 = \tfrac{3}{4} - 0.
$$

Thus

$$
U_2 = \tfrac{21}{16}, \qquad U_3 = \tfrac{3}{2}.
$$

	A	B	C	D	E	F	G	H	I	J	K	L	M	N
6	K_overall													F_overall
7	10	-10	0	0	0	0	0	0	0	0	0			0.1
8	-10	20	-10	0	0	0	0	0	0	0	0			0.2
9	0	-10	20	-10	0	0	0	0	0	0	0			0.2
10	0	0	-10	20	-10	0	0	0	0	0	0			0.2
11	0	0	0	-10	20	-10	0	0	0	0	0			0.2
12	0	0	0	0	-10	20	-10	0	0	0	0			0.2
13	0	0	0	0	0	-10	20	-10	0	0	0			0.2
14	0	0	0	0	0	0	-10	20	-10	0	0			0.2
15	0	0	0	0	0	0	0	-10	20	-10	0			0.2
16	0	0	0	0	0	0	0	0	-10	20	-10			0.2
17	0	0	0	0	0	0	0	0	0	-10	10			0.1
18	K													F
19	20	-10	0	0	0	0	0	0	0	0				0.2
20	-10	20	-10	0	0	0	0	0	0	0				0.2
21	0	-10	20	-10	0	0	0	0	0	0				0.2
22	0	0	-10	20	-10	0	0	0	0	0				0.2
23	0	0	0	-10	20	-10	0	0	0	0				0.2
24	0	0	0	0	-10	20	-10	0	0	0				0.2
25	0	0	0	0	0	-10	20	-10	0	0				0.2
26	0	0	0	0	0	0	-10	20	-10	0				0.2
27	0	0	0	0	0	0	0	-10	20	-10				0.2
28	0	0	0	0	0	0	0	0	-10	10				0.1
29														
30	K_inverse													U
31	0.1	0.1	0.1	0.1	0.1	0.1	0.1	0.1	0.1	0.1				0.19
32	0.1	0.2	0.2	0.2	0.2	0.2	0.2	0.2	0.2	0.2				0.36
33	0.1	0.2	0.3	0.3	0.3	0.3	0.3	0.3	0.3	0.3				0.51
34	0.1	0.2	0.3	0.4	0.4	0.4	0.4	0.4	0.4	0.4				0.64
35	0.1	0.2	0.3	0.4	0.5	0.5	0.5	0.5	0.5	0.5				0.75
36	0.1	0.2	0.3	0.4	0.5	0.6	0.6	0.6	0.6	0.6				0.84
37	0.1	0.2	0.3	0.4	0.5	0.6	0.7	0.7	0.7	0.7				0.91
38	0.1	0.2	0.3	0.4	0.5	0.6	0.7	0.8	0.8	0.8				0.96
39	0.1	0.2	0.3	0.4	0.5	0.6	0.7	0.8	0.9	0.9				0.99
40	0.1	0.2	0.3	0.4	0.5	0.6	0.7	0.8	0.9	1				1

Fig. 3.35 Spreadsheet for ten-element solution to Exercise 3.3.

(ii) Take the four nodes at $x = 0, \frac{1}{4}, \frac{1}{2}, 1$; then the overall matrices may be obtained in a similar manner to that in (i). In this case

$$
\mathbf{K} = 4 \begin{bmatrix} \overset{1}{4} & \overset{2}{-4} & \overset{3}{0} & \overset{4}{0} \\ & 8 & -4 & 0 \\ & & 6 & -2 \\ \text{sym} & & & 2 \end{bmatrix} \begin{matrix} 1 \\ 2 \\ 3 \\ 4 \end{matrix}, \qquad \mathbf{F} = \frac{1}{4} \begin{bmatrix} 1 \\ 2 \\ 3 \\ 2 \end{bmatrix} \begin{matrix} 1 \\ 2 \\ 3 \\ 4 \end{matrix}.
$$

Enforcing the non-homogeneous Dirichlet boundary condition $U_1 = 1$ and solving the resulting system of equations gives

$$
U_2 = \tfrac{23}{16}, \qquad U_3 = \tfrac{7}{4}, \qquad U_4 = 2.
$$

The exact solution at $x = \frac{1}{4}$ is $\frac{23}{16}$.

Solution 3.5 (i) The element stiffness matrices are given in Example 3.1 as

$$
\mathbf{k}^e = \frac{1}{h} \begin{bmatrix} 1 & -1 \\ -1 & 1 \end{bmatrix},
$$

where h is the element length. Using the notation of Fig. 3.7, the element force vector is given by

$$
\mathbf{f}^e = \int_{-1}^{1} \frac{1}{2} [(1-\xi) \ (1+\xi)]^T \ e^{x_m + h\xi/2} \frac{h}{2} \, d\xi
$$

$$
= e^{x_m} \left[\left(1 + \frac{2}{h}\right) \sinh \frac{h}{2} - \cosh \frac{h}{2} \quad \left(1 - \frac{2}{h}\right) \sinh \frac{h}{2} + \cosh \frac{h}{2} \right]^T
$$

$$
= e^{x_m} [0.12011 \ 0.13054]^T, \quad \text{since } h = 0.25.
$$

Thus

$$
\mathbf{f}^1 = \begin{bmatrix} 0.13610 \\ 0.14792 \end{bmatrix}, \quad \mathbf{f}^2 = \begin{bmatrix} 0.17476 \\ 0.18994 \end{bmatrix}, \quad \mathbf{f}^3 = \begin{bmatrix} 0.22439 \\ 0.24389 \end{bmatrix}, \quad \mathbf{f}^4 = \begin{bmatrix} 0.28813 \\ 0.31315 \end{bmatrix}.
$$

The overall stiffness matrix is given in Example 3.1, so that the equations for the unknown nodal values U_2, U_3, U_4 (see Fig. 3.10) are, after enforcing the essential Dirichlet boundary conditions,

$$
\begin{aligned}
8U_2 - 4U_3 \qquad\quad &= 0.32268, \\
-4U_2 + 8U_3 - 4U_4 &= 0.41433, \\
-4U_3 + 8U_4 &= 0.53202.
\end{aligned}
$$

Solving yields $U_2 = 0.1455, U_3 = 0.2104, U_4 = 0.1717$. These solutions may be compared with the exact solution and the classical Rayleigh–Ritz solution given in Table 2.5 and shown in Fig. 3.36, from which it is seen that the finite element

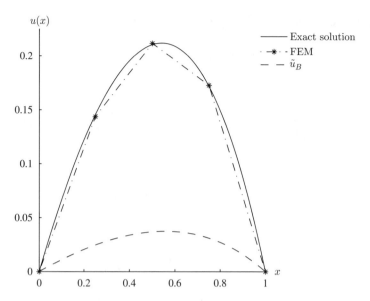

Fig. 3.36 Comparison of the finite element solution and the classical Rayleigh–Ritz solution with the exact solution for Exercise 3.5.

solution is far superior. Once again we observe superconvergence, with the finite element solution being exact at the nodes.

(ii) Using the result of part (i),

$$\mathbf{f}^e = e^{x_m} \, [0.049187 \quad 0.050854]^T .$$

$\mathbf{K}$ is the tridiagonal matrix

$$\lceil 20 \quad -10; -10 \quad 20 \quad -10; \ldots; -10 \quad 20 \quad -10; -10 \quad 20 \rfloor$$

and, enforcing the essential boundary condition,

$$\mathbf{F} = [\, 0.110609 \quad 0.122242 \quad 0.135098$$
$$0.149307 \quad 0.165010 \quad 0.182364$$
$$0.201543 \quad 0.222740 \quad 0.246165 \,]^T,$$

so that

$$\mathbf{U} = [\, 0.066657 \quad 0.122254 \quad 0.165626$$
$$0.195488 \quad 0.210420 \quad 0.208850$$
$$0.189045 \quad 0.149085 \quad 0.086851 \,]^T.$$

Again, these nodal values are exact.

Solution 3.6 In element $[e]$, approximate e^x by

$$e^x \approx \frac{1}{2}\left[(1-\xi)\ (1+\xi)\right]\left[e^{x_m-h/2}\ e^{x_m+h/2}\right]^T.$$

Then

$$\mathbf{f}^e = \left(\int_{-1}^{1}\frac{1}{4}\begin{bmatrix}1-\xi\\1+\xi\end{bmatrix}\left[(1-\xi)\ \ (1+\xi)\right]\frac{h}{2}d\xi\right)\begin{bmatrix}e^{x_m-h/2}\\e^{x_m+h/2}\end{bmatrix}$$

$$= \frac{h}{6}e^{x_m}\left[2e^{-h/2}+e^{h/2}\ \ e^{-h/2}+2e^{h/2}\right]^T$$

$$= e^{x_m}\left[0.12076\ \ 0.13120\right]^T,\qquad \text{since } h = 0.25.$$

Thus

$$\mathbf{f}^1 = \begin{bmatrix}0.13683\\0.14860\end{bmatrix},\quad \mathbf{f}^2 = \begin{bmatrix}0.17570\\0.19089\end{bmatrix},\quad \mathbf{f}^3 = \begin{bmatrix}0.22560\\0.24511\end{bmatrix},\quad \mathbf{f}^4 = \begin{bmatrix}0.28968\\0.31473\end{bmatrix}.$$

It follows, as in Exercise 3.5, that the equations for the unknowns U_2, U_3, U_4 are

$$\begin{aligned}
8U_2 -4U_3\qquad\quad &= 0.32437,\\
-4U_2 +8U_3 -4U_4 &= 0.41649,\\
-4U_3 +8U_4 &= 0.53479.
\end{aligned}$$

Solving yields $U_2 = 0.1463$, $U_3 = 0.2115$, $U_4 = 0.1726$. This solution is compared with the solution of Exercise 3.5, in which the exponential term is integrated exactly, in Table 3.4.

Solution 3.7 For the generalized Poisson equation

$$-\mathrm{div}(\boldsymbol{\kappa}\ \mathrm{grad}\ u) = f,$$

we use Galerkin's method just as in Section 3.6:

$$\iint_D (-\mathrm{div}(\boldsymbol{\kappa}\ \mathrm{grad}\ \tilde{u}) - f)\, w_i\, dx\, dy = 0,\quad i = 1, 2, \ldots, n.$$

Using the first form of Green's theorem, this becomes

$$\iint_D \mathrm{grad}\ w_i \cdot (\boldsymbol{\kappa}\ \mathrm{grad}\ \tilde{u})\, dx\, dy - \oint_C w_i(\boldsymbol{\kappa}\ \mathrm{grad}\ \tilde{u}) \cdot \mathbf{n}\, ds$$

$$- \iint_D fw_i\, dx\, dy = 0,\qquad i = 1, 2, \ldots, n.$$

Therefore the element stiffness matrix is

$$\mathbf{k}^e = \frac{1}{2}\iint_{[e]} \boldsymbol{\alpha}^T(\boldsymbol{\kappa} + \boldsymbol{\kappa}^T)\boldsymbol{\alpha}\, dx\, dy.$$

Table 3.4 A comparison of the solutions to Exercises 3.5 and 3.6

x	0	0.25	0.5	0.75	1
Interpolation of e^x	0	0.1463	0.2115	0.1726	0
Exact integration of e^x	0	0.1455	0.2104	0.1717	0
Exact solution	0	0.1455	0.2104	0.1717	0

When κ is symmetric,

$$\mathbf{k}^e = \iint_{[e]} \boldsymbol{\alpha}^T \boldsymbol{\kappa} \boldsymbol{\alpha} \, dx \, dy.$$

Using the Robin boundary condition gives the contribution from the boundary integral as

$$\int_{C_2^e} w_i (h - \sigma w_j) \, ds,$$

and it follows that the matrices $\bar{\mathbf{k}}^e$ and $\bar{\mathbf{f}}^e$ are unchanged.

Solution 3.8 In element $[e]$, the field variable is related to the 'potential' by

$$\tilde{\mathbf{q}}^e = -\boldsymbol{\kappa} \operatorname{grad} \tilde{u}^e$$

$$= -\boldsymbol{\kappa} \begin{bmatrix} \partial/\partial x \\ \partial/\partial y \end{bmatrix} \mathbf{N}^e \mathbf{U}^e$$

$$= -\boldsymbol{\kappa} \boldsymbol{\alpha}^e \mathbf{U}^e.$$

Thus

$$\mathbf{S}^e = -\boldsymbol{\kappa} \boldsymbol{\alpha}^e.$$

Solution 3.9 The shape function matrix is

$$\mathbf{N}^e = \frac{1}{2} \left[(1 - \xi) \quad (1 + \xi) \right];$$

thus

$$\boldsymbol{\alpha} = \frac{1}{h} \begin{bmatrix} -1 & 1 \end{bmatrix}.$$

Now $\kappa = [k(x)]$, so that, using Exercise 3.6,

$$\mathbf{k}^e = \frac{1}{2h} \int_{-1}^{1} k(x) \begin{bmatrix} 1 & -1 \\ -1 & 1 \end{bmatrix} d\xi$$

$$= \frac{1}{2h} \begin{bmatrix} 1 & -1 \\ -1 & 1 \end{bmatrix} \left(\int_{1}^{\xi_0} k_1 \, d\xi + \int_{\xi_0}^{1} k_2 \, d\xi \right)$$

$$= \frac{1}{2h} \{(k_1 + k_2) + \xi_0(k_1 - k_2)\} \begin{bmatrix} 1 & -1 \\ -1 & 1 \end{bmatrix},$$

where $\xi_0 = 2(x_0 - x_m)/h$. We notice that if $k_1 = k_2 = k$ (constant), then we recover the form developed in Example 3.1.

Solution 3.10 The element stiffness and force matrices are given in the solution to Exercise 3.3(i). If $\mathbf{U}$ is the vector of overall nodal values, then the vector of element nodal values $\mathbf{U}^e$ is given by eqn (3.49),

$$\mathbf{U}^3 = \mathbf{a}^3 \mathbf{U}.$$

Now,

$$\mathbf{U} = [U_1 \ U_2 \ U_3 \ U_4 \ U_5]^T$$

and

$$\mathbf{U}^1 = [U_1 \ U_2]^T, \quad \mathbf{U}^2 = [U_2 \ U_3]^T, \quad \mathbf{U}^3 = [U_3 \ U_4]^T, \quad \mathbf{U}^4 = [U_4 \ U_5]^T.$$

Thus

$$\mathbf{a}^1 = \begin{matrix} & 1 \ \ 2 \ 3 \ 4 \ 5 \\ \begin{bmatrix} 1 & & & & \\ & 1 & & & \end{bmatrix} \end{matrix}, \quad \mathbf{a}^2 = \begin{matrix} 1 & 2 \ \ 3 \ 4 \ 5 \\ \begin{bmatrix} & 1 & & & \\ & & 1 & & \end{bmatrix} \end{matrix},$$

$$\mathbf{a}^3 = \begin{matrix} 1 \ 2 & 3 & 4 \ \ 5 \\ \begin{bmatrix} & & 1 & & \\ & & & 1 & \end{bmatrix} \end{matrix}, \quad \mathbf{a}^4 = \begin{matrix} 1 \ 2 \ \ 3 & 4 & 5 \\ \begin{bmatrix} & & & 1 & \\ & & & & 1 \end{bmatrix} \end{matrix}.$$

Equations (3.50) and (3.51) give the overall stiffness and nodal force matrices as

$$\mathbf{K} = \sum_e \mathbf{a}^{e^T} \mathbf{k}^e \mathbf{a}^e \quad \text{and} \quad \mathbf{F} = \sum_e \mathbf{a}^{e^T} \mathbf{f}^e.$$

Now,

$$\mathbf{a}^{1^T} \mathbf{k}^1 \mathbf{a}^1 = \begin{bmatrix} 1 & \\ & 1 \\ & \\ & \\ & \end{bmatrix} \begin{bmatrix} 8 & -8 \\ -8 & 8 \end{bmatrix} \begin{bmatrix} 1 & \\ & 1 \end{bmatrix}$$

$$= \begin{matrix} 1 & 2 & 3 \ 4 \ 5 \\ \begin{bmatrix} 8 & -8 & \\ -8 & 8 & \\ & & \\ & & \\ & & \end{bmatrix} \begin{matrix} 1 \\ 2 \\ 3 \\ 4 \\ 5 \end{matrix} \end{matrix}$$

and

$$
\mathbf{a^1}^T \mathbf{f^1} = \begin{bmatrix} 1 & & & \\ & 1 & & \\ & & & \\ & & & \\ & & & \end{bmatrix} \begin{bmatrix} \frac{1}{8} \\ \frac{1}{8} \end{bmatrix} = \begin{bmatrix} \frac{1}{8} \\ \frac{1}{8} \\ \\ \\ \end{bmatrix} \begin{matrix} 1 \\ 2 \\ 3 \\ 4 \\ 5 \end{matrix} .
$$

Similarly,

$$
\mathbf{a^2}^T \mathbf{k^2 a^2} = \begin{matrix} 1 \quad 2 \quad\; 3 \quad 4\;5 \end{matrix} \begin{bmatrix} & & & \\ & 8 & -8 & \\ & -8 & 8 & \\ & & & \\ & & & \end{bmatrix} \begin{matrix} 1 \\ 2 \\ 3 \\ 4 \\ 5 \end{matrix} , \qquad \mathbf{a^2}^T \mathbf{f^2} = \begin{bmatrix} \\ \frac{1}{8} \\ \frac{1}{8} \\ \\ \end{bmatrix} \begin{matrix} 1 \\ 2 \\ 3 \\ 4 \\ 5 \end{matrix} ,
$$

$$
\mathbf{a^3}^T \mathbf{k^3 a^3} = \begin{matrix} 1 \quad\; 2 \quad\; 3 \quad\;\, 4 \quad\;\, 5 \end{matrix} \begin{bmatrix} & & & & \\ & & & & \\ & & 4 & -4 & \\ & & -4 & 4 & \\ & & & & \end{bmatrix} \begin{matrix} 1 \\ 2 \\ 3 \\ 4 \\ 5 \end{matrix} , \qquad \mathbf{a^3}^T \mathbf{f^3} = \begin{bmatrix} \\ \\ \frac{1}{4} \\ \frac{1}{4} \\ \end{bmatrix} \begin{matrix} 1 \\ 2 \\ 3 \\ 4 \\ 5 \end{matrix} ,
$$

$$
\mathbf{a^4}^T \mathbf{k^4 a^4} = \begin{matrix} 1 \quad\; 2 \quad 3 \quad\;\, 4 \quad\;\;\; 5 \end{matrix} \begin{bmatrix} & & & & \\ & & & & \\ & & & & \\ & & & 2 & -2 \\ & & & -2 & 2 \end{bmatrix} \begin{matrix} 1 \\ 2 \\ 3 \\ 4 \\ 5 \end{matrix} , \qquad \mathbf{a^4}^T \mathbf{f^4} = \begin{bmatrix} \\ \\ \\ \frac{1}{2} \\ \frac{1}{2} \end{bmatrix} \begin{matrix} 1 \\ 2 \\ 3 \\ 4 \\ 5 \end{matrix} .
$$

Thus

$$
\mathbf{K} = \begin{matrix} 1 \quad\;\; 2 \quad\;\; 3 \quad\;\; 4 \quad\;\; 5 \end{matrix} \begin{bmatrix} 8 & -8 & & & \\ & 16 & -8 & & \\ & & 12 & -4 & \\ & & & 6 & -2 \\ \text{sym} & & & & 2 \end{bmatrix} \begin{matrix} 1 \\ 2 \\ 3 \\ 4 \\ 5 \end{matrix} , \qquad \mathbf{F} = \begin{bmatrix} \frac{1}{8} \\ \frac{1}{4} \\ \frac{3}{8} \\ \frac{3}{4} \\ \frac{1}{2} \end{bmatrix} \begin{matrix} 1 \\ 2 \\ 3 \\ 4 \\ 5 \end{matrix} .
$$

Solution 3.11 Interpolating $f(x, y)$ throughout the element gives

$$
f(x, y) \approx [L_1 \; L_2 \; L_3] \begin{bmatrix} f(x_1, y_1) \\ f(x_2, y_2) \\ f(x_3, y_3) \end{bmatrix}
$$

$$
= [L_1 \; L_2 \; L_3] \, \mathbf{s}, \quad \text{say.}
$$

Thus, using eqn (3.70),

$$\mathbf{f}^e = \int\!\!\int_A [L_1 \; L_2 \; L_3]^T \, f(x,y) \, dx \, dy$$

$$\approx \left(\int\!\!\int_A \begin{bmatrix} L_1^2 & L_1 L_2 & L_1 L_3 \\ L_2 L_1 & L_2^2 & L_2 L_3 \\ L_3 L_1 & L_3 L_2 & L_3^2 \end{bmatrix} dx \, dy \right) \mathbf{s},$$

i.e.

$$\mathbf{f}^e = \Lambda \mathbf{s},$$

where

$$\Lambda = \frac{A}{12} \begin{bmatrix} 2 & 1 & 1 \\ 1 & 2 & 1 \\ 1 & 1 & 2 \end{bmatrix}.$$

Solution 3.12 The maximum difference in node numbers for any element is 7; this occurs in elements 2 and 11. Thus $d = 7 + 1 = 8$. There are many ways to reduce this value; the minimum is obtained by numbering across the region as shown in Fig. 3.37. In this case the maximum difference in node numbers for any element is 4. Thus $d = 4 + 1 = 5$. This exercise illustrates the general strategy that nodes should be numbered across the narrowest part of the region.

Solution 3.13 Suppose that the square is divided into four square elements as shown in Fig. 3.38. Using eqn (3.54), the element stiffness matrices are of the form

$$\mathbf{k}^e = \frac{1}{6} \begin{bmatrix} 4 & -1 & -2 & -1 \\ & 4 & -1 & -2 \\ & & 4 & -1 \\ \text{sym} & & & 4 \end{bmatrix}.$$

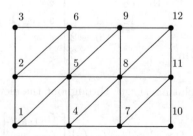

Fig. 3.37 Node-numbering system which yields the minimum semi-bandwidth for the region of Fig. 3.29.

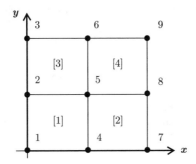

Fig. 3.38 Four-element discretization of the region in Exercise 3.13.

Using the notation of Fig. 3.19, the element force vectors are given by

$$\mathbf{f}^e = \int_{-1}^{1}\int_{-1}^{1}\left[2\left(x_m + \frac{a\xi}{2}\right)\left(y_m + \frac{b\eta}{2}\right) - 4\right]\frac{1}{4}$$

$$\times \left[(1-\xi)(1-\eta)\ (1+\xi)(1-\eta)\ (1-\xi)(1+\eta)\ (1+\xi)(1+\eta)\right]^T \frac{ab}{4}\, d\xi\, d\eta.$$

Since $a = b = \frac{1}{2}$,

$$\mathbf{f}^e = \frac{1}{8}(x_m y_m - 2)\,[1\ 1\ 1\ 1]^T$$

$$+\frac{y_m}{96}\,[-1\ 1\ -1\ 1]^T$$

$$+\frac{x_m}{96}\,[-1\ -1\ 1\ 1]^T$$

$$+\frac{1}{1152}\,[1\ -1\ -1\ 1]^T.$$

Now, node 5 is the only node where a Dirichlet boundary condition does not act; thus only the contributions to the equation corresponding to node 5 need be assembled. Row 5 of $\mathbf{K}$ is

$$\frac{1}{3}[-2\quad -1-1\quad -2\quad -1-1\quad 4+4+4+4\quad -1-1\quad -2\quad -1-1\quad -2].$$

Also, $F_5 = f_3^1 + f_4^2 + f_2^3 + f_1^4$, where the subscripts refer to the local node numbering given in Fig. 3.19:

$$f_3^1 = -35/144,\quad f_4^2 = -31/144,\quad f_2^3 = -2/9,\quad f_1^4 = -7/8.$$

Thus

$$F_5 = -\frac{7}{8}.$$

Table 3.5 A comparison of finite element solutions in Exercises 3.13 and 3.15

(x, y)	$\left(\frac{1}{4}, \frac{1}{4}\right)$	$\left(\frac{1}{2}, \frac{1}{2}\right)$	$\left(\frac{3}{4}, \frac{3}{4}\right)$	$\left(\frac{1}{4}, \frac{1}{2}\right)$	$\left(\frac{1}{4}, \frac{3}{4}\right)$	$\left(\frac{1}{2}, \frac{3}{4}\right)$
4 rectangular elements (Exercise 3.13)	0.193	0.273	0.318	0.262	0.506	0.387
4 triangular elements (Exercise 3.15)	0.125	0.25	0.25	0.25	0.395	0.375
Exact solution	0.094	0.25	0.281	0.219	0.438	0.344

Since

$$U_1 = 0, \qquad U_2 = \tfrac{1}{4}, \qquad U_3 = 1, \qquad U_4 = \tfrac{1}{4},$$

$$U_6 = \tfrac{1}{2}, \qquad U_7 = 1, \qquad U_8 = \tfrac{1}{2}, \qquad U_9 = 0,$$

it follows that the equation for U_5 is

$$\tfrac{16}{3}U_5 = -\tfrac{7}{8} + \tfrac{1}{3}.7,$$

which yields $U_5 = 0.27344$.

By virtue of the interpolation (3.53), the function value at the midpoint of each element, i.e. at the point where $\xi = \eta = 0$, is just the mean of the nodal values, and values along interelement boundaries are found by linear interpolation from the nodal values. These values are compared with the corresponding values obtained using triangular elements with and the exact solution, $u_0(x, y) = (1 - x)y^2 + (1 - y)x^2$, in Table 3.5.

Solution 3.14 On side 3,1, $\sigma = \sigma_1 L_1 + \sigma_3(1 - L_1)$. Then, using eqn (3.72),

$$\bar{\mathbf{k}} = \left(b_2^2 + c_2^2\right)^{1/2}$$

$$\times \int_0^1 \begin{bmatrix} \sigma_1 L_1^3 + \sigma_3 \left(L_1^2 - L_1^3\right) & 0 & \sigma_1 \left(L_1^2 - L_1^3\right) + \sigma_3 L_1 \left(1 - L_1^2\right) \\ & 0 & 0 \\ \text{sym} & & \sigma_1 L_1(1 - L_1)^2 + \sigma_3(1 - L_1)^3 \end{bmatrix} dL_1$$

$$= \left(b_2^2 + c_2^2\right)^{1/2} \begin{bmatrix} 3\sigma_1 + \sigma_3 & 0 & \sigma_1 + \sigma_3 \\ & 0 & 0 \\ \text{sym} & & \sigma_1 + 3\sigma_3 \end{bmatrix}.$$

Similarly, using eqn (3.73),

$$\bar{\mathbf{f}} = \frac{\left(b_2^2 + c_2^2\right)^{1/2}}{6} \begin{bmatrix} 2h_1 + h_3 & 0 & h_1 + 2h_3 \end{bmatrix}^T.$$

Solution 3.15 Notice that the problem has symmetry about the line $y = x$; thus, consider only the region shown in Fig. 3.39. Along the line $y = x$, $\partial u/\partial n = 0$.

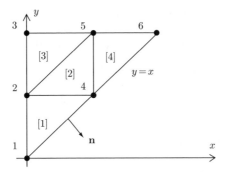

Fig. 3.39 Finite element idealization for the problem of Exercise 3.15.

The element stiffness matrices may be obtained directly from eqn (3.69) or by using the results of Example 3.4; in either case, they are found to be given by

$$
k^1 = \begin{array}{c c} & \begin{matrix} 1 & 4 & 2 \end{matrix} \\ \begin{bmatrix} 2 & 0 & -2 \\ 0 & 2 & -2 \\ -2 & -2 & 4 \end{bmatrix} & \begin{matrix} 1 \\ 4 \\ 2 \end{matrix} \end{array},
\qquad
k^2 = \begin{array}{c c} & \begin{matrix} 5 & 2 & 4 \end{matrix} \\ \begin{bmatrix} 2 & 0 & -2 \\ 0 & 2 & -2 \\ -2 & -2 & 4 \end{bmatrix} & \begin{matrix} 5 \\ 2 \\ 4 \end{matrix} \end{array},
$$

$$
k^3 = \begin{array}{c c} & \begin{matrix} 2 & 5 & 3 \end{matrix} \\ \begin{bmatrix} 2 & 0 & -2 \\ 0 & 2 & -2 \\ -2 & -2 & 4 \end{bmatrix} & \begin{matrix} 2 \\ 5 \\ 3 \end{matrix} \end{array},
\qquad
k^4 = \begin{array}{c c} & \begin{matrix} 4 & 6 & 5 \end{matrix} \\ \begin{bmatrix} 2 & 0 & -2 \\ 0 & 2 & -2 \\ -2 & -2 & 4 \end{bmatrix} & \begin{matrix} 4 \\ 6 \\ 5 \end{matrix} \end{array}.
$$

The element force vector is given by eqn (3.70) as

$$
f_i = \iint_A L_i \left[2(x+y) - 4 \right] dx\, dy, \qquad i = 1, 2, 3,
$$

where 1, 2, 3 is the local nodal numbering defined in Fig. 3.24. Then, using eqn (3.60) and (3.61),

$$
f_i = \iint_A L_i \left[2L_1(x_1+y_1) + 2L_2(x_2+y_2) + 2L_3(x_3+y_3) - 4 \right] dx\, dy
$$

$$
= \tfrac{1}{24}(x_i+y_i) + \tfrac{1}{48}(x_j+y_j) + \tfrac{1}{48}(x_k+y_k) - \tfrac{1}{6}.
$$

The only node without an essential boundary condition is node 4; thus only the equation associated with node 4 is assembled. This equation is $\sum_{j=1}^{6} K_{4j}U_j = F_4$, where **K** and **F** are the overall stiffness and force matrices, respectively. Row 4 of **K** is $[0 \ -1 \ 2 \ -1 \ 0]$, and $F_4 = f_2^1 + f_3^2 + f_1^4$; the subscripts refer to the local nodal numbers in Fig. 3.24. We obtain

$$f_2^1 = \tfrac{1}{24}\left(\tfrac{1}{2}+\tfrac{1}{2}\right) + \tfrac{1}{48}\left(0+\tfrac{1}{2}\right) + \tfrac{1}{48}(0+0) - \tfrac{1}{16} = -\tfrac{11}{96},$$

$$f_3^2 = \tfrac{1}{24}\left(\tfrac{1}{2}+\tfrac{1}{2}\right) + \tfrac{1}{48}\left(\tfrac{1}{2}+1\right) + \tfrac{1}{48}\left(0+\tfrac{1}{2}\right) - \tfrac{1}{6} = -\tfrac{8}{96},$$

$$f_1^4 = \tfrac{1}{24}\left(\tfrac{1}{2}+\tfrac{1}{2}\right) + \tfrac{1}{48}(1+1) + \tfrac{1}{48}\left(\tfrac{1}{2}+1\right) - \tfrac{1}{6} = -\tfrac{5}{96};$$

therefore $F_4 = -\tfrac{1}{4}$.

The essential boundary conditions give

$$U_1 = 0, \qquad U_2 = \tfrac{1}{4}, \qquad U_3 = 1, \qquad U_5 = \tfrac{1}{2}, \qquad U_6 = 0,$$

so that U_4 is given by

$$-\tfrac{1}{4} + 2U_4 - \tfrac{1}{2} = -\tfrac{1}{4},$$

which gives $U_4 = \tfrac{1}{4}$.

The solution at $\left(\tfrac{1}{4},\tfrac{1}{4}\right)$, $\left(\tfrac{1}{4},\tfrac{3}{4}\right)$, $\left(\tfrac{1}{4},\tfrac{1}{2}\right)$, $\left(\tfrac{1}{2},\tfrac{3}{4}\right)$ and $\left(\tfrac{3}{4},\tfrac{3}{4}\right)$ is found by linear interpolation between nodes 1 and 4, 2 and 5, 2 and 4, and 4 and 6. The results are compared with the corresponding results obtained using rectangular elements in Table 3.5.

Solution 3.16 The element stiffness matrices are given by eqn (3.69) as

$$
\mathbf{k}^1 =
\begin{array}{c}
\quad\; 1,5 \qquad\quad 2,4 \qquad\quad 4,2 \\
\begin{bmatrix}
1.0021 & -0.5333 & -0.4687 \\
 & 0.5333 & 0 \\
\text{sym} & & 0.4687
\end{bmatrix}
\begin{array}{c} 1,5 \\ 2,4 = \mathbf{k}^2, \\ 4,2 \end{array}
\end{array}
$$

$$
\mathbf{k}^4 =
\begin{array}{c}
\quad\; 4 \qquad\quad 5 \qquad\quad 6 \\
\begin{bmatrix}
1.1333 & -0.3 & -0.8333 \\
 & 0.3 & 0 \\
\text{sym} & & 0.8333
\end{bmatrix}
\begin{array}{c} 4 \\ 5. \\ 6 \end{array}
\end{array}
$$

The only node at which U is unknown is node 4, and node 4 is not in element 3; thus $\mathbf{k}^3$ is not obtained. Since the governing differential equation is Laplace's equation, $\mathbf{f}^e = \mathbf{0}$.

There is a contribution to the stiffness and force matrices due to the non-homogeneous Robin boundary condition on the y-axis. Since $\sigma = 2$ and $h = 2y - 7$ are linear functions, these matrices may be written down using the results of Exercise 3.14:

$$
\bar{\mathbf{k}}^1 =
\begin{array}{c}
\;\; 1 \quad\;\; 2 \quad\;\; 4 \\
\begin{bmatrix}
0.4267 & 0 & 0.2133 \\
 & 0 & 0 \\
\text{sym} & & 0.4267
\end{bmatrix}
\begin{array}{c} 1 \\ 2, \\ 4 \end{array}
\end{array}
\qquad
\bar{\mathbf{f}}^1 =
\begin{bmatrix}
-2.1035 \\
0 \\
-1.9669
\end{bmatrix}
\begin{array}{c} 1 \\ 2, \\ 4 \end{array}
$$

$$\bar{k}^4 = \begin{matrix} & 4 & 5 & 6 \\ \begin{bmatrix} 0.24 & 0 & 0.12 \\ & 0 & 0 \\ \text{sym} & & 0.24 \end{bmatrix} & \begin{matrix} 4 \\ 5 \\ 6 \end{matrix} \end{matrix}, \qquad \bar{f}^4 = \begin{bmatrix} -0.9864 \\ 0 \\ -0.9432 \end{bmatrix} \begin{matrix} 4 \\ 5 \\ 6 \end{matrix}.$$

Thus, assembling the equation for the unkown U_4,

$$-0.2554U_1 + 0U_2 + 0U_3 + 2.8021U_4 - 0.8333U_5 - 0.7133U_6 = -2.9533.$$

The essential boundary conditions give

$$U_1 = -2, \qquad U_2 = -0.2, \qquad U_3 = 1, \qquad U_5 = 0.44, \qquad U_6 = -1.$$

Therefore it follows that $U_4 = -1.36$. This agrees with the exact solution $u_0(x, y) = y + 3x - 2$ at $(0, 0.64)$, which is to be expected since the exact solution is itself linear.

Solution 3.17 The element stiffness matrices and the force vectors due to the Robin boundary conditions are identical with those given in Exercise 3.16. The element force vectors due to $f(x, y) \equiv 2$ are given by

$$\mathbf{f}^1 = \begin{bmatrix} 0.128 \\ 0.128 \\ 0.128 \end{bmatrix} \begin{matrix} 1,5 \\ 2,4 \\ 4,2 \end{matrix} = \mathbf{f}^2, \qquad \bar{\mathbf{f}}^4 = \begin{bmatrix} 0.072 \\ 0.072 \\ 0.072 \end{bmatrix} \begin{matrix} 4 \\ 5 \\ 6 \end{matrix}.$$

Thus the process of assembling the equation for the unknown U_4 follows just as in Exercise 3.16. In this case the right-hand side equals -2.6258. Enforcing the essential boundary conditions $U_1 = -2, U_2 = -0.56, U_3 = 0, U_5 = 0.08, U_6 = -1$ yields the solution $U_4 = -1.35$.

Solution 3.18 The element stiffness and force matrices may be obtained from Example 3.4 and Exercise 3.11 as

$$\mathbf{k}^e = \frac{1}{2} \begin{matrix} & 1 & 2 & 3 \\ \begin{bmatrix} 1 & -1 & 0 \\ & 2 & -1 \\ \text{sym} & & 1 \end{bmatrix} & \begin{matrix} 1 \\ 2 \\ 3 \end{matrix} \end{matrix}$$

and

$$\mathbf{f}^e = \frac{h^2}{3} \begin{bmatrix} 1 \\ 1 \\ 1 \end{bmatrix} \begin{matrix} 1 \\ 2 \\ 3 \end{matrix},$$

where the local nodal numbering is defined in Fig. 3.40.

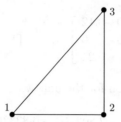

Fig. 3.40 Local node numbering for the triangles in Exercise 3.18.

For the idealization of Fig. 3.31(a), row 5 of the overall system of equations is

$$\tfrac{1}{2}[-1+(-1) \quad 0+0 \quad -1+(-1) \quad 1+1+2+1+1+2$$

$$-1+(-1) \quad 0+0 \quad -1+(-1)][U_2 \ldots U_8]^T$$

$$= \tfrac{1}{3}h^2(1+1+1+1+1+1),$$

i.e.

$$\frac{1}{h^2}(U_2 + U_4 + U_6 + U_8 - 4U_5) = -2.$$

This is the usual five-point finite-difference replacement for Poisson's equation (Smith 1985).

For the idealization of Fig. 3.31(b), row 5 of the overall system of equations is

$$\tfrac{1}{2}[0+0 \quad -1+(-1) \quad 0+0 \quad -1+(-1) \quad 1+1+1+1+1+1+1+1$$

$$-1+(-1) \quad 0+0 \quad -1+(-1) \quad 0+0][U_1,\ldots,U_9]^T$$

$$= \tfrac{1}{3}h^2(1+1+1+1+1+1),$$

i.e.

$$\frac{1}{h^2}(U_2 + U_4 + U_6 + U_8 - 4U_5) = -\tfrac{8}{3}.$$

Solution 3.19 Poisson's equation (3.74) may be rewritten as

$$-\frac{\partial}{\partial r}\left(r\frac{\partial u}{\partial r}\right) - \frac{\partial}{\partial z}\left(r\frac{\partial u}{\partial z}\right) = rf(r,z)$$

and, comparing with eqn (3.1), it follows that

$$k_{ij} = \iint_A r\frac{(b_ib_j + c_ic_j)}{4A^2}2\pi r\,dr\,dz.$$

In this case the integral arising in k_{ij} is a volume integral in which $dV = 2\pi r \, dr \, dz$. Now, $r = r_1 L_1 + r_2 L_2 + r_3 L_3$; thus

$$r^2 = [r_1 \quad r_2 \quad r_3] \begin{bmatrix} L_1^2 & L_1 L_2 & L_1 L_3 \\ L_2 L_1 & L_2^2 & L_2 L_3 \\ L_3 L_1 & L_3 L_2 & L_3^2 \end{bmatrix} \begin{bmatrix} r_1 \\ r_2 \\ r_3 \end{bmatrix}.$$

Using the result of Exercise 3.11, it follows that

$$\iint_A r^2 \, dr \, dz = \mathbf{x}^T \mathbf{\Lambda} \mathbf{x},$$

where $\mathbf{x} = [r_1 \; r_2 \; r_3]^T$ and $\mathbf{\Lambda}$ is defined in Exercise 3.11. Defining

$$R_0^2 = \frac{1}{A} \mathbf{x}^T \mathbf{\Lambda} \mathbf{x} = \frac{1}{6} \left(r_1^2 + r_2^2 + r_3^2 + r_1 r_2 + r_2 r_3 + r_3 r_1 \right),$$

it follows that

$$k_{ij} = \frac{\pi}{2A} R_0^2 (b_i b_j + c_i c_j).$$

In practice, in the integration for k_{ij}, the value of r is often replaced by $\bar{r} = \frac{1}{3}(r_1 + r_2 + r_3)$, which is the radial coordinate of the centroid of the triangle. In this case

$$k_{ij} = \frac{\pi}{2A} \bar{r}^2 (b_i b_j + c_i c_j).$$

The difference between the stiffness coefficients in the two cases arises only from the difference between $\bar{r}^2$ and R_0^2. This difference is usually small; for example, if $r_1 = 2$, $r_2 = 4$ and $r_3 = 3$, $\bar{r}^2 = 9$ and $R_0^2 = 9.167$; the relative difference is thus approximately 1.8 per cent.

Solution 3.20 By reference to eqn (3.70), the element force vector may be written as

$$f_i = \iint_A r f(r, z) L_i 2\pi r \, dr \, dz.$$

In the case $f = $ constant, eqn (3.73) gives

$$f_i = 2\pi f \iint_A r^2 L_i \, dr \, dz.$$

Replacing r^2 by $\bar{r} r$ (see Exercise 3.19),

$$f_i = 2\pi \bar{r} f \iint_A (L_1 r_1 + L_2 r_2 + L_3 r_3) L_i \, dr \, dz$$

$$= \frac{1}{6} \pi \bar{r} A f (2r_i + r_j + r_k).$$

Notice that in this case the source term f is not evenly distributed between the nodes as in the case of two-dimensional problems, see Exercise 3.11, with $f(x, y) = f$.

Solution 3.21 By comparison with eqns (3.46) and (3.48),

$$\bar{k}^e = \int_{S_2^e} \sigma(S) \mathbf{N}^{e^T} \mathbf{N}^e \, dS$$

and

$$\bar{f}^e = \int_{S_2^e} h(S) \mathbf{N}^{e^T} \, dS,$$

where the integrals are taken over the boundary S_2^e of element $[e]$, where the Robin boundary condition holds. Thus if σ is a constant, it follows that

$$\bar{k}^e = 2\pi\sigma \left(b_2^2 + c_2^2\right)^{1/2} \int_0^1 \begin{bmatrix} L_1^2 & 0 & L_1 - L_1^2 \\ 0 & 0 & 0 \\ L_1 - L_1^2 & 0 & (1 - L_1^2) \end{bmatrix} [r_1 L_1 + r_3(1 - L_1)] \, dL_1$$

(*cf.* eqn (3.72)). Therefore

$$\bar{k}^e = \frac{\pi\sigma}{6} \left(b_2^2 + c_2^2\right)^{1/2} \begin{bmatrix} 3r_1 + r_3 & 0 & r_1 + r_3 \\ 0 & 0 & 0 \\ r_1 + r_3 & 0 & r_1 + 3r_3 \end{bmatrix}.$$

Similarly, if h is a constant,

$$\bar{f}^e = \frac{\pi h}{3} \left(b_2^2 + c_2^2\right)^{1/2} \begin{bmatrix} 2r_1 + r_3 \\ 0 \\ r_1 + 2r_3 \end{bmatrix}.$$

Solution 3.22 The shape functions are given by

$$N_i(\xi, \eta, \zeta) = \tfrac{1}{8}(1 + \xi_0)(1 + \eta_0)(1 + \zeta_0), \qquad i = 1, \dots, 8,$$

and the element stiffness matrix is given by

$$k_{ij} = \int\!\!\int\!\!\int_V \left(\frac{\partial N_i}{\partial x} \frac{\partial N_j}{\partial x} + \frac{\partial N_i}{\partial y} \frac{\partial N_j}{\partial y} + \frac{\partial N_i}{\partial z} \frac{\partial N_j}{\partial z} \right) dx \, dy \, dz.$$

Now,

$$\frac{\partial N_i}{\partial x} \frac{\partial N_j}{\partial x} = \frac{\xi_i \xi_j}{16a^2} [1 + \eta_i \eta_j \eta^2 + (\eta_i + \eta_j)\eta][1 + \zeta_i \zeta_j \zeta^2 + (\zeta_i + \zeta_j)\zeta].$$

Thus

$$\int\!\!\int\!\!\int_V \frac{\partial N_i}{\partial x} \frac{\partial N_j}{\partial x} \, dx \, dy \, dz = \frac{bc}{144a} \xi_i \xi_j (3 + \eta_i \eta_j)(3 + \zeta_i \zeta_j).$$

The other two terms in the integral for k_{ij} may be obtained by cyclic permutation.

Solution 3.23 The volume coordinates may be written in terms of the global coordinates as follows:

$$L_i = (a_i + b_i x + c_i y + d_i z)/6V,$$

where

$$a_i = \begin{vmatrix} x_j & y_j & z_j \\ x_k & y_k & z_k \\ x_l & y_l & z_l \end{vmatrix}, \qquad b_i = - \begin{vmatrix} 1 & y_j & z_j \\ 1 & y_k & z_k \\ 1 & y_l & z_l \end{vmatrix},$$

$$c_i = - \begin{vmatrix} x_j & 1 & z_j \\ x_k & 1 & z_k \\ x_l & 1 & z_l \end{vmatrix}, \qquad d_i = - \begin{vmatrix} x_j & y_j & 1 \\ x_k & y_k & 1 \\ x_l & y_l & 1 \end{vmatrix}.$$

The shape function matrix is $\mathbf{N}^e = [L_1 \ L_2 \ L_3 \ L_4]$. Hence

$$k_{ij} = \iiint_V \left(\frac{\partial L_i}{\partial x} \frac{\partial L_j}{\partial x} + \frac{\partial L_i}{\partial y} \frac{\partial L_j}{\partial y} + \frac{\partial L_i}{\partial z} \frac{\partial L_j}{\partial z} \right) dx \, dy \, dz$$

$$= \frac{1}{36V} (b_i b_j + c_i c_j + d_i d_j).$$

Solution 3.24 The development of the element matrices follows exactly as in Section 3.6, with the element force vector in this case given by (see eqn (3.47))

$$\iint_D \mathbf{N}^{e^T} \lambda \tilde{u}^e \, dx \, dy = \lambda \left(\iint_D \mathbf{N}^{e^T} \mathbf{N}^e \, dx \, dy \right) \mathbf{U}^e$$

$$= \lambda \mathbf{m}^e \mathbf{U}^e.$$

Thus assembling the overall system yields the generalized eigenvalue problem

$$\mathbf{KU} = \lambda \mathbf{MU}.$$

The eigenvalues are given by $\det[\mathbf{K} - \lambda \mathbf{M}] = 0$. For the eigenvalue problem $-u'' = \lambda u$, $u(0) = u(1) = 0$, use the linear element of Example 3.1; see Fig. 3.7. The element stiffness matrix is

$$\mathbf{k}^e = \frac{1}{h} \begin{bmatrix} 1 & -1 \\ -1 & 1 \end{bmatrix}.$$

The element mass matrix is given by

$$\mathbf{m}^e = \frac{h}{8} \int_{-1}^{1} \begin{bmatrix} 1-\xi \\ 1+\xi \end{bmatrix} [(1-\xi)(1+\xi)] \, d\xi$$

$$= \frac{h}{6} \begin{bmatrix} 2 & 1 \\ 1 & 2 \end{bmatrix}.$$

(i) Using two elements, the generalized eigenvalue problem is, after enforcing the essential homogeneous boundary condition,

$$4U_2 = \lambda \tfrac{1}{3} U_2.$$

Thus

$$\lambda_1 = 12.$$

(ii) Using three elements,

$$\begin{bmatrix} 6 & -3 \\ -3 & 6 \end{bmatrix} \begin{bmatrix} U_2 \\ U_3 \end{bmatrix} = \frac{\lambda}{18} \begin{bmatrix} 4 & 1 \\ 1 & 4 \end{bmatrix} \begin{bmatrix} U_2 \\ U_3 \end{bmatrix}.$$

Thus

$$\begin{vmatrix} 6 - 2\lambda/9 & -3 - \lambda/18 \\ -3 - \lambda/18 & 6 - 2\lambda/9 \end{vmatrix} = 0.$$

Expanding the determinant and then solving the resulting quadratic equation yields $\lambda_1 = 10.8$, $\lambda_2 = 54$. The exact value of λ_1 is $\pi^2 \approx 9.87$.

(iii) Using ten elements, the overall matrices are 9×9 tridiagonal matrices given by

$$\mathbf{K} = 10 \lceil 2 \quad -1; \quad -1 \quad 2 \quad -1; \quad \dots; \quad -1 \quad 2 \quad -1; \quad -1 \quad 2 \rfloor,$$

$$\mathbf{M} = \tfrac{1}{60} \lceil 4 \quad 1; \quad 1 \quad 4 \quad 1; \quad \dots; \quad 1 \quad 4 \quad 1; \quad 1 \quad 4 \rfloor.$$

Then, using any suitable eigenvalue routine, we find

$$\lambda_1 = 10.0, \quad \lambda_2 = 40.8, \quad \lambda_3 = 95.6, \quad \dots.$$

The exact values of λ_n are $n^2 \pi^2$, so that λ_1 is computed with an error of about one per cent.

4   Higher-order elements: the isoparametric concept

4.1   A two-point boundary-value problem

In Chapter 3, the finite element philosophy was developed and illustrated by means of examples in which the interpolation functions were linear polynomials. There is no reason to choose linear functions only; indeed, higher-order polynomials are frequently used and in principle cause no more difficulty. We shall illustrate the approach using the Galerkin finite element method of Sections 3.2–3.5.

Example 4.1

$$-u'' = f(x), \quad 0 < x < 1, \quad u(0) = 1, \quad u'(1) + 2u(1) = 3.$$

Suppose that the region $0 \leq x \leq 1$ is divided into E elements, each element having three nodes; a typical element is shown in Fig. 4.1. In terms of its three nodal values, the variable u may be uniquely interpolated as a quadratic polynomial, the interpolation being given by

$$u^e = \mathbf{N}^e \mathbf{U}^e,$$

where $\mathbf{U}^e = [U_1 \ U_2 \ U_3]^T$ and the shape function matrix is

$$\mathbf{N}^e = \tfrac{1}{2} \left[\xi^2 - \xi \quad 2(1 - \xi^2) \quad \xi^2 + \xi \right].$$

The elements of $\mathbf{N}^e$ are the usual quadratic Lagrange interpolation polynomials; see Fig. 4.2.

The development of the method follows in exactly the same way as in Section 3.5, and we use the notation of Section 3.6 to obtain the relevant element matrices.

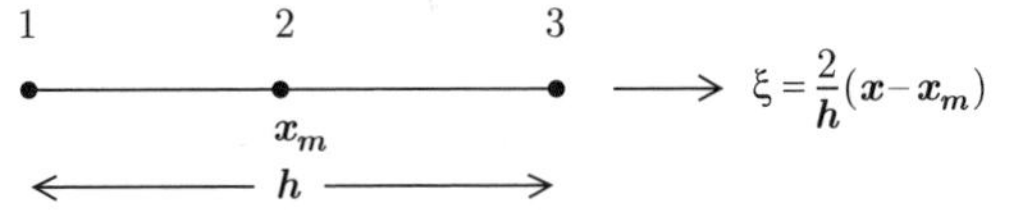

Fig. 4.1 The three-node element for the problem of Example 4.1.

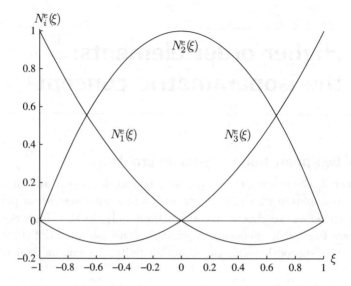

Fig. 4.2 Lagrange quadratic interpolation polynomials.

Thus, in this case,

$$\boldsymbol{\alpha}^e = [d/dx]\,\mathbf{N}^e$$

$$= \frac{1}{h}\,[2\xi - 1 \quad -4\xi \quad 2\xi + 1].$$

Therefore

$$(4.1) \qquad \mathbf{k}^e = \frac{1}{h^2} \int_{-1}^{1} \begin{bmatrix} 1 - 4\xi + 4\xi^2 & 4\xi - 8\xi^2 & -1 + 4\xi^2 \\ & 16\xi^2 & -4\xi - 8\xi^2 \\ \text{sym} & & 1 + 4\xi - 8\xi^2 \end{bmatrix} \frac{h}{2}\,d\xi,$$

i.e.

$$\mathbf{k}^e = \frac{1}{3h} \begin{array}{ccc} & \begin{smallmatrix}1 & 2 & 3\end{smallmatrix} & \\ \begin{bmatrix} 7 & -8 & 1 \\ & 16 & -8 \\ \text{sym} & & 7 \end{bmatrix} & \begin{smallmatrix}1\\2\\3\end{smallmatrix} \end{array}.$$

The element force vector is given by

$$\bar{\mathbf{f}}^e = \int_{-1}^{1} f\mathbf{N}^{e^T} \frac{h}{2}\,d\xi$$

$$= \frac{fh}{6}\,[1 \quad 4 \quad 1]^T$$

when $f(x) \equiv f$, a constant.

For boundary elements, there are also contributions to the stiffness and force matrices due to the Robin boundary conditions. These are given by

$$\bar{k}^e = 2N^{e^T}N^e\big|_{\xi=1}$$

and

$$\bar{f}^e = 3N^{e^T}\big|_{\xi=1}.$$

Suppose that $f = 1$ and the problem is to be solved using two elements with five nodes at $x = 0, \frac{1}{4}, \frac{1}{2}, \frac{3}{4}, 1$. Then

$$
k^1 = \begin{matrix} & \begin{matrix} 1,3 & \quad 2,4 & \quad 3,5 \end{matrix} & \\ \begin{bmatrix} \frac{14}{3} & -\frac{16}{3} & \frac{2}{3} \\ & \frac{32}{3} & -\frac{16}{3} \\ \text{sym} & & \frac{14}{3} \end{bmatrix} & \begin{matrix} 1,3 \\ 2,4 \\ 3,5 \end{matrix} \end{matrix} = k^2,
$$

$$
f^1 = \begin{matrix} \begin{matrix} 1,3 & 2,4 & 3,5 \end{matrix} \\ \begin{bmatrix} \frac{1}{12} & \frac{1}{3} & \frac{1}{12} \end{bmatrix} \end{matrix} = f^2,
$$

$$
\bar{k}^2 = \begin{matrix} & \begin{matrix} 3 & \;\; 4 & \;\; 5 \end{matrix} & \\ \begin{bmatrix} 0 & 0 & 0 \\ & 0 & 0 \\ \text{sym} & & 2 \end{bmatrix} & \begin{matrix} 3 \\ 4 \\ 5 \end{matrix} \end{matrix},
$$

$$
\bar{f}^2 = \begin{matrix} \begin{matrix} 3 & 4 & 5 \end{matrix} \\ \begin{bmatrix} 0 & 0 & 3 \end{bmatrix} \end{matrix}^T.
$$

The overall stiffness and force matrices are thus

$$
K = \begin{matrix} & \begin{matrix} 1 & \quad 2 & \quad 3 & \quad 4 & \quad 5 \end{matrix} & \\ \begin{bmatrix} \frac{14}{3} & -\frac{16}{3} & \frac{2}{3} & 0 & 0 \\ & \frac{32}{3} & -\frac{16}{3} & 0 & 0 \\ & & \frac{28}{3} & -\frac{16}{3} & \frac{2}{3} \\ & & & \frac{23}{3} & -\frac{16}{3} \\ \text{sym} & & & & \frac{20}{3} \end{bmatrix} & \begin{matrix} 1 \\ 2 \\ 3 \\ 4 \\ 5 \end{matrix} \end{matrix},
$$

$$
F = \begin{matrix} \begin{matrix} 1 & \;\; 2 & \;\; 3 & \;\; 4 & \;\; 5 \end{matrix} \\ \begin{bmatrix} \frac{1}{12} & \frac{1}{3} & \frac{1}{6} & \frac{1}{3} & \frac{37}{12} \end{bmatrix} \end{matrix}^T.
$$

Enforcing the essential boundary condition $U_1 = 1$, the unknown nodal values $U_2, \ldots, U_5$ are given by

$$
\begin{aligned}
32U_2 - 16U_3 &= 17, \\
-16U_2 + 28U_3 - 16U_4 + 2U_5 &= -\tfrac{3}{2}, \\
-16U_3 + 32U_4 - 16U_5 &= 1, \\
2U_3 - 16U_4 + 20U_5 &= \tfrac{37}{4}.
\end{aligned}
$$

Solving these equations yields

$$U_2 = \tfrac{39}{32}, \quad U_3 = \tfrac{11}{8}, \quad U_4 = \tfrac{47}{32}, \quad U_5 = \tfrac{3}{2}.$$

These agree with the exact solution $u_0(x) = 1 + x - \tfrac{1}{2}x^2$, as would be expected since the approximation is quadratic. In a similar manner, cubic, quartic and high-order elements may be developed using suitable Lagrange interpolation polynomials.

4.2 Higher-order rectangular elements

The shape functions for the bilinear rectangular element, discussed in Section 3.7, were obtained by taking products of the Lagrange linear interpolation polynomials. This idea may be extended to develop higher-order rectangular elements; for example, the shape functions for the elements in Fig. 4.3 are obtained by taking products of Lagrange quadratic, cubic and quartic polynomials.

Although the Lagrange family is convenient to use and easy to set up, it does have two drawbacks:

1. The elements have many internal nodes and, in the overall system, these contribute to one element only. This may, however, be overcome by the 'condensation' procedure of Section 4.5.

2. The expressions for the shape functions contain relatively high-order terms, while some lower-order terms are missing.

Probably the most useful family of rectangular elements is one in which the nodes are placed on the element boundary. Unfortunately, simply removing the internal nodes from elements in the Lagrange family does not yield suitable elements; see Exercise 4.2.

The most frequently used quadratic and cubic elements are shown in Fig. 4.4. The local coordinates for these elements are just the same as those for the bilinear element shown in Fig. 3.19. Here, the notation $\xi_0 = \xi\xi_i$, $\eta_0 = \eta\eta_i$ will be used; see Exercise 3.22.

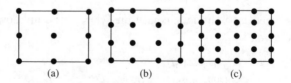

(a) (b) (c)

Fig. 4.3 Rectangular elements whose shape functions are Lagrange interpolation polynomials: (a) quadratic; (b) cubic; (c) quartic.

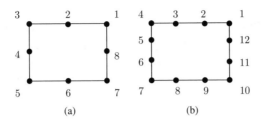

(a)   (b)

Fig. 4.4 Rectangular elements containing only boundary nodes: (a) quadratic; (b) cubic.

The shape functions are as follows:

(a) *Quadratic element.*
   Corner nodes:

(4.2) $$N_i = \tfrac{1}{4}(1+\xi_0)(1+\eta_0)(\xi_0 + \eta_0 - 1).$$

Mid-side nodes:

(4.3) $\xi_i = 0, \quad N_i = \tfrac{1}{2}(1-\xi^2)(1+\eta_0); \quad \eta_i = 0, \quad N_i = \tfrac{1}{2}(1+\xi_0)(1-\eta^2).$

(b) *Cubic element.*
   Corner nodes:

$$N_i = \tfrac{1}{32}(1+\xi_0)(1+\eta_0)[9(\xi^2 + \eta^2) - 10].$$

Mid-side nodes:

$$N_i = \tfrac{9}{32}(1+\xi_0)(1-\eta^2)(1+9\eta_0)$$

when $\xi_i = \pm 1$ and $\eta_i = \pm\tfrac{1}{3}$; the expression for the other mid-side nodes is obtained by interchanging ξ and η.

These elements are conforming elements when the rectangles have sides parallel to the coordinate axes. Whenever this is not the case, the elements may still conform provided that along the common side the function values are expressed in terms of the local coordinates, just as in the case of the bilinear element of Section 3.7.

4.3   Higher-order triangular elements

The quadratic, cubic and quartic triangular elements are shown in Fig. 4.5. The shape functions are very easy to generate in terms of the area coordinates given by eqn (3.62).

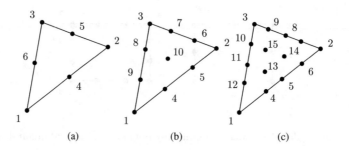

Fig. 4.5 Triangular elements: (a) quadratic; (b) cubic; (c) quartic.

(a) *Quadratic element.*
   Corner nodes:

(4.4) $$N_i = L_i(2L_i - 1).$$

   Mid-side nodes:

(4.5) $$N_4 = 4L_1L_2, \text{ etc.}$$

(b) *Cubic element.*
   Corner nodes:

$$N_i = \tfrac{1}{2}(3L_i - 1)(3L_i - 2).$$

   Side nodes:

$$N_4 = \tfrac{9}{2}L_1L_2(3L_1 - 1),$$
$$N_5 = \tfrac{9}{2}L_1L_2(3L_1 - 2), \text{ etc.}$$

   Node at centroid:

$$N_{10} = 27L_1L_2L_3.$$

(c) *Quartic element.* The shape functions are obtained in Exercise 4.5.

   For all these triangular elements, the function values along a side are uniquely determined by the nodal values on that side. Consequently, these elements are conforming elements. The element matrices for both triangular and rectangular elements are obtained in the usual way; however, it is not difficult to see that the integrals involved make hand calculation impracticable for the cubic and higher-order elements. For this reason, it is usual to use numerical integration to obtain the matrices.

   The matrices for the quadratic rectangle and triangle are obtained in closed form in Exercises 4.3 and 4.4, respectively. It is worth remembering here that the reason for the introduction of higher-order elements is to get a better polynomial approximation for a given number of elements. Improvements may also be made by refining the finite element mesh. Which procedure is the 'better' from a practical point of view is not known. This point was discussed by Desai and Abel (1972), who cited an example due to Clough (1969) in which mesh refinement was found to be better for one problem, whereas higher-order elements provided a better approximation for another.

It is beyond the scope of this text to deal in any detail with the subject of which is the better method. Indeed, there is no specific criterion which states conditions under which one process is better than the other. However, both processes may be used in an adaptive manner, i.e. either the mesh is refined, which is called h-refinement, or higher-order polynomials are used, which is called p-refinement (Zienkiewicz *et al.* 2005).

4.4   Two degrees of freedom at each node

So far, in the generalization of the method to field problems, the nodal variables have simply been the function values at the nodes. There is no reason why derivatives may not also be used.

Example 4.2 Consider again the problem of Example 4.1:

$$-u'' = f(x), \qquad 0 < x < 1,$$

$$u(0) = 1, \qquad u'(1) + 2u(1) = 3.$$

Suppose that in this case the element has two nodes with two degrees of freedom at each node, viz. u and u'; see Fig. 4.6.

In the usual way, u is interpolated throughout the element, in terms of its nodal values, by

$$\tilde{u}^e = [N_1^e \ \ N_2^e \ \ N_3^e \ \ N_4^e][U_1 \ \ U_1' \ \ U_2 \ \ U_2']^T.$$

The Hermite interpolation polynomials are given by (see Fig. 4.7)

(4.6) $H_1(\xi) = \frac{1}{4}(2 - 3\xi + \xi^3), \qquad H_2(\xi) = \frac{1}{4}(1 - \xi - \xi^2 + \xi^3),$

(4.7) $H_3(\xi) = \frac{1}{4}(2 + 3\xi - \xi^3), \qquad H_4(\xi) = \frac{1}{4}(-1 - \xi + \xi^2 + \xi^3).$

Fig. 4.6 Element with two degrees of freedom at each node for the solution of the problem of Example 4.2.

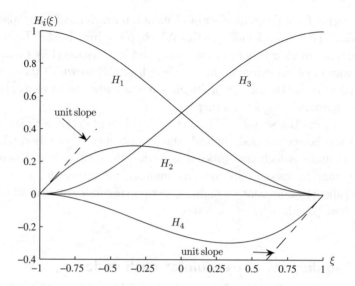

Fig. 4.7 Hermite interpolation polynomials

The shape functions are given by

$$(4.8) \qquad N_1^e(\xi) = H_1(\xi), \qquad N_2^e(\xi) = \frac{h}{2} H_2(\xi),$$

$$(4.9) \qquad N_3^e(\xi) = H_3(\xi), \qquad N_4^e(\xi) = \frac{h}{2} H_4(\xi).$$

Notice the factor $h/2$ in N_2^e and N_4^e; this is due to the fact that these shape functions must give a unit value to du/dx at the nodes and the corresponding Hermite interpolation polynomials give unit values to $dH_i/d\xi$.

It is not difficult to show that these functions have the necessary properties associated with the shape functions; see Exercise 4.8. In the usual way, the element stiffness matrix is given by

$$\mathbf{k}^e = \int_{-1}^{1} \boldsymbol{\alpha}^T \boldsymbol{\alpha} \frac{h}{2} \, d\xi,$$

where

$$(4.10) \quad \boldsymbol{\alpha} = \frac{1}{4h} \left[2(-3 + 3\xi^2) \quad h(-1 - 2\xi + 3\xi^2) \quad 2(3 - 3\xi^2) \quad h(-1 + 2\xi + 3\xi^2) \right].$$

Thus

$$\mathbf{k}^e = \frac{1}{480h} \begin{bmatrix} 576 & 48h & -576 & 48h \\ & 64h^2 & -48h & -16h^2 \\ & & 576 & -48h^2 \\ \text{sym} & & & 64h^2 \end{bmatrix}.$$

The element force vector is given by

$$\mathbf{f}^e = \int_{-1}^1 f \mathbf{N}^{eT} \frac{h}{2} \, d\xi.$$

In the case $f(x) = x$,

$$\mathbf{f}^e = \frac{h}{120} \left[12(5x_m - h) \quad h(10x_m - h) \quad 12(5x_m + h) \quad -h(10x_m + h) \right]^T.$$

There are also contributions to the overall stiffness and force matrices due to the non-homogeneous Robin boundary condition at $x = 1$; these contributions are

$$\bar{\mathbf{k}}^e = 2\mathbf{N}^{eT}\mathbf{N}^e \Big|_{x=1}$$

$$= \begin{bmatrix} 0 & 0 & 0 & 0 \\ & 0 & 0 & 0 \\ & & 2 & 0 \\ \text{sym} & & & 2 \end{bmatrix}$$

and

$$\bar{\mathbf{f}}^e = 3\mathbf{N}^{eT} \Big|_{x=1}$$

$$= [0 \quad 0 \quad 3 \quad 0]^T.$$

Thus in this case, where $f(x) = x$, a one-element solution yields the overall matrices as

$$\mathbf{K} = \frac{1}{480} \begin{bmatrix} 576 & 48 & -576 & 48 \\ & 64 & -48 & -16 \\ & & 1536 & -48 \\ \text{sym} & & & 64 \end{bmatrix},$$

$$\mathbf{F} = \tfrac{1}{60} [9 \quad 2 \quad 21 \quad -3].$$

Thus the overall system of equations is, enforcing the essential boundary condition $U_1 = 1$,

$$64U_1' \quad -48U_2 -16U_2' = 16 - 48,$$
$$-48U_1' +1536U_2 -48U_2' = 1608 + 576,$$
$$-16U_1' \quad -48U_2 +64U_2' = -24 - 48.$$

Solving yields $U_1' = \frac{11}{8}$, $U_2 = \frac{13}{8}$, $U_2' = \frac{1}{9}$, which agrees with the exact solution $u_0(x) = \frac{1}{6}x(1 - x^2)$.

In a similar manner, rectangular elements may be derived in which the nodal variables are u, $\partial u/\partial x$, $\partial u/\partial y$ and the shape functions are products of Hermite polynomials; compare the Lagrange family of elements in Section 4.2. Unlike the linear element, this representation does not give continuity of nodal variables across interelement boundaries; see Exercise 5.1. In order that a compatible rectangular element is obtained, it is necessary to have $\partial^2 u/\partial x \partial y$ as a nodal variable in addition to u, $\partial u/\partial x$ and $\partial u/\partial y$ (Irons and Draper 1965). The rectangular element is, as has already been mentioned in Section 3.8, of limited use because of its special shape; triangles are frequently preferred.

If derivatives at nodes are allowed to be nodal variables, a whole variety of elements is available, and these are described in detail by Zienkiewicz *et al.* (2005). Only one is described here, since it is one of the few two-dimensional elements whch has complete continuity of the first derivatives across interelement boundaries.

Consider the triangle with six nodes shown in Fig. 4.5(a). The element has 21 degrees of freedom throughout the element. The nodal variables at the corner nodes are u, $\partial u/\partial x$, $\partial u/\partial y$, $\partial^2 u/\partial x^2$, $\partial^2 u/\partial y^2$ and $\partial^2 u/\partial x \partial y$. At the mid-side nodes there is just one degree of freedom $\partial u/\partial n$, the normal derivative.

Suppose that s is the distance along the side 3,1. Along this side, u may be taken as a function of the single variable s; also, du/ds and $d^2 u/ds^2$ may be expressed in terms of the first and second partial derivatives of u. Thus, at nodes 3 and 1, there are six quantities U_3, U_3', U_3'', U_1, U_1', U_1'' which uniquely determine a quintic variation of u along this side. Also, $\partial u/\partial n$ may be determined by the five quantities U_{n3}, U_{n6}, U_{n1}, U_{nn3}, U_{nn1}, giving a unique quartic variation of $u_n (\equiv \partial u/\partial n)$ along this side. It follows, then, that u and $\partial u/\partial n$ are continuous across interelement boundaries. For this element, it is more convenient to use Cartesian coordinates and to set up the stiffness matrix using the generalized coordinate formulation of Section 3.4. In practice, the matrix $\mathbf{C}$ obtained is inverted numerically (Zienkiewicz *et al.* 2005).

This 21 degree-of-freedom element is sometimes reduced to an 18 degree-of-freedom element by removing the mid-side nodes, thus allowing a cubic variation of $\partial u/\partial n$ along this side; this of course still maintains compatibility of u and its derivatives.

The elements presented in this section are a very small sample of those available, and the interested reader is recommended to follow up the references. These elements were first developed for solving plate-bending problems, in which the governing differential equation is fourth-order and trial functions must have continuity of the first derivative. For problems governed by second-order equations, it is permissible for the trial functions to have discontinuities in the first derivatives. In these cases the user has a choice: either use function values only as nodal variables, develop simple elements and then find the field variables using the stress matrix, or use function values and derivatives as nodal variables, develop relatively sophisticated elements and then find the field variables directly.

4.5 Condensation of internal nodal freedoms

For elements with internal nodes, there is a procedure which may be adopted at the element stage, i.e. before assembly, which has the effect of removing from the overall system of equations the freedoms at those nodes. This is possible because there is no coupling between these freedoms and freedoms in other elements. It is desirable to do this, since savings in computational cost come from the resulting reduction in overall equation size.

For the sake of generality, we have deliberately avoided reference to the structural origins of the finite element method. However, an analogy is very helpful here. The system of equations

$$(4.11) \qquad\qquad \mathbf{KU} = \mathbf{F}$$

represents the *equilibrium* of the structural system; we have already explained the terms 'stiffness matrix' and 'force vector'. The vector $\mathbf{U}$, in structural terms, describes the system displacements. In terms of the element matrices, eqn (4.11) can be considered to be an amalgam, via the Boolean selection matrices of Section 3.6, of the element equilibrium equations

$$\mathbf{k}^e \mathbf{U}^e = \mathbf{f}^e.$$

We shall use this relationship as follows.

Suppose that the element equilibrium equation may be partitioned in the form

$$\begin{bmatrix} \mathbf{k}_{11} & \mathbf{k}_{12} \\ \mathbf{k}_{21} & \mathbf{k}_{22} \end{bmatrix} \begin{bmatrix} \mathbf{U}_1 \\ \mathbf{U}_2 \end{bmatrix} = \begin{bmatrix} \mathbf{f}_1 \\ \mathbf{f}_2 \end{bmatrix},$$

where $\mathbf{U}_2$ is the vector of nodal variables at the internal nodes and $\mathbf{f}_2$ is the corresponding nodal force vector. Thus

$$(4.12) \qquad\qquad \mathbf{k}_{11}\mathbf{U}_1 + \mathbf{k}_{12}\mathbf{U}_2 = \mathbf{f}_1$$

and

$$\mathbf{k}_{21}\mathbf{U}_1 + \mathbf{k}_{22}\mathbf{U}_2 = \mathbf{f}_2.$$

Hence

$$(4.13) \qquad\qquad \mathbf{U}_2 = \mathbf{k}_{22}^{-1}\mathbf{f}_2 - \mathbf{k}_{22}^{-1}\mathbf{k}_{21}\mathbf{U}_1.$$

Equation (4.12) then gives

$$\hat{\mathbf{k}}\mathbf{U}_1 = \hat{\mathbf{f}},$$

where

$$\hat{k} = k_{11} - k_{12}k_{22}^{-1}k_{21}$$

and

$$\hat{f} = f_1 - k_{12}k_{22}^{-1}f_2.$$

The overall matrices are then assembled in the usual manner.

Example 4.3 Consider again the three-node element of Example 4.1; then

$$k^e = \frac{1}{3h}\left[\begin{array}{cc|c} \overset{1}{7} & \overset{3}{1} & \overset{2}{-8} \\ \hline & 7 & -8 \\ \hline \text{sym} & & 16 \end{array}\begin{array}{c} 1 \\ 3 \\ 2 \end{array}\right]$$

and

$$f^e = \frac{h}{6}\left[\begin{array}{cc|c} \overset{1}{1} & \overset{3}{1} & \overset{2}{4} \end{array}\right]^T.$$

Thus

$$\hat{k} = \frac{1}{3h}\begin{bmatrix} 7 & 1 \\ 1 & 7 \end{bmatrix} - \begin{bmatrix} -\frac{8h}{3} \\ -\frac{8h}{3} \end{bmatrix}\begin{bmatrix} \frac{3h}{16} \end{bmatrix}\begin{bmatrix} -\frac{8h}{3} & -\frac{8h}{3} \end{bmatrix}$$

$$= \frac{1}{h}\begin{bmatrix} 1 & -1 \\ -1 & 1 \end{bmatrix}$$

and

$$\hat{f} = \frac{h}{6}\begin{bmatrix} 1 \\ 1 \end{bmatrix} - \begin{bmatrix} \frac{8h}{3} \\ \frac{8h}{3} \end{bmatrix}\begin{bmatrix} \frac{3h}{16} \end{bmatrix}\begin{bmatrix} \frac{4h}{6} \end{bmatrix}$$

$$= \frac{h}{2}[1 \ \ 1]^T.$$

The contributions to the stiffness and force matrices due to the non-homogeneous boundary condition are given in Example 4.2. It thus follows that the overall stiffness and force matrices for the two-element idealization are

$$K = \left[\begin{array}{ccc} \overset{1}{2} & \overset{3}{-2} & \overset{5}{0} \\ & 4 & -2 \\ \text{sym} & & 4 \end{array}\begin{array}{c} 1 \\ 3 \\ 5 \end{array}\right], \quad F = \left[\begin{array}{c} \frac{1}{4} \\ \frac{1}{2} \\ \frac{13}{4} \end{array}\begin{array}{c} 1 \\ 3 \\ 5 \end{array}\right].$$

Enforcing the essential boundary condition $U_1 = 1$ and solving the equations yields $U_3 = \frac{11}{8}$ and $U_5 = \frac{3}{2}$, as given in Example 4.1.

We now find the remaining nodal values from eqn (4.13) on an element-by-element basis:

$$[U_2] = \left[\tfrac{3h}{16}\right]\left[\tfrac{4h}{6}\right] - \left[\tfrac{3h}{16}\right]\left[-\tfrac{8}{3h} - \tfrac{8}{3h}\right][U_1 \quad U_3]^T,$$

so that $U_2 = 39/32$. Similarly, $U_4 = 47/32$ as given in Example 4.1.

4.6 Curved boundaries and higher-order elements: isoparametric elements

One advantage of higher-order elements is that the better polynomial approximation allows larger elements to be used. Unfortunately, this also means that curved boundaries are not approximated so well; see Fig. 4.8. This can be overcome by placing nodes 2 and 3 on the boundary and fitting a curve through the nodes. This distorts the element and approximates the boundary by a curve, see Fig. 4.9.

The distortion of the element is accomplished by working with a set of local curvilinear coordinates (ξ, η). The global coordinates are expressed in terms of these cordinates by

$$x = x(\xi, \eta), \quad y = y(\xi, \eta).$$

The terms in the integrals for the element stiffness matrices contain x and y derivatives. These derivatives may be expressed in terms of ξ and η derivatives by

Fig. 4.8 Boundary approximation by the side of a high-order element, compared with that by the sides of three linear elements.

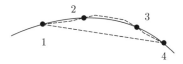

Fig. 4.9 Boundary approximation by a curve, cubic in this case, obtained by placing nodes 2 and 3 on the boundary and thus distorting the side of the element.

$$\begin{bmatrix} \partial/\partial\xi \\ \partial/\partial\eta \end{bmatrix} = \mathbf{J} \begin{bmatrix} \partial/\partial x \\ \partial/\partial y \end{bmatrix},$$

where $\mathbf{J}$ is the Jacobian matrix, given by

$$(4.14) \qquad\qquad \mathbf{J} = \begin{bmatrix} \partial x/\partial\xi & \partial y/\partial\xi \\ \partial x/\partial\eta & \partial y/\partial\eta \end{bmatrix}.$$

Thus, provided that $\mathbf{J}$ is non-singular, it follows that

$$\begin{bmatrix} \partial/\partial x \\ \partial/\partial y \end{bmatrix} = \mathbf{J}^{-1} \begin{bmatrix} \partial/\partial\xi \\ \partial/\partial\eta \end{bmatrix}.$$

In the *isoparametric* approach, the shape functions which interpolate the nodal variables are also used to transform the coordinates. Thus, if the function u is interpolated in element $[e]$ by

$$\tilde{u}^e = N_1 U_1 + N_2 U_2 + \ldots + N_m U_m,$$

then the coordinate transformation is given by

$$(4.15) \qquad\qquad x = N_1 x_1 + N_2 x_2 + \ldots + N_m x_m$$

and

$$(4.16) \qquad\qquad y = N_1 y_1 + N_2 y_2 + \ldots + N_m y_m.$$

Notice that the isoparametric transformation (4.15), (4.16) will always yield the constant-derivative condition (see Exercise 3.2), provided that $\sum N_i = 1$.

The principle of the isoparametric concept is to map a 'parent' element in the $\xi\eta$ plane to a curvilinear element in the xy plane, the sides of which pass through the chosen nodes; see, for example, Figs. 4.10 and 4.11. It is important that no gaps should occur between adjacent distorted elements. This will be the case if the two adjacent elements are generated from parent elements in which the shape functions satisfy the condition necessary for the continuity of u (Zienkiewicz *et al.* 2005).

The terms in the element matrices are treated as follows. Using eqns (4.15) and (4.16),

$$\mathbf{J} = \beta\mathbf{X},$$

where

$$\beta = \begin{bmatrix} \dfrac{\partial\mathbf{N}^e}{\partial\xi} & \dfrac{\partial\mathbf{N}^e}{\partial\eta} \end{bmatrix}^T$$

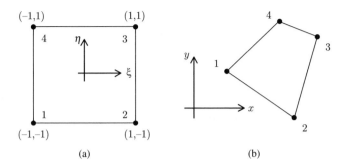

Fig. 4.10 The linear isoparametric quadrilateral: (a) parent element; (b) distorted element.

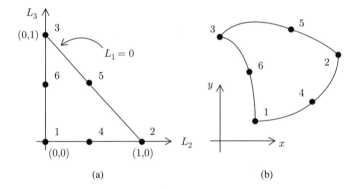

Fig. 4.11 The quadratic isoparametric triangle: (a) parent element; (b) distorted element.

and

$$\mathbf{X} = \begin{bmatrix} x_1 & y_1 \\ x_2 & y_2 \\ \vdots & \vdots \\ x_m & y_m \end{bmatrix}.$$

Since

$$\boldsymbol{\alpha} = \begin{bmatrix} \dfrac{\partial \mathbf{N}^e}{\partial x} & \dfrac{\partial \mathbf{N}^e}{\partial y} \end{bmatrix}^T,$$

it follows that

$$\boldsymbol{\alpha} = \mathbf{J}^{-1}\boldsymbol{\beta}.$$

The region of integration in the xy plane comes from a region in the $\xi\eta$ plane, and the element of area transforms as

$$dx\,dy = |\det \mathbf{J}|\,d\xi\,d\eta.$$

Also, the presence of a Robin boundary condition requires the evaluation of curvilinear integrals. These are obtained by noting that

$$ds^2 = (dx^2 + dy^2),$$

i.e.

(4.17) $ds = \pm\left\{(J_{11}\,d\xi + J_{21}\,d\eta)^2 + (J_{12}\,d\xi + J_{22}\,d\eta)^2\right\}^{1/2}.$

Since the boundary curve has an equation in local coordinates of the form $\eta = \eta(\xi)$ (or $\xi = \xi(\eta)$), the contour integrals are easily transformed to local coordinates. In eqn (4.17), it should be noted that a positive or negative sign must be associated with ds according as ξ (or η) is increasing or decreasing.

Thus, in the evaluation of the element matrices, it is necessary to obtain integrals of the form

$$\int_{\eta_1}^{\eta_2}\int_{\xi_1}^{\xi_2} f(\xi,\eta)\,d\xi\,d\eta \quad \text{and} \quad \int_{\xi_1}^{\xi_2} g(\xi)\,d\xi.$$

These integrals are not at all convenient for analytical evaluation, and Gauss quadrature is used to obtain the results numerically.

Example 4.4 *The linear isoparametric quadrilateral for the solution of the generalized Poisson equation.* One of the disadvantages of the rectangular element described in Section 3.7 is the fact that rectangles rarely provide good approximations to the geometry under consideration. However, quadrilaterals can be used to provide a piecewise straight-line approximation which is arbitrarily close to a given curve.

The parent and distorted elements are shown in Fig. 4.10. The shape function matrix is given by eqn (3.53); thus the matrices necessary for the evaluation of $\mathbf{k}^e$ are

$$\boldsymbol{\beta} = \frac{1}{4}\begin{bmatrix} -(1-\eta) & (1-\eta) & (1+\eta) & -(1+\eta) \\ -(1-\xi) & -(1+\xi) & (1+\xi) & (1-\xi) \end{bmatrix},$$

$$\mathbf{X} = \begin{bmatrix} x_1 & y_1 \\ x_2 & y_2 \\ x_3 & y_3 \\ x_4 & y_4 \end{bmatrix}$$

and

$$\mathbf{J} = \boldsymbol{\beta}\mathbf{X}.$$

It then follows from the result of Exercise 3.7 that

(4.18) $$\mathbf{k}^e = \int_{-1}^{1}\int_{-1}^{1} \boldsymbol{\beta}^T\mathbf{J}^{-T}\boldsymbol{\kappa}\mathbf{J}^{-1}\boldsymbol{\beta}\,|\det\mathbf{J}|\,d\xi\,d\eta.$$

The element force vector is

(4.19) $$\mathbf{f}^e = \int_{-1}^{1}\int_{-1}^{1} f\mathbf{N}^{e^T}|\det\mathbf{J}|\,d\xi\,d\eta.$$

A simple problem which is amenable to hand calculation is given in Exercise 4.10.

Example 4.5 *The quadratic isoparametric triangle.* The parent and distorted elements are shown in Fig. 4.11, the local coordinates being the area coordinates (L_1, L_2, L_3). The shape functions are given by eqn (4.4) and (4.5); thus the matrices necessary for the evaluation of $\mathbf{k}^e$ are

$$\boldsymbol{\beta} = \begin{bmatrix} 4(L_2 + L_3) - 3\ 4L_2 - 1 & 0 & 4(1 - 2L_2 - L_3)\ 4L_3 & -4L_3 \\ 4(L_2 + L_3) - 3 & 0 & 4L_3 - 1 & -4L_2 & 4L_2\ 4(1 - L_2 - 2L_3) \end{bmatrix},$$

$$\mathbf{X} = \begin{bmatrix} x_1\ y_1 \\ \vdots\ \vdots \\ x_6\ y_6 \end{bmatrix}$$

and

$$\mathbf{J} = \boldsymbol{\beta}\mathbf{X}.$$

The element stiffness matrix is then given by

(4.20) $$\mathbf{k}^e = \int_{0}^{1}\int_{0}^{1-L_3} \boldsymbol{\beta}^T\mathbf{J}^{-T}\boldsymbol{\kappa}\mathbf{J}^{-1}\boldsymbol{\beta}\,|\det\mathbf{J}|\,dL_2\,dL_3.$$

The isoparametric concept may also be used to distort three-dimensional elements; see Fig. 4.12.

It is for three-dimensional problems that the concept is at its most useful, since relatively few isoparametric elements can be used to obtain an acceptable boundary approximation. The isoparametric transformation approximates curved boundaries with curved arcs; for example, the quadratic transformation approximates boundaries with parabolic arcs. While this is a big improvement on piecewise linear approximation, there is still in general a difference between the original and the approximating geometry. If essential Dirichlet boundary conditions are enforced only at the nodes, then there will be an error introduced, since these conditions will not hold everywhere on the approximate boundary.

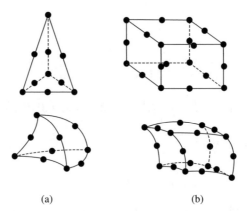

(a)   (b)

Fig. 4.12 Some three-dimensional isoparametric elements: (a) tetrahedron; (b) rectangular brick.

This difficulty may be overcome by using blending functions which interpolate the boundary conditions exactly along the boundary in the $\xi\eta$ plane (Gordon and Hall 1973). To illustrate the procedure, consider the following example.

Example 4.6 Suppose that u is given on the boundary of the four-node rectangle shown in Fig. 3.19. Let $L_1(t) = \frac{1}{2}(1-t)$ and $L_2(t) = \frac{1}{2}(1+t)$ be the usual Lagrange interpolation polynomials; then

$$\bar{u}^e(x,y) = L_1(\eta)u(\xi,-1) + L_2(\eta)u(\xi,1) + L_1(\xi)u(-1,\eta) + L_2(\xi)u(1,\eta)$$

(4.21) $$- L_1(\xi)L_1(\eta)U_1 - L_2(\xi)L_1(\eta)U_2 - L_2(\xi)L_2(\eta)U_3 - L_1(\xi)L_2(\eta)U_4$$

interpolates u throughout the rectangle and gives the exact value of u on its boundaries.

A boundary element will usually have only one or two sides as boundary sides, and in these cases not all the terms in eqn (4.21) are retained. The technique is incorporated into the finite element procedure by replacing eqn (3.5) by (Wait and Mitchell 1985)

$$\tilde{u}^e(x,y) = \sum_e \{u^e(x,y) + \bar{u}^e(x,y)\},$$

where $\bar{u}^e(x,y)$ is non-zero in element $[e]$ only if that element is a boundary element where a Dirichlet boundary condition is specified. With this approximating function $\tilde{u}$, the finite element procedure of Section 3.6 follows through in an identical fashion. The forms of the element matrices for elements other than those adjacent to a boundary with a Dirichlet condition are exactly the same as given by eqns (3.45)–(3.48). For a Dirichlet boundary element, eqn (4.21) may be written in the form

$$\tilde{u}^e = v(\xi, \eta; x, y) - w_1 U_1 - w_2 U_2 - w_3 U_3 - w_4 U_4.$$

It follows, then, that in the expression for the element matrices, the shape functions are replaced by $N_1 - w_1$, etc. There are also contributions to $\mathbf{f}^e$ and $\bar{\mathbf{f}}^e$ given by

$$f_i = \iint_{[e]} \left(f N_i^e - \frac{\partial v}{\partial x}\frac{\partial N_i^e}{\partial x} - \frac{\partial v}{\partial y}\frac{\partial N_i^e}{\partial y} \right) dx\, dy$$

and

$$\bar{f}_i = \int_{C_2^e} (h N_i^e - \sigma N_i^e v)\, ds.$$

Example 4.7 Solve Laplace's equation $\nabla^2 u = 0$ in the square with vertices at $(0,0), (1,0), (1,1), (0,1)$, subject to the boundary conditions $u(0,y) = y$, $u(x,0) = x$, $\partial u/\partial n = y$ on $x = 1$, $\partial u/\partial n = x$ on $y = 1$.

Using one bilinear blended element, eqn (4.21) is modified to give

$$\tilde{u}^e = \tfrac{1}{2}(1 - \eta)u(\xi, -1) + \tfrac{1}{2}(1 - \xi)u(-1, \eta) - \tfrac{1}{4}(1 - \xi)(1 - \eta)U_1.$$

In this case $w_1 = N_1$, $w_2 = w_3 = w_4 = 0$, so that the shape function associated with node 1 is identically zero and the overall system is immediately reduced to a 3×3 set. The element stiffness matrix is obtained from Example 3.4 as

$$\mathbf{k} = \frac{1}{6}\begin{array}{c}\begin{array}{ccc} 2 & 3 & 4 \end{array} \\ \left[\begin{array}{rrr} 4 & -1 & -2 \\ -1 & 4 & -1 \\ -2 & -1 & 4 \end{array}\right]\begin{array}{c} 2 \\ 3 \\ 4 \end{array}\end{array},$$

$$v = \tfrac{1}{2}(1 - \eta)x + \tfrac{1}{2}(1 - \xi)y;$$

thus

$$\frac{\partial v}{\partial x} = \tfrac{1}{2}(1 - \eta) - y = -\eta$$

and

$$\frac{\partial v}{\partial y} = -x + \tfrac{1}{2}(1 - \xi) = -\xi.$$

Thus

$$\mathbf{f} = \int_{-1}^{1}\int_{-1}^{1}\frac{2}{4}\left[\begin{array}{c} (1 - \eta)\eta - \xi(1 + \xi) \\ (1 + \eta)\eta + \xi(1 + \xi) \\ -(1 + \eta)\eta + \xi(1 - \xi) \end{array}\right]\frac{1}{4}\, d\xi\, d\eta$$

$$= \tfrac{1}{3}\begin{array}{c}\begin{array}{ccc} 2 & 3 & 4 \end{array} \\ \left[\begin{array}{ccc} -1 & 1 & 1 \end{array}\right]^T\end{array}.$$

using eqn (3.53) for the shape functions, and

$$\bar{\mathbf{f}} = \int_{-1}^{1} \tfrac{1}{2}(1+\eta)\tfrac{1}{4} \begin{bmatrix} 2(1-\eta) \\ 2(1+\eta) \\ 0 \end{bmatrix} \tfrac{1}{2}\,d\eta + \int_{-1}^{1} \tfrac{1}{2}(1+\xi)\tfrac{1}{4} \begin{bmatrix} 0 \\ 2(1+\xi) \\ 2(1-\xi) \end{bmatrix} \left(-\tfrac{1}{2}\,d\xi\right)$$

$$= \tfrac{1}{6} \begin{bmatrix} \overset{2}{1} & \overset{3}{4} & \overset{4}{1} \end{bmatrix}^{T}.$$

Thus, assembling the equation for the only unknown, U_3,

$$-\tfrac{1}{6}U_2 + \tfrac{2}{3}U_3 - \tfrac{1}{6}U_4 = \tfrac{1}{3} + \tfrac{2}{3}.$$

Now, $U_2 = U_4 = 1$ so that $U_3 = 2$. Blending functions may be used with triangular elements (Barnhill *et al.* 1973) and may also be used to develop elements which satisfy natural boundary conditions exactly (Hall and Heinrich 1978).

4.7   Exercises and solutions

Exercise 4.1 Obtain the matrix $\boldsymbol{\alpha}^e$ for a cubic element to be used in the solution of $-u'' = f(x)$.

Exercise 4.2 For the quadratic rectangular element of the Lagrange family (Fig. 4.3(a)), obtain the shape functions. If the central node is removed, show that the remaining shape functions are not a suitable set.

Exercise 4.3 Obtain the stiffness matrix for the quadratic rectangular element shown in Fig. 4.4(a).

Exercise 4.4 Obtain the shape function for the rectangular brick element shown in Fig. 4.13.

Exercise 4.5 Obtain the shape functions for the quartic triangular element shown in Fig. 4.5(c).

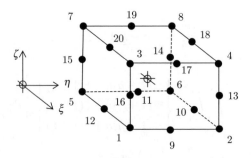

Fig. 4.13 Twenty-node rectangular brick element.

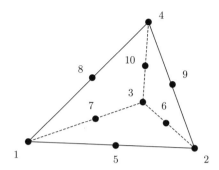

Fig. 4.14 Ten-node tetrahedral element.

Exercise 4.6 Obtain the stiffness matrix for the quadratic triangular element shown in Fig. 4.5(a).

Exercise 4.7 Obtain the shape functions for the tetrahedral element shown in Fig. 4.14. Use the volume coordinates as defined in Exercise 3.23.

Exercise 4.8 Show that the shape functions given by eqn (4.8) and (4.9), defined in terms of the Hermite interpolation polynomials, satisfy the usual necessary conditions for the shape functions.

Exercise 4.9 The rectangular element with four corner nodes shown in Fig. 3.19 is to be used with three degrees of freedom at each node, these being u, $\partial u/\partial x$ and $\partial u/\partial y$. Show that in the resulting element, $\partial^2 u/\partial x \partial y \neq \partial^2 u/\partial y \partial x$ at the nodes, so that the element is not a fully compatible element.

Exercise 4.10 u satisfies Laplace's equation in the square with vertices at A (0, 0), B (1, −1), C (2, 0), D (1, 1) as shown in Fig. 4.15. On sides BC and CD, $u = 2x - 3$; on sides AB and AD, $\partial u/\partial n = \sqrt{2}(1 - 2x)$. Using the isoparametric approach with one element, obtain the stiffness and force matrices and hence find $u(x, y)$.

Exercise 4.11 The generalized Poisson equation $-\text{div}(\kappa \; \text{grad} \, u) = f$ is to be solved using quadratic isoparametric triangles with numerical integration. u satisfies the Dirichlet boundary condition $u = g(s)$ on some part, C_1, of the boundary, and the Robin boundary condition $(\kappa \; \text{grad} \, u) \cdot \mathbf{n} + \sigma u = h(s)$ on the remainder, C_2. Obtain expressions for the element matrices necessary for the solution of the problem.

Exercise 4.12 Obtain the matrices necessary for the computation of the element matrices for the isoparametric quadratic rectangle.

Solution 4.1 Using the usual local coordinate $\xi = 2(x - x_m)/h$, the shape function matrix is

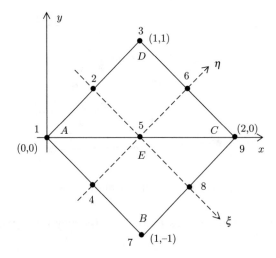

Fig. 4.15 Square region for the problem of Exercise 4.11. (ξ, η) are the usual local coordinates.

$$\mathbf{N}^e = \frac{1}{16} \begin{bmatrix} 9(\xi^2 - 1)(1 - \xi) \\ 9(1 - 3\xi)(1 - \xi^2) \\ 9(1 + 3\xi)(\xi^2 - 1) \\ (9\xi^2 - 1)(1 + \xi) \end{bmatrix}^T .$$

Thus

$$\boldsymbol{\alpha}^e = \frac{1}{8h} \begin{bmatrix} -27\xi^2 + 18\xi + 1 \\ 81\xi^2 - 18\xi - 27 \\ 81\xi^2 + 18\xi - 27 \\ 27\xi^2 + 18\xi - 1 \end{bmatrix}^T .$$

Solution 4.2 Taking (ξ, η) as local coordinates at the centre of the element, the shape functions are

$$\tfrac{1}{4}(\xi^2 - \xi)(\eta^2 - \eta), \quad \tfrac{1}{2}(1 - \xi^2)(\eta^2 - \eta), \quad \tfrac{1}{4}(\xi^2 + \xi)(\eta^2 - \eta),$$
$$\tfrac{1}{2}(\xi^2 + \xi)(1 - \eta^2), \quad \tfrac{1}{4}(\xi^2 + \xi)(\eta^2 + \eta), \quad \tfrac{1}{2}(1 - \xi^2)(\eta^2 + \eta),$$
$$\tfrac{1}{4}(\xi^2 - \xi)(\eta^2 + \eta), \quad \tfrac{1}{2}(\xi^2 - \xi)(1 - \eta^2), \quad \tfrac{1}{4}(1 - \xi^2)(1 - \eta^2).$$

If the node at the centroid is removed, then the remaining shape functions satisfy

$$\sum N_i = \xi^2 + \eta^2 - \xi^2\eta^2 \neq 1.$$

Thus they do not satisfy one of the conditions necessary to recover an arbitrary linear form, see Exercise 3.2.

Solution 4.3 The shape functions are given by eqn (4.2) as

$$N_i = \tfrac{1}{4}(1 + \xi\xi_i)(1 + \eta\eta_i)(\xi\xi_i + \eta\eta_i - 1).$$

Thus the matrix $\boldsymbol{\alpha}^e$ is given by

$$
\left.
\begin{aligned}
\alpha_{1j} &= \frac{\partial N_j}{\partial x} = \frac{1}{2a}(1 + \eta\eta_j)\xi_j(\eta\eta_j + 2\xi\xi_j), \\
\alpha_{2j} &= \frac{\partial N_j}{\partial y} = \frac{1}{2b}(1 + \xi\xi_j)\eta_j(2\eta\eta_j + \xi\xi_j),
\end{aligned}
\right\} \quad j = 1, 3, 5, 7,
$$

$$\alpha_{1j} = -\frac{\xi}{2a}(1 + \eta\eta_j), \quad \alpha_{2j} = \frac{\eta_j}{4b}(1 - \xi^2), \quad j = 2, 6,$$

$$\alpha_{1j} = \frac{\xi_j}{4a}(1 - \eta^2), \quad \alpha_{2j} = -\frac{\eta}{2b}(1 + \xi\xi_j), \quad j = 4, 8.$$

Now,

$$k_{ij} = \int_{-1}^{1} \int_{-1}^{1} (\alpha_{1i}\alpha_{1j} + \alpha_{2i}\alpha_{2j}) \frac{ab}{4} d\xi\, d\eta.$$

Thus, after some algebra, it follows that $\mathbf{k}^e = \frac{1}{90} \times$

$$
\begin{array}{cccccccc}
1 & 2 & 3 & 4 & 5 & 6 & 7 & 8 \\
\end{array}
$$

$$
\begin{bmatrix}
52(r+1/r) & 6r-80/r & 17r+28/r & -40r-6/r & 23(r+1/r) & -6r-40/r & 28r+17/r & -80r+6/r \\
 & 48r+160/r & 6r-80/r & 0 & -6r-40/r & -48r+80/r & -6r-40/r & 0 \\
 & & 52(r+1/r) & -80r+6/r & 28r+17/r & -6r-40/r & 23(r+1/r) & -40r-6/r \\
 & & & 160r+48/r & -80r+6/r & 0 & -40r-6/r & 80r-48/r \\
 & & & & 52(r+1/r) & 6r-80/r & 17r+28/r & -40r-6/r \\
 & & & & & 48r+160/r & 6r-80/r & 0 \\
 & & & & & & 52(r+1/r) & -80r+6/r \\
\text{sym} & & & & & & & 160r+48/r \\
\end{bmatrix}
\begin{array}{c}
1 \\ 2 \\ 3 \\ 4 \\ 5 \\ 6 \\ 7 \\ 8
\end{array}
$$

Solution 4.4 Following the notation of Zienkiewicz *et al.* (2005), the shape functions are as follows.

Corner nodes:

$$N_i = \tfrac{1}{8}(1 + \xi\xi_i)(1 + \eta\eta_i)(1 + \zeta\zeta_i)(\xi\xi_i + \eta\eta_i + \zeta\zeta_i - 2).$$

Mid-side nodes $9, 17, 19, 11$ (i.e. $\eta_i = 0$) :

$$N_i = \tfrac{1}{4}(1 - \eta^2)(1 + \xi\xi_i)(1 + \zeta\zeta_i).$$

The shape functions for the other mid-side nodes follow in a similar manner.

Solution 4.5 The shape functions are as follows.

Corner nodes:

$$N_i = \tfrac{1}{6}L_i(4L_i - 1)(4L_i - 2)(4L_i - 3).$$

Side nodes:

$$N_4 = \tfrac{8}{3}L_1L_2(4L_1 - 1)(4L_1 - 2),$$
$$N_5 = 4L_1L_2(4L_1 - 1)(4L_2 - 1),$$
$$N_6 = \tfrac{8}{3}L_1L_2(4L_2 - 1)(4L_2 - 2), \text{ etc.}$$

Internal nodes:

$$N_{13} = 32L_1L_2L_3(4L_2 - 1), \text{ etc.}$$

Solution 4.6 The shape functions are given by eqns (4.4) and (4.5); the local coordinates (L_1, L_2, L_3) are related to the global coordinates (x, y) by eqn (3.62). Using these results, the matrix $\boldsymbol{\alpha}$ may be written as

$$\boldsymbol{\alpha} = \frac{1}{2A}\boldsymbol{\omega}\mathbf{B},$$

where

$$\boldsymbol{\omega} = \begin{bmatrix} L_1 & L_2 & L_3 & 0 & 0 & 0 \\ 0 & 0 & 0 & L_1 & L_2 & L_3 \end{bmatrix}$$

and

$$\mathbf{B} = \begin{bmatrix} -3b_1 & -b_2 & -b_3 & 4b_2 & 0 & 4b_3 \\ -b_1 & 3b_2 & -b_3 & 4b_1 & 4b_3 & 0 \\ -b_1 & -b_2 & 3b_3 & 0 & 4b_2 & 4b_1 \\ 3c_1 & -c_2 & -c_3 & 4c_2 & 0 & 4c_3 \\ -c_1 & 3c_2 & -c_3 & 4c_1 & 4c_3 & 0 \\ -c_1 & -c_2 & -3c_3 & 0 & 4c_2 & 4c_1 \end{bmatrix}.$$

Thus

$$\mathbf{k}^e = \frac{1}{4A^2}\mathbf{B}^T\left(\iint_A \boldsymbol{\omega}^T\boldsymbol{\omega}\,dx\,dy\right)\mathbf{B}.$$

Now,

$$\iint_A \boldsymbol{\omega}^T\boldsymbol{\omega}\,dx\,dy = \begin{bmatrix} \boldsymbol{\Lambda} & \mathbf{0} \\ \mathbf{0} & \boldsymbol{\Lambda} \end{bmatrix}$$

$$= \mathbf{D}, \text{ say,}$$

where

$$\boldsymbol{\Lambda} = \frac{A}{12} \begin{bmatrix} 2 & 1 & 1 \\ 1 & 2 & 1 \\ 1 & 1 & 2 \end{bmatrix}; \quad \text{see Exercise 3.11.}$$

Therefore

$$\mathbf{k}^e = \frac{1}{4A^2} \mathbf{B}^T \mathbf{D} \mathbf{B}.$$

Solution 4.7 The shape functions are as follows.

Corner nodes:

$$N_i = (2L_i - 1)L_i, \quad i = 1, \ldots, 4.$$

Mid-side nodes:

$$N_i = 4L_k L_j, \quad i = 5, \ldots, 10,$$

where node i lies on the edge joining corner nodes k and j.

Solution 4.8 (i) Suppose that the nodal points are at $x = x_1$; see Fig. 4.6. Using eqn (3.10), the shape functions must satisfy the following.

At node 1:

$$N_1 = \frac{dN_2}{dx} = 1, \quad \frac{dN_1}{dx} = N_2 = N_3 = \frac{dN_3}{dx} = N_4 = \frac{dN_4}{dx} = 0.$$

At node 2:

$$N_1 = \frac{dN_1}{dx} = N_2 = \frac{dN_2}{dx} = \frac{dN_3}{dx} = N_4 = 0, \quad N_3 = \frac{dN_4}{dx} = 1.$$

These results are easily verified using eqn (4.6)–(4.9), remembering that

$$\frac{dN_i}{dx} = \frac{2}{h} \frac{dN_i}{d\xi}.$$

(ii) The shape functions must be able to recover an arbitrary linear form. Thus solutions $u^e = 1$ and $u^e = x$ must be realizable. It follows then that the shape functions must satisfy $N_1 + N_2 = 1$ and $N_1 x_1 + N_2 + N_3 x_2 + N_4 = x$. The first of these is easily verified; the second can be verified as follows:

$$N_1 x_1 + N_2 + N_3 x_2 + N_4 = \frac{1}{2}(x_1 + x_2) + \frac{3\xi}{4}(x_2 - x_1) + \frac{\xi^3}{4}(x_1 - x_2) - \frac{h\xi}{4} + \frac{h\xi^3}{4}$$

$$= x_m + \frac{h\xi}{2} = x.$$

Solution 4.9 (Irons and Draper 1965). The element has 12 degrees of freedom to specify u, four to specify $\partial u/\partial x$ and four to specify $\partial u/\partial y$. Thus

$$u = a_0 + a_1 x + a_2 y + \ldots,$$

$$\frac{\partial u}{\partial x} = b_0 + b_1 x + b_2 y + b_3 xy,$$

$$\frac{\partial u}{\partial y} = c_0 + c_1 x + c_2 y + c_3 xy.$$

Along side 3,4, on which y is constant,

$$u = \alpha_1 + \alpha_2 x + \alpha_3 x^2 + \alpha_4 x^3,$$

and there are exactly four nodal parameters, u and $\partial u/\partial x$ at each node, to define uniquely the value of u along this side. It thus follows that u is continuous across this boundary. Similarly, it may be shown that u is continuous across the other three boundaries. Now, along side 3,4, $\partial u/\partial y$ will be interpolated linearly between its nodal values, so that on this side

$$\frac{\partial u}{\partial y} = \tfrac{1}{2}(1 - \xi)(u_y)_4 + \tfrac{1}{2}(1 + \xi)(u_y)_3.$$

Thus

$$\frac{\partial^2 u}{\partial x \partial y} = \frac{1}{a}\left\{(u_y)_3 - (u_y)_4\right\}.$$

Similarly, along side 2,3,

$$\frac{\partial u}{\partial x} = \tfrac{1}{2}(1 - \eta)(u_x)_2 + \tfrac{1}{2}(1 + \eta)(u_x)_3.$$

Thus

$$\frac{\partial^2 u}{\partial y \partial x} = \frac{1}{b}\left\{(u_x)_3 - (u_x)_2\right\}.$$

In general, then, it follows that at node 3,

$$\frac{\partial^2 u}{\partial x \partial y} \neq \frac{\partial^2 u}{\partial y \partial x},$$

and so the necessary conditions for a continuous first derivative are not satisfied, i.e. the element is not fully compatible.

Solution 4.10 For Laplace's equation,

$$\kappa = \begin{bmatrix} 1 & 0 \\ 0 & 1 \end{bmatrix}.$$

The shape function matrix is

$$\mathbf{N}^e = \tfrac{1}{4}\left[(1-\xi)(1-\eta) \quad (1+\xi)(1-\eta) \quad (1+\xi)(1+\eta) \quad (1-\xi)(1+\eta)\right].$$

The isoparametric transformation gives

$$x = \mathbf{N}^e \begin{bmatrix} 0 & 1 & 2 & 1 \end{bmatrix}^T, \qquad y = \mathbf{N}^e \begin{bmatrix} 0 & -1 & 0 & 1 \end{bmatrix}^T.$$

Using Example 4.4,

$$\boldsymbol{\beta} = \frac{1}{4}\begin{bmatrix} -(1-\eta) & (1-\eta) & (1+\eta) & -(1+\eta) \\ -(1-\xi) & -(1+\xi) & (1+\xi) & (1-\xi) \end{bmatrix},$$

$$\mathbf{X} = \begin{bmatrix} 0 & 0 \\ 1 & -1 \\ 2 & 0 \\ 1 & 1 \end{bmatrix}.$$

Thus

$$\mathbf{J} = \begin{bmatrix} \tfrac{1}{2} & -\tfrac{1}{2} \\ \tfrac{1}{2} & \tfrac{1}{2} \end{bmatrix}, \qquad |\det \mathbf{J}| = \tfrac{1}{2}$$

and

$$\mathbf{J}^{-1} = \begin{bmatrix} 1 & 1 \\ -1 & 1 \end{bmatrix}.$$

Therefore

$$\mathbf{J}^{-T}\boldsymbol{\kappa}\mathbf{J}^{-1} = \begin{bmatrix} 2 & 0 \\ 0 & 2 \end{bmatrix}.$$

Thus, using eqn (4.18),

$$\mathbf{k}^e = \int_{-1}^{1}\int_{-1}^{1} \boldsymbol{\beta}^T \begin{bmatrix} 2 & 0 \\ 0 & 2 \end{bmatrix} \boldsymbol{\beta}\tfrac{1}{2}\,d\xi\,d\eta$$

$$= \frac{1}{6}\begin{bmatrix} 4 & -1 & -2 & -1 \\ & 4 & -1 & -2 \\ & & 4 & -1 \\ \text{sym} & & & 4 \end{bmatrix}.$$

The force vector has contributions from the non-homogeneous Neumann boundary condition.

$$\text{On side } DA, \quad 1 - 2x = -\eta \quad \text{and} \quad ds = -\frac{1}{\sqrt{2}} d\eta.$$

$$\text{On side } AB, \quad 1 - 2x = -\xi \quad \text{and} \quad ds = \frac{1}{\sqrt{2}} d\xi.$$

Thus

$$\mathbf{f} = \int_{-1}^{1} \tfrac{1}{4} [2(1 - \eta) \quad 0 \quad 0 \quad 2(1 + \eta)]^T \, \eta \, d\eta$$

$$- \int_{-1}^{1} \tfrac{1}{4} [2(1 - \xi) \quad 2(1 + \xi) \quad 0 \quad 0]^T \, \xi \, d\xi$$

$$= \tfrac{1}{3} [2 \quad -1 \quad 0 \quad 1]^T.$$

Thus assembling the equation for the one unknown u_A, enforcing the essential boundary conditions $u_B = -1, u_C = 1, u_D = -1$ and solving yields $u_A = 1$.
Interpolating through the element gives

$$\tilde{u}^e(x, y) = \mathbf{N} [1 \quad -1 \quad 1 \quad -1]^T$$

$$= \xi \eta$$

$$= x^2 - y^2 - 2x + 1.$$

This is the exact solution, which has been recovered since it is a linear combination of $1, \xi, \eta$ and $\xi\eta$. In general, an arbitrary quadratic function cannot be expressed as a linear combination of $1, \xi, \eta$ and $\xi\eta$.

Solution 4.11 Using eqn (4.20),

$$\mathbf{k}^e = \int_0^1 \int_0^{1-L_3} \boldsymbol{\beta}^T \mathbf{J}^{-T} \kappa \mathbf{J}^{-1} \boldsymbol{\beta} \, |\det \mathbf{J}| \, dL_2 \, dL_3.$$

Thus, using a Gauss quadrature formula for the integral,

$$\mathbf{k}^e \approx \sum_{g=1}^{p_1} w_g \boldsymbol{\beta}_g^T \mathbf{J}_g^{-T} \kappa_g \mathbf{J}_g^{-1} \boldsymbol{\beta}_g \, |\det \mathbf{J}_g|,$$

where the subscript g means we evaluate at Gauss point g and a quadrature of order p_1 is chosen; w_g is the corresponding weight; see Appendix D.

The force vector is given by

$$\mathbf{f}^e = \int_0^1 \int_0^{1-L_3} f \mathbf{N}^{eT} |\det \mathbf{J}| \, dL_2 \, dL_3$$

$$\approx \sum_{g=1}^{p_1} w_g f_g \mathbf{N}_g^{eT} |\det \mathbf{J}_g| .$$

For a boundary element which coincides with a part of C_2, there are contributions to the stiffness and force given by eqns (3.46) and (3.48) as

$$\bar{\mathbf{k}}^e = \int_{C_2^e} \sigma \mathbf{N}^{eT} \mathbf{N}^e \, ds,$$

$$\bar{\mathbf{f}}^e = \int_{C_2^e} h \mathbf{N}^{eT} \, ds,$$

where ds is given by eqn (4.17). Thus, using a one-dimensional Gauss quadrature formula of order p_2,

$$\bar{\mathbf{k}}^e \approx \sum_{g=1}^{p_2} w_g \sigma_g \mathbf{N}_g^{eT} \mathbf{N}_g^e \, ds_g,$$

$$\bar{\mathbf{f}}^e \approx \sum_{g=1}^{p_2} w_g h_g \mathbf{N}_g^{eT} \, ds_g,$$

where w_g is the weight associated with Gauss point g.

Solution 4.12 The shape functions are given by eqn (4.2) as

$$\boldsymbol{\beta} = \begin{bmatrix} \partial \mathbf{N}^e / \partial \xi \\ \partial \mathbf{N}^e / \partial \eta \end{bmatrix},$$

and these partial derivatives have already been obtained in Exercise 4.3, remembering that in the $\xi\eta$ plane, $a = b = 2$:

$$\left. \begin{aligned} \beta_{1j} &= \frac{1}{4}(1 + \eta\eta_i)\xi_i(\eta\eta_i + 2\xi\xi_i), \\ \beta_{2i} &= \frac{1}{4}(1 + \xi\xi_i)\eta_i(2\eta\eta_i + \xi\xi_i), \end{aligned} \right\} j = 1, 3, 5, 7,$$

$$\beta_{1j} = -\frac{\xi}{4}(1 + \eta\eta_j), \quad \beta_{2j} = \frac{\eta_j}{8}(1 - \xi^2), \quad j = 2, 6,$$

$$\beta_{1j} = \frac{\xi_j}{8}(1 - \eta^2), \quad \beta_{2j} = -\frac{\eta}{4}(1 + \xi\xi_j), \quad j = 4, 8,$$

$$\mathbf{X} = \begin{bmatrix} x_1 & y_1 \\ x_2 & y_2 \\ \vdots & \vdots \\ x_8 & y_8 \end{bmatrix}.$$

The Jacobian may now be evaluated using $\mathbf{J} = \beta\mathbf{X}$. Using these matrices, eqns (4.18) and (4.19) give the element matrices.

5 Further topics in the finite element method

So far, elliptic problems only have been considered, and we shall see in Section 5.1 how, with reference to Poisson problems, the variational approach is equivalent to Galerkin's method. Of course, for many problems of practical interest, such variational principles may not exist, or where they do, a suitable functional may not be known. In this chapter we shall consider procedures for a wide variety of problems, including parabolic, hyperbolic and non-linear problems. It is not intended to be any more than an introduction, and the ideas are presented by way of particular problems. The reader with a specific interest in any one subject area will find the references useful for further detail.

5.1 The variational approach

We seek a finite element solution of the problem given by eqns (3.30)–(3.32), viz.

$$(5.1) \qquad\qquad -\mathrm{div}(k\,\mathrm{grad}\,u) = f(x,y) \quad \text{in } D,$$

with the Dirichlet boundary condition

$$(5.2) \qquad\qquad u = g(s) \quad \text{on } C_1$$

and the Robin boundary condition

$$(5.3) \qquad\qquad k(s)\frac{\partial u}{\partial n} + \sigma(s)u = h(s) \quad \text{on } C_2.$$

The functional for this problem is found from eqn (2.44) as

$$I[u] = \iint_D \left\{ k\left(\frac{\partial u}{\partial x}\right)^2 + k\left(\frac{\partial u}{\partial y}\right)^2 - 2uf \right\} dx\, dy + \int_{C_2} (\sigma u^2 - 2uh)\, ds$$

and the solution, u, of eqns (5.1)–(5.3) is that function, u_0, which minimizes $I[u]$ subject to the essential boundary condition $u_0 = g(s)$ on C_1.

We follow exactly the finite element philosophy of Section 3.4, writing

$$(5.4) \qquad\qquad \tilde{u}(x,y) = \sum_e \tilde{u}^e(x,y)$$

with

(5.5) $$\tilde{u}^e(x, y) = \mathbf{N}^e(x, y)\mathbf{U}^e.$$

Then

$$I[\tilde{u}] = \int\int_D \left\{ k\left(\frac{\partial}{\partial x}\sum_e \tilde{u}^e\right)^2 + k\left(\frac{\partial}{\partial y}\sum_e \tilde{u}^e\right)^2 - 2\sum_e \tilde{u}^e f \right\} dx\, dy$$

$$+ \int_{C_2}\left\{ \sigma\left(\sum_e \tilde{u}^e\right)^2 - 2\sum_e \tilde{u}^e h \right\} ds.$$

Now, since $\tilde{u}^e$ is zero outside element $[e]$, the only non-zero contribution to $I[\tilde{u}]$ from $\tilde{u}^e$ comes from integration over the element itself. Thus

$$I[\tilde{u}] = \sum_e \int\int_{[e]} \left\{ k\left(\frac{\partial \tilde{u}^e}{\partial x}\right)^2 + k\left(\frac{\partial \tilde{u}^e}{\partial y}\right)^2 - 2\tilde{u}^e f \right\} dx\, dy$$

$$+ \sum_e \int_{C_2}\left\{ \sigma(\tilde{u}^e)^2 - 2\tilde{u}^e h \right\} ds$$

$$= \sum_e I^e, \quad \text{say.}$$

The second term applies only if the element has a boundary coincident with C_2; see Fig. 3.15.

Using eqns (5.4) and (5.5),

$$I[\tilde{u}] = I(U_1, U_2, \ldots, U_n).$$

Then, using the Rayleigh–Ritz procedure to minimize I with respect to the variational parameters U_i gives

$$\frac{\partial I}{\partial U_i} = 0, \quad i = 1, \ldots, n,$$

i.e.

(5.6) $$\sum_e \frac{\partial I^e}{\partial U_i} = 0, \quad i = 1, \ldots, n.$$

Before developing the element matrices, it is helpful to express the equations (5.6) as a single matrix equation.

Define

$$\frac{\partial I^e}{\partial \mathbf{U}} = \left[\frac{\partial I^e}{\partial U_1} \quad \frac{\partial I^e}{\partial U_2} \quad \cdots \quad \frac{\partial I^e}{\partial U_n} \right]^T.$$

Suppose that element $[e]$ has nodes $p, q, \ldots, i, \ldots, s$; see Fig. 3.16. Then

$$\frac{\partial I^e}{\partial \mathbf{U}} = [\overset{1}{0} \; \overset{2}{0} \; \cdots \; \overset{p}{\frac{\partial I^e}{\partial U_p}} \; 0 \; \cdots \; 0 \; \overset{q}{\frac{\partial I^e}{\partial U_q}} \; 0 \; \cdots \; 0 \; \overset{i}{\frac{\partial I^e}{\partial U_i}} \; 0 \; \cdots$$

(5.7)
$$\cdots \; 0 \; \overset{s}{\frac{\partial I^e}{\partial U_s}} \; 0 \; \cdots \; 0 \; \overset{n}{0} \;]^T \, ,$$

and the equations (5.6) become

$$\sum_e \frac{\partial I^e}{\partial \mathbf{U}} = \mathbf{0}.$$

Now,

$$\frac{\partial I^e}{\partial U_i} = \int\int_{[e]} \left\{ k \frac{\partial}{\partial U_i} \left(\frac{\partial \tilde{u}^e}{\partial x} \right)^2 + k \frac{\partial}{\partial U_i} \left(\frac{\partial \tilde{u}^e}{\partial y} \right)^2 - 2 \frac{\partial \tilde{u}^e}{\partial U_i} f \right\} dx \, dy$$

(5.8)
$$+ \int_{C_2} \left\{ \sigma \frac{\partial}{\partial U_i} (\tilde{u}^e)^2 - -2 \frac{\partial \tilde{u}^e}{\partial U_i} h \right\} ds.$$

If node i is not associated with element $[e]$, then $\partial I^e / \partial U_i = 0$; a non-zero contribution to $\partial I^e / \partial U_i$ will occur only if node i is associated with element $[e]$. This is shown in eqn (5.7). For element $[e]$, as shown in Fig. 3.16,

$$\tilde{u}^e(x, y) = N_p^e U_p + N_q^e U_q + \ldots + N_i^e U_i + \ldots + N_s^e U_s = \sum_{j \in [e]} N_j^e U_j.$$

Then

$$\frac{\partial}{\partial U_i} \left(\frac{\partial \tilde{u}^e}{\partial x} \right)^2 = 2 \frac{\partial \tilde{u}^e}{\partial x} \frac{\partial}{\partial U_i} \left(\frac{\partial \tilde{u}^e}{\partial x} \right)$$

$$= 2 \frac{\partial \tilde{u}^e}{\partial x} \frac{\partial}{\partial x} \left(\frac{\partial \tilde{u}^e}{\partial U_i} \right)$$

$$= 2 \frac{\partial}{\partial x} (\mathbf{N}^e \mathbf{U}^e) \frac{\partial N_i^e}{\partial x}$$

$$= 2 \left[\frac{\partial N_i^e}{\partial x} \frac{\partial N_p^e}{\partial x} \; \cdots \; \frac{\partial N_i^e}{\partial x} \frac{\partial N_s^e}{\partial x} \right] [U_p \ldots U_s]^T .$$

Similarly,

$$\frac{\partial}{\partial U_i} \left(\frac{\partial \tilde{u}^e}{\partial y} \right)^2 = 2 \left[\frac{\partial N_i^e}{\partial y} \frac{\partial N_p^e}{\partial y} \; \cdots \; \frac{\partial N_i^e}{\partial y} \frac{\partial N_s^e}{\partial y} \right] [U_p \ldots U_s]^T .$$

Now,

$$\frac{\partial \tilde{u}^e}{\partial U_i} = N_i^e$$

and

$$\frac{\partial}{\partial U_i} (\tilde{u}^e)^2 = 2 \left[N_i^e N_p^e \dots N_i^e N_s^e \right] \left[U_p \dots U_s \right]^T.$$

Thus eqn (5.8) becomes

$$\frac{\partial I^e}{\partial U_i} = 2 \iint_{[e]} k \left[\left(\frac{\partial N_i^e}{\partial x} \frac{\partial N_p^e}{\partial x} + \frac{\partial N_i^e}{\partial y} \frac{\partial N_p^e}{\partial y} \right) \right.$$

$$\left. \dots \left(\frac{\partial N_i^e}{\partial x} \frac{\partial N_s^e}{\partial x} + \frac{\partial N_i^e}{\partial y} \frac{\partial N_s^e}{\partial y} \right) \right] \begin{bmatrix} U_p \\ \vdots \\ U_s \end{bmatrix} dx \, dy$$

$$- 2 \iint_{[e]} f N_i^e \, dx \, dy + 2 \int_{C_2^e} \sigma \left[N_i^e N_p^e \dots N_i^e N_s^e \right] \begin{bmatrix} U_p \\ \vdots \\ U_s \end{bmatrix} ds$$

$$- 2 \int_{C_2^e} h N_i^e \, ds,$$

i.e.

$$\frac{\partial I^e}{\partial U_i} = 2 \sum_{j \in [e]} k_{ij}^e U_j + 2 \sum_{j \in [e]} \bar{k}_{ij}^e U_j - 2 f_i^e - 2 \bar{f}_i^e,$$

where

(5.9) $$k_{ij}^e = \iint_{[e]} k \left(\frac{\partial N_i^e}{\partial x} \frac{\partial N_j^e}{\partial x} + \frac{\partial N_i^e}{\partial y} \frac{\partial N_j^e}{\partial y} \right) dx \, dy,$$

(5.10) $$\bar{k}_{ij}^e = \int_{C_2^e} \sigma N_i^e N_j^e \, ds,$$

(5.11) $$f_i^e = \iint_{[e]} f N_i^e \, dx \, dy,$$

(5.12) $$\bar{f}_i^e = \int_{C_2^e} h N_i^e \, ds,$$

which are exactly eqns (3.40)–(3.43), developed using Galerkin's method in Chapter 3, and these lead as before to the matrix form given by eqns (3.45)–(3.48). The Galerkin approach of Chapter 3 is more general, since it is applicable in cases where a variational principle does not exist. However, the variational procedure ensures that the resulting stiffness matrix, reduced by enforcing the essential boundary condition, is positive definite and hence non-singular, provided that the differential operator $\mathcal{L}$ is positive definite.

Example 5.1 Consider the differential operator $\mathcal{L}$ given by

$$\mathcal{L}u = -\text{div}(\boldsymbol{\kappa} \text{ grad } u) \, dx \, dy.$$

Then

$$\iint_D u\mathcal{L}u \, dx \, dy = -\iint_D u \text{ div}(\boldsymbol{\kappa} \text{ grad } u) \, dx \, dy$$

$$= \iint_D \text{grad } u \cdot (\boldsymbol{\kappa} \text{ grad } u) \, dx \, dy - \oint_C u \, (\boldsymbol{\kappa} \text{ grad } u) \cdot \mathbf{n} \, ds$$

using the generalized first form of Green's theorem (2.6). For homogeneous Dirichlet boundary conditions, the boundary integral vanishes, and hence $\mathcal{L}$ is positive definite provided that $\boldsymbol{\kappa}$ is positive definite. For a homogeneous Robin boundary condition of the form

$$(\boldsymbol{\kappa} \text{ grad } u) \cdot \mathbf{n} + \sigma \mathbf{u} = \mathbf{0},$$

it is also necessary that $\sigma > 0$ in order that $\mathcal{L}$ is positive definite (*cf.* Example 2.2).

Suppose that $v = \sum_j w_j v_j$, where v_j is arbitrary and w_j is the nodal function associated with a node at which a Dirichlet boundary condition is not specified. Then

$$\iint_D v\mathcal{L}v \, dx \, dy = \iint_D -v \text{ div} (\boldsymbol{\kappa} \text{ grad } v) \, dx \, dy$$

$$= \iint_D \text{grad } v \cdot (\boldsymbol{\kappa} \text{ grad } v) \, dx \, dy - \int_C v \, (\boldsymbol{\kappa} \text{ grad } v) \cdot \mathbf{n} \, ds$$

$$= \mathbf{v}^T \left(\iint_D \begin{bmatrix} \partial w_1/\partial x & \partial w_1/\partial y \\ \partial w_2/\partial x & \partial w_2/\partial y \\ \vdots & \vdots \end{bmatrix} \boldsymbol{\kappa} \begin{bmatrix} \partial w_1/\partial x & \partial w_1/\partial y \dots \\ \partial w_2/\partial x & \partial w_2/\partial y \dots \end{bmatrix} dx \, dy \right) \mathbf{v}$$

$$+ \mathbf{v}^T \left(\int_{C_2} \sigma \begin{bmatrix} w_1 \\ w_2 \\ \vdots \end{bmatrix} [w_1 \quad w_2 \quad \dots] \, ds \right) \mathbf{v},$$

where C_2 is that part of the boundary on which a homogeneous mixed boundary condition holds. Thus it may be seen, by comparison with Example 3.3, that

$$\iint_D v\mathcal{L}v \, dx \, dy = \mathbf{v}^T \mathbf{K} \mathbf{v},$$

where $\mathbf{K}$ is the reduced overall stiffness matrix. Now, provided that κ is positive definite and $\sigma > 0$, $\mathcal{L}$ is positive definite; consequently, it follows that

$$\mathbf{v}^T \mathbf{K} \mathbf{v} > 0,$$

i.e. $\mathbf{K}$ is positive definite.

When a variational principle exists, it is always equivalent to a weighted residual procedure. However, the converse is not true, since weighted residual methods are applied directly to the boundary-value problem under consideration, irrespective of whether a variational principle exists or not.

To establish this result, consider the functional

$$I[u] = \iint_D F\left(x, y, u, \frac{\partial u}{\partial x}, \frac{\partial u}{\partial y}, \ldots\right) dx\, dy + \oint_C G\left(x, y, u, \frac{\partial u}{\partial x}, \frac{\partial u}{\partial y}, \ldots\right) ds,$$

(5.13)

which is stationary when $u = u_0$.

Suppose that $u = u_0 + \alpha v$; then the stationary point occurs when $(dI/d\alpha)|_{\alpha=0} = 0$; see Section 2.6. This yields an equation of the form

$$(5.14) \qquad\qquad \iint_D v \mathcal{L}_E(u)\, dx\, dy + \oint_C v \mathcal{B}_E(u)\, ds = 0,$$

which holds for arbitrary v; thus it follows that

$$(5.15) \qquad\qquad\qquad\qquad \mathcal{L}_E(u) = 0 \quad \text{in } D$$

and

$$(5.16) \qquad\qquad\qquad\qquad \mathcal{B}_E(u) = 0 \quad \text{on } C.$$

Equation (5.15) is the so-called Euler equation for the functional (5.13).

If eqns (5.15) and (5.16) are precisely the differential equation and boundary conditions under consideration, then the variational principle is said to be a natural principle and it follows immediately that eqn (5.14) gives the corresponding Galerkin method, the weighting function being the trial function v. However, not all differential equations are Euler equations of an appropriate functional; nevertheless, it is always possible to apply a weighted residual method.

Thus, if the Euler equations of the variational principle are identical with the differential equations of the problem, then the Galerkin and Rayleigh–Ritz methods yield the same system of equations. In particular, it follows from Section 2.3 that, since the variational principle associated with a linear self-adjoint operator is a natural one, the Galerkin and Rayleigh–Ritz methods yield identical results.

It is worth concluding this section with a note on the terminology, since the method described here is often associated with the name 'Bubnov–Galerkin method'. When piecewise weighting functions other than the nodal functions are used, then the name 'Petrov–Galerkin' is associated with the procedure.

5.2 Collocation and least squares methods

Recall the weighted residual method (Section 2.3) for the solution of

$$(5.17) \qquad \mathcal{L}u = f \quad \text{in } D$$

subject to the boundary condition

$$(5.18) \qquad \mathcal{B}u = b \quad \text{on } C.$$

Define the residual

$$r_1(\tilde{u}) = \mathcal{L}\tilde{u} - f$$

and the boundary residual

$$r_2(\tilde{u}) = \mathcal{B}\tilde{u} - b;$$

then eqn (2.23) suggests the following general weighted residual equations:

$$(5.19) \qquad \iint_D r_1 v_i \, dx \, dy + \oint_{C_2} r_2 v_i \, ds = 0, \quad i = 1, \dots, n,$$

where $\{v_i\}$ is a set of linearly independent weighting functions which satisfy $v_i \equiv 0$ on C_1, that part of C on which an essential boundary condition applies. The trial functions $\tilde{u}$ are defined in the usual piecewise sense by eqn (5.4) as

$$\tilde{u} = \sum_e \tilde{u}^e,$$

with $\tilde{u}^e$ interpolated through element $[e]$ in terms of the nodal values. The equations (5.19) then yield a set of algebraic equations for these nodal values. Notice that no restriction is placed on the operator $\mathcal{L}$; it may be non-linear, in which case the resulting set of equations is a non-linear algebraic set; see Section 5.3. Very often, the equations (5.19) are transformed by the use of an integration-by-parts formula, Green's theorem, so that the highest-order derivative occurring in the integrand is reduced, thus reducing the continuity requirement for the chosen trial function.

The point collocation method requires that the boundary-value problem be satisfied exactly at n points in the domain; this is accomplished by choosing

$v_i(x, y) = \delta(x - x_i, y - y_i)$, the usual Dirac delta function. In practice, the collocation is usually performed at m points in the domain $(m \gg n)$ and the resulting overdetermined system is solved by the method of least squares; see Exercise 5.3.

In the subdomain collocation method, the region is divided into N subdomains (elements) D_j, and the weighting function is given by

$$v_j(x, y) = \begin{cases} 1, & (x, y) \in D_j, \\ 0, & \text{otherwise.} \end{cases}$$

In the least squares method, the integral

$$I = \iint_D r_1^2 \, dx \, dy + \int_C r_2^2 \, ds$$

is minimized with respect to the nodal variables U_j, which leads to the set of equations

(5.20) $$\frac{\partial I}{\partial U_i} = 0, \qquad i = 1, \ldots n.$$

In the case where the trial functions are chosen to satisfy the boundary conditions, eqn (5.20) yields

(5.21) $$\iint_D r_1 \frac{\partial r_1}{\partial U_i} \, dx \, dy = 0, \qquad i = 1, \ldots n,$$

so that the weighting functions are given by $\partial r_1 / \partial U_i$.

Example 5.2 Consider Poisson's equation,

(5.22) $$-\nabla^2 u = f.$$

The usual finite element approximation is written in the form

$$\tilde{u} = \sum_e \left(\sum_{j \in [e]} N_j^e U_j \right),$$

which gives the residual

$$r(\tilde{u}) = -\sum_e \left\{ \sum_{j \in [e]} (\nabla^2 N_j^e) U_j \right\} - f.$$

Thus it follows from eqn (5.21) that

$$\sum_e \iint_{[e]} \left\{ \sum_{j \in e} (\nabla^2 N_j^e) U_j + f \right\} \nabla^2 N_i^e \, dx \, dy = 0.$$

Thus element stiffness and force matrices may be obtained, given by

$$k_{ij}^e = \iint_{[e]} \left(\nabla^2 N_j^e\right) \nabla^2 N_i^e \, dx \, dy$$

and

$$f_i^e = -\iint_{[e]} f \, \nabla^2 N_i^e \, dx \, dy.$$

Unfortunately, these integrals contain second derivatives, which means that the trial functions must have continuous first derivatives. For this reason, the least squares method has not been very attractive. However, if the governing partial differential equation (5.22) is replaced by a set of first-order equations (Lynn and Arya 1973, 1974), then the continuity requirement may be relaxed. Let

(5.23)
$$\xi = \frac{\partial U}{\partial x}, \quad \eta = \frac{\partial U}{\partial y};$$

then eqn (5.22) becomes

(5.24)
$$\frac{\partial \xi}{\partial x} + \frac{\partial \eta}{\partial y} = -f,$$

and the system of equations (5.23) and (5.24) is used instead of the original equation (5.22).

The least squares approach to minimizing the residual errors then leads to three integral expressions. The usual finite element representation for the unknowns U, ξ, η is then substituted into these expressions to obtain the necessary stiffness and force matrices; see Exercise 5.4.

5.3 Use of Galerkin's method for time-dependent and non-linear problems

When the finite element method is applied to time-dependent problems, the time variable is usually treated in one of two ways:

(1) Time is considered as an extra dimension, and shape functions in space and time are used. This is illustrated in Example 5.3.

(2) The nodal variables are considered as functions of time, and the space variables are used in the finite element analysis. This leads to a system of ordinary differential equations, which may be solved by a finite difference or weighted residual method. This approach is illustrated in Example 5.4.

A Laplace transform approach is also possible, and this is considered in Section 5.5.

Example 5.3 Consider the diffusion equation

$$\nabla^2 u = \frac{1}{\alpha}\frac{\partial u}{\partial t} \quad \text{in } D$$

subject to the boundary conditions

$$u = g(s,t) \quad \text{on } C_1,$$

$$\frac{\partial u}{\partial n} + \sigma(s,t)u = h(s,t) \quad \text{on } C_2.$$

Suppose that the approximation in xyt space is given by

(5.25) $$\tilde{u} = \sum_e \tilde{u}^e,$$

where

(5.26) $$\tilde{u}^e = \mathbf{N}^e(x,y,t)\mathbf{U}^e.$$

The Galerkin procedure involves choosing the nodal functions as weighting functions and setting the integrals of the weighted residuals to zero, just as in Chapters 2 and 3:

$$\int_0^T \iint_D \left(-\nabla^2\tilde{u} + \frac{1}{\alpha}\frac{\partial\tilde{u}}{\partial t}\right) w_i \, dx \, dy \, dt + \int_0^T \int_{C_2} \left(\frac{\partial\tilde{u}}{\partial n} + \sigma\tilde{u} - h\right) w_i \, ds \, dt = 0,$$

where the nodal functions are chosen such that $w_i \equiv 0$ on C_1. The term in the first integral is written in this form to be consistent with the notation of Chapters 2 and 3, where, for Poisson's equation, the differential operator was written as $-\nabla^2$.

The first integral may be transformed using Green's theorem for the space variables to give

$$\int_0^T \iint_D \left(\text{grad } w_i \cdot \text{grad } \tilde{u} + \frac{w_i}{\alpha}\frac{\partial\tilde{u}}{\partial t}\right) dx \, dy \, dt + \int_0^T \int_{C_2} (\sigma\tilde{u} - h) w_i \, ds \, dt.$$

Then, using eqns (5.25) and (5.26), the following system of equations may be obtained just as before:

(5.27) $$\mathbf{KU} = \mathbf{F},$$

where the element stiffness and forces are given by

(5.28) $$k_{ij}^e = \int_0^T \iint_{[e]} \left(\frac{\partial N_i^e}{\partial x}\frac{\partial N_j^e}{\partial x} + \frac{\partial N_i^e}{\partial y}\frac{\partial N_j^e}{\partial y} + \frac{1}{\alpha}N_i^e\frac{\partial N_j^e}{\partial x}\right) dx \, dy \, dt,$$

(5.29)
$$\bar{k}_{ij}^e = \int_0^T \int_{C_2^e} \sigma N_i^e N_j^e \, ds \, dt$$

and

(5.30)
$$\bar{f}_i^e = \int_0^T \int_{C_2^e} h N_i^e \, ds \, dt.$$

Notice that the stiffness matrix is no longer symmetric; this is due to the fact that the parabolic operator $-\nabla^2 + (1/\alpha)(\partial/\partial t)$ is not self-adjoint.

The essential boundary condition must be enforced in eqn (5.27) together with the nodal values along the initial plane $t = 0$. The equations (5.27) then give the solution at time T. This solution may then be used to step forward in time again, and the whole time development of the solution may be obtained in a stepping manner very similar to that used in the finite difference method (Smith 1985).

To illustrate the procedure, consider the one-dimensional problem

$$\frac{\partial^2 u}{\partial x^2} = \frac{\partial u}{\partial t}$$

subject to the boundary conditions

$$u(0, t) = t, \quad u(2, t) = 2 + t$$

and the initial condition

$$u(x, 0) = \tfrac{1}{2} x^2.$$

Suppose that eight triangular elements are used in the xt plane, where the time step and the mesh size are each one unit, as shown in Fig. 5.1.

The element matrices for this problem may be obtained by using the results and notation of Section 3.8 with y replaced by t, as follows. The shape function matrix is

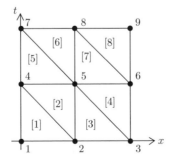

Fig. 5.1 Eight-element discretization for the problem of Example 5.3.

$$\mathbf{N}^e = \begin{bmatrix} L_1 & L_2 & L_3 \end{bmatrix}$$

so that, using eqn (5.28),

$$k_{ij}^e = \frac{b_i b_j}{4A} + \int\int_A L_i \frac{c_j}{2A}\, dx\, dt,$$

i.e.

$$k_{ij}^e = \frac{b_i b_j}{4A} + \frac{c_i}{6}.$$

Thus, using the results of Exercise 5.6,

$$
\mathbf{k}^1 = \begin{bmatrix} \frac{1}{3} & -\frac{1}{2} & \frac{1}{6} \\ -\frac{2}{3} & \frac{1}{2} & \frac{1}{2} \\ -\frac{1}{6} & 0 & \frac{1}{6} \end{bmatrix} \begin{matrix} 1 \\ 2 \\ 4 \end{matrix}, \quad
\mathbf{k}^2 = \begin{bmatrix} \frac{1}{2} & -\frac{1}{6} & -\frac{1}{3} \\ 0 & -\frac{1}{6} & \frac{1}{6} \\ -\frac{1}{2} & -\frac{1}{6} & \frac{2}{3} \end{bmatrix} \begin{matrix} 4 \\ 2 \\ 5 \end{matrix},
$$

(top labels: $\mathbf{k}^1$: 1, 2, 4; $\mathbf{k}^2$: 4, 2, 5)

$$
\mathbf{k}^3 = \begin{bmatrix} \frac{1}{3} & -\frac{1}{2} & \frac{1}{2} \\ -\frac{2}{3} & \frac{1}{2} & \frac{1}{6} \\ -\frac{1}{6} & 0 & \frac{1}{6} \end{bmatrix} \begin{matrix} 2 \\ 3 \\ 5 \end{matrix}, \quad
\mathbf{k}^4 = \begin{bmatrix} \frac{1}{2} & -\frac{1}{6} & -\frac{1}{3} \\ 0 & -\frac{1}{6} & \frac{1}{6} \\ -\frac{1}{2} & -\frac{1}{6} & \frac{2}{3} \end{bmatrix} \begin{matrix} 5 \\ 3 \\ 6 \end{matrix}.
$$

(top labels: $\mathbf{k}^3$: 2, 3, 5; $\mathbf{k}^4$: 5, 3, 6)

In the overall system of equations, it is only the equation for U_5 that need be set up. This equation is

$$\begin{bmatrix} K_{51} & K_{52} & K_{53} & K_{54} & K_{55} & K_{56} \end{bmatrix} \begin{bmatrix} U_1 & \cdots & U_6 \end{bmatrix}^T = 0,$$

i.e.

$$\left(-\tfrac{1}{6} - \tfrac{1}{6}\right) U_2 + \left(0 - \tfrac{1}{6}\right) U_3 - \tfrac{1}{2} U_4 + \left(\tfrac{2}{3} + \tfrac{1}{6} + \tfrac{1}{2}\right) U_5 - \tfrac{1}{3} U_6 = 0.$$

Enforcing the essential boundary conditions

$$U_3 = 2, \quad U_4 = 1, \quad U_6 = 3$$

and the initial condition

$$U_2 = \tfrac{1}{2},$$

it follows that $U_5 = \tfrac{3}{2}$.

To step forward to the next time level, $t = 2$, it is clear that

$$\mathbf{k}^{n+4} = \mathbf{k}^n, \quad n = 1, 2, 3, 4,$$

and the difference equation at node 8 thus becomes

$$-\tfrac{1}{3} U_5 - \tfrac{1}{6} U_6 - \tfrac{1}{2} U_7 + \tfrac{4}{3} U_8 - \tfrac{1}{3} U_9 = 0.$$

Enforcing the boundary conditions and the previously calculated value of U_5, it follows that $U_8 = \frac{5}{2}$.

The procedure may be repeated until the required time level is reached. The solution at points inside an element may be found from the nodal values using the usual linear interpolation polynomials.

The wave equation may be treated in a similar manner; see Exercise 5.8.

Example 5.4 Consider again the initial-boundary-value problem of Example 5.3 but in this case suppose that the approximation in each element is, instead of eqn (5.26), given by

$$(5.31) \qquad \tilde{u}^e = \mathbf{N}^e(x, y) \mathbf{U}^e(t),$$

where the nodal values are now assumed to be functions of time. Then, using Galerkin's method, the weighted residual equations become

$$\iint_D \left(-\nabla^2 \tilde{u} + \frac{1}{\alpha} \frac{\partial \tilde{u}}{\partial t} \right) w_i \, dx \, dy + \int_{C_2} \left(\frac{\partial \tilde{u}}{\partial n} + \sigma \tilde{u} - h \right) w_i \, ds = 0.$$

Thus, substituting the approximation given by eqns (5.25) and (5.31) and using Green's theorem, these become

$$\sum_e \iint_{[e]} \sum_{j \in [e]} (\text{grad } w_i \cdot \text{grad } N_j^e) \, U_j \, dx \, dy + \sum_e \iint_{[e]} \frac{1}{\alpha} \sum_{j \in [e]} w_i N_j^e \frac{dU_j}{dt} \, dx \, dy$$

$$+ \sum_e \int_{C_2^e} \sigma \sum_{j \in [e]} w_i N_j^e U_j \, ds - \sum_e \int_{C_2^e} h w_i \, ds = 0.$$

This system is seen to be a set of ordinary differential equations of the form

$$(5.32) \qquad \mathbf{C}\dot{\mathbf{U}} + \mathbf{K}\mathbf{U} = \mathbf{F},$$

where $\mathbf{K}$ and $\mathbf{F}$ are the usual overall stiffness and force matrices and the matrix $\mathbf{C}$ is often referred to as the overall *conductivity matrix*, since the diffusion equation models heat conduction. Since $\mathbf{c}^e = \iint_{[e]} \mathbf{N}^{e^T} \mathbf{N}^e \, dx \, dy$ (*cf.* the mass matrix $\mathbf{m}^e$ in Exercise 3.24), it is clear that $\mathbf{C}$ is symmetric. Some authors, by virtue of the structural origins of the finite element method, refer to $\mathbf{C}$ as the *damping matrix*.

A finite difference approach to the solution of this system of equations is to take a sequence of time steps of length Δt from time level j to $j + 1$.

Using a forward difference scheme given by

$$\dot{\mathbf{U}} \approx (\mathbf{U}_{j+1} - \mathbf{U}_j)/\Delta t,$$

eqn (5.32) becomes approximated by

$$(5.33) \qquad \frac{1}{\Delta t} \mathbf{C}\mathbf{U}_{j+1} + \left[\mathbf{K} - \frac{1}{\Delta t}\mathbf{C} \right] \mathbf{U}_j = \mathbf{F}_j.$$

Knowing the initial value $\mathbf{U}_0$, the time-stepping procedure can start and the solution be developed by marching forward to the next time level, etc.

To illustrate the method, return to the one-dimensional problem of Example 5.3. In this case, two elements are used in space and these are taken to be simple linear elements of the type discussed in Section 3.5. Thus, using the results of that section,

$$\mathbf{K} = \begin{bmatrix} 1 & -1 & 0 \\ -1 & 2 & -1 \\ 0 & -1 & 1 \end{bmatrix}.$$

The element conductivity matrix is given by

$$\mathbf{c}^e = \int_{-1}^{1} \begin{bmatrix} \frac{1}{2}(1-\xi) \\ \frac{1}{2}(1+\xi) \end{bmatrix} \begin{bmatrix} \frac{1}{2}(1-\xi) & \frac{1}{2}(1+\xi) \end{bmatrix} \frac{h}{2} d\xi,$$

where h is the length of the element as in Section 3.5. For the problem under consideration, $h = 1$, and thus

$$\mathbf{c}^e = \frac{1}{6} \begin{bmatrix} 2 & 1 \\ 1 & 2 \end{bmatrix}.$$

The overall conductivity matrix is, therefore,

$$\mathbf{C} = \frac{1}{6} \begin{bmatrix} 2 & 1 & 0 \\ 1 & 4 & 1 \\ 0 & 1 & 2 \end{bmatrix}.$$

If the three nodes are situated at $x = 0, 1, 2$ and if $\Delta t = 1$, it follows that, since $\mathbf{F}_j = \mathbf{0}$, eqn (5.33) yields the set of equations

$$\frac{1}{6} \begin{bmatrix} 2 & 1 & 0 \\ 1 & 4 & 1 \\ 0 & 1 & 2 \end{bmatrix} \begin{bmatrix} U_1 \\ U_2 \\ U_3 \end{bmatrix}_{j+1} + \frac{1}{6} \begin{bmatrix} 4 & -7 & 0 \\ -7 & 8 & -7 \\ 0 & -7 & 4 \end{bmatrix} \begin{bmatrix} U_1 \\ U_2 \\ U_3 \end{bmatrix}_j = \begin{bmatrix} 0 \\ 0 \\ 0 \end{bmatrix}.$$

Initially, $\mathbf{U}_0 = \begin{bmatrix} 0 & \frac{1}{2} & 2 \end{bmatrix}^T$ and the boundary conditions give

$$U_{1j} = j\,\Delta t = j, \qquad U_{3j} = 2 + j\,\Delta t = 2 + j;$$

thus, at time level 1,

$$1 + 4U_{21} + 3 + 8 \times \tfrac{1}{2} - 7 \times 2 = 0,$$

which gives $U_{21} = \frac{3}{2}$. Similarly, at time level 2,

$$2 + 4U_{22} + 4 - 7 + 8 \times \tfrac{3}{2} - 7 \times 3 = 0,$$

which gives $U_{22} = \frac{5}{2}$.

Other difference schemes could of course be used instead of forward differences in eqn (5.32); this approach actually replaces U_j by a suitably truncated Taylor series. If, instead, a weighted residual approach in time is used, a more general recurrence relation can be set up (Zienkiewicz and Taylor 2000a).

Suppose that between time levels j and $j+1$, $U_i(t)$ is interpolated by

$$U_i(t) = [1 - \tau \quad \tau][U_{ij} \quad U_{ij+1}]^T,$$

where $t = (j + \tau)\,\Delta t$ with $0 \leq \tau \leq 1$; see Fig. 5.2.

Then

$$\dot{U}_i(t) = \frac{1}{\Delta t}[-1 \quad 1]\,[U_{ij} \quad U_{ij+1}]^T.$$

Using the same interpolation for F, eqn (5.32) may be written as

$$(5.34) \qquad \frac{1}{\Delta t}C[U_{j+1} - U_j] + K\,[\tau U_{j+1} + (1 - \tau)U_j] = \tau F_{j+1} + (1 - \tau)F_j.$$

Thus, multiplying by a weighting function v, and integrating with respect to τ from 0 to 1, eqn (5.34) becomes

$$(5.35) \qquad \left[\frac{1}{\Delta t}C + rK\right]U_{j+1} + \left[-\frac{1}{\Delta t}C + (1 - r)K\right]U_j = rF_{j+1} + (1 - r)F_j,$$

where $r = \int_0^1 \tau v\,d\tau / \int_0^1 v\,d\tau$.

Equation (5.35) is a recurrence formula for the nodal values at two time levels. By choosing different values for the parameter r, well-known difference formulae are recovered; for example, $r = 0, \frac{1}{2}$ and 1 give the forward, Crank–Nicolson and backward difference formulae, respectively. Recurrence formulae for nodal values at three time levels may be obtained by integrating over two consecutive time intervals in the weighted residual equation; see Exercise 5.10. Finally, the procedure may also be adopted to solve second-order equations of the form

$$M\ddot{U} + C\dot{U} + KU = F,$$

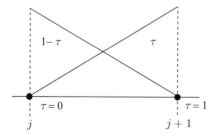

Fig. 5.2 Node i at times levels j and $j+1$, showing the linear interpolation polynomials in time.

which arise in the finite element solution of propagation problems; see Exercises 5.11 and 5.12.

This discussion of time-dependent problems has been necessarily brief; for further details, there are two excellent chapters in the book by Zienkiewicz *et al.* (2005).

As well as time-dependent problems, the weighted residual method may also be used to tackle non-linear problems as illustrated in Example 5.5. In these cases, however, the resulting system of algebraic equations is no longer linear, and iterative techniques are usually necessary for their solution.

Example 5.5 Consider the equation

$$-\mathrm{div}(k \ \mathrm{grad} \ u) = f \quad \text{in } D$$

subject to the boundary conditions

$$u = g \quad \text{on } C_1,$$

$$k\frac{\partial u}{\partial n} + \sigma u = h \quad \text{on } C_2.$$

In this case suppose that k, f, σ and h all depend on u as well as on position. The weighted residual method may be applied in the usual manner, and if the weighting functions are the nodal functions, then the element matrices become

(5.36)
$$\mathbf{k}^e(\mathbf{U}^e) = \iint_{[e]} k\boldsymbol{\alpha}^T \boldsymbol{\alpha} \, dx \, dy,$$

$$\bar{\mathbf{k}}^e(\mathbf{U}^e) = \int_{C_2^e} \sigma \mathbf{N}^{e^T} \mathbf{N}^e \, ds,$$

$$\mathbf{f}^e(\mathbf{U}^e) = \iint_{[e]} f\mathbf{N}^{e^T} \, dx \, dt,$$

$$\bar{\mathbf{f}}^e(\mathbf{U}^e) = \int_{C_2^e} h\mathbf{N}^e \, ds$$

(*cf.* eqns (3.45)–(3.48)). However, in this case the unknown u occurs under the integral sign, so that the matrices themselves are functions of the nodal variables. The overall system is assembled in the usual way to yield a set of equations of the form

(5.37)
$$\mathbf{K}(\mathbf{U})\mathbf{U} = \mathbf{F}(\mathbf{U}).$$

To illustrate the idea, consider the one-dimensional problem

$$-(uu')' + 1 = 0, \qquad 0 < x < 1,$$

$$u(0) = 1, \qquad u(1) = 0.$$

Suppose that two linear elements of the type discussed in Section 3.5 are used. Then

$$\mathbf{k}^e(u) = \int_{-1}^{1} u \begin{bmatrix} -1/h \\ 1/h \end{bmatrix} [-1/h \quad 1/h] \frac{h}{2} \, d\xi$$

$$= \frac{1}{2h} \begin{bmatrix} 1 & -1 \\ -1 & 1 \end{bmatrix} \int_{-1}^{1} [\tfrac{1}{2}(1-\xi) \quad \tfrac{1}{2}(1+\xi)] [u_A \quad u_B]^T d\xi,$$

where the notation of Fig. 3.7 is used. Thus

$$\mathbf{k}^e(\mathbf{U}^e) = \frac{U_A + U_B}{2h} \begin{bmatrix} 1 & -1 \\ -1 & 1 \end{bmatrix}.$$

With the two-element idealization of Fig. 3.6, it follows that, with $h = \tfrac{1}{2}$,

$$\mathbf{K(U)} = \begin{bmatrix} \overset{1}{U_1 + U_2} & \overset{2}{-U_1 - U_2} & \overset{3}{0} \\ & U_1 + 2U_2 + U_3 & -U_2 - U_3 \\ \text{sym} & & U_2 + U_3 \end{bmatrix} \begin{matrix} 1 \\ 2 \\ 4 \end{matrix}.$$

Since $f \equiv -1$, it follows from Example 3.1 that

$$\mathbf{F} = -\tfrac{1}{4} \begin{bmatrix} \overset{1}{1} & \overset{2}{2} & \overset{3}{1} \end{bmatrix}^T.$$

The only node without an essential boundary condition is node 2; thus assembling only the equation for node 2 yields

$$(-U_1 - U_2)U_1 + (U_1 + 2U_2 + U_3)U_2 + (-U_2 - U_3)U_3 = -\tfrac{1}{2}.$$

Since $U_1 = 1$ and $U_3 = 0$, it follows that

$$U_2^2 = \tfrac{1}{4}$$

and hence that

$$U_2 = \pm\tfrac{1}{2}.$$

The non-linear algebraic system has led to two possible solutions, and some knowledge of the behaviour of the solution to the original boundary-value problem is necessary in order that the correct value may be chosen. In this case, it may be shown that the solution $U_2 = -\tfrac{1}{2}$ is not possible, since this would imply that there is a point $x_p \in (0,1)$ with the properties $U(x_p) < 0, U(x_p) = 0, U''(x_p) > 0$. These properties are not consistent with the original differential equation. Thus the solution required is $U_2 = \tfrac{1}{2}$. This situation is typical of non-linear problems, where it is extremely useful to have some intuitive idea of the behaviour of the solution or the physics of the problem.

The problem considered here is of course very simple, leading to a non-linear algebraic equation whose solution is amenable to hand calculation. In practice, this does not occur, and numerical methods of solution are necessary. In general, either an iterative or incremental approach is adopted. Write eqn (5.37) as

(5.38) $$\mathbf{A}(\mathbf{U}) \equiv \mathbf{K}(\mathbf{U})\mathbf{U} - \mathbf{F}(\mathbf{U}) = \mathbf{0};$$

then the Newton–Raphson method gives the following iterative scheme:

(5.39) $$\mathbf{U}_{n+1} = \mathbf{U}_n - \mathbf{G}_n^{-1}\mathbf{A}(\mathbf{U}_n),$$

where

$$\mathbf{G}_n = [\partial A_i/\partial U_j]_n.$$

The solution proceeds by assuming an initial value $\mathbf{U}_0$ and then iterating with eqn (5.39) until sufficient accuracy is obtained.

Unfortunately, an iterative approach will not always converge; however, incremental methods will always converge. Suppose that, for some given value $\mathbf{F}_0$ of $\mathbf{F}$, eqn (5.38) has the known solution $\mathbf{U}_0$; then the solution proceeds by adding small increments to $\mathbf{F}$ and finding the corresponding increments in $\mathbf{U}$. Write eqn (5.37) as

$$\mathbf{H}(\mathbf{U}) - \lambda\mathbf{F}^* = \mathbf{0},$$

where

$$\mathbf{H}(\mathbf{U}) = \mathbf{K}(\mathbf{U})\mathbf{U} \quad \text{and} \quad \lambda\mathbf{F}^* = \mathbf{F}(\mathbf{U}).$$

Then, differentiating with respect to λ,

$$\mathbf{J}\frac{d\mathbf{U}}{d\lambda} - \mathbf{F}^* = \mathbf{0},$$

where

(5.40) $$\mathbf{J} = [\partial H_i/\partial U_j];$$

thus

(5.41) $$\frac{d\mathbf{U}}{d\lambda} = \mathbf{J}^{-1}\mathbf{F}^*.$$

Using a forward difference to approximate the derivative in eqn (5.41),

$$(\mathbf{U}_{n+1} - \mathbf{U}_n)/\Delta\lambda_n = \mathbf{J}_n^{-1}\mathbf{F}^*,$$

where

$$\mathbf{J}_n = \mathbf{J}(\mathbf{U}_n).$$

Thus the incremental scheme is

$$(5.42) \qquad \mathbf{U}_{n+1} = \mathbf{U}_n + \mathbf{J}_n^{-1}\Delta\mathbf{F}_n,$$

where $\Delta\mathbf{F}_n = \Delta\lambda_n\mathbf{F}^*$ is the nth increment in $\mathbf{F}$. Simple problems illustrating the use of these methods may be found in Exercises 5.13–5.16.

It has been the intention of this section to illustrate the way in which the Galerkin finite element approach may be used to solve time-dependent and non-linear problems. Only simple examples have been introduced, the general area of such problems being well beyond the scope of this text. Much research has been conducted in these areas, and the interested reader is recommended to consult the books by Zienkiewicz and Taylor (2000a,b) for references to specific subject areas such as plasticity, electromagnetic theory and viscous flow.

5.4 Time-dependent problems using variational principles which are not extremal

In Section 2.9, variational principles for time-dependent problems were introduced. In this section, they are used to develop finite element solutions to the wave and heat equations. Some authors feel that a direct approach to the problem at hand is better than using such variational problems, either because the solutions do not yield extrema for the functionals involved (Finlayson and Scriven 1967) or because a weighted residual approach is more general, since there is always a Galerkin method equivalent to a given variational principle (Zienkiewicz and Taylor 2000a). However, it is a possible approach when such principles exist and a suitable functional is known, and it gives users another weapon in their finite element armoury.

Example 5.6 (Noble 1973). Consider the wave equation

$$(5.43) \qquad \frac{\partial^2 u}{\partial x^2} = \frac{1}{c^2}\frac{\partial^2 u}{\partial t^2}, \quad 0 < x < 1, \quad t > 0,$$

subject to the boundary conditions

$$(5.44) \qquad u(0,t) = u(l,t) = 0$$

and the initial conditions

$$(5.45) \qquad u(x,0) = f(x),$$

$$(5.46) \qquad \frac{\partial u}{\partial t}(x,0) = g(x).$$

From Section 2.9, it follows that the functional (2.75) is stationary at the solution of eqn (5.43) subject to eqn (5.44) and (5.45) and a condition at time Δt of the form

$$u(x, \Delta t) = h(x),$$

which replaces eqn (5.46).

Of course, $h(x)$ is not known; however, it is treated as if it is known and then, at the end of the analysis, eqn (5.46) is used to eliminate this unknown function.

Suppose that $l = 1$ and that $[0, 1]$ is divided into E elements. Consider the usual finite element approximation

$$\tilde{u} = \sum_e \tilde{u}^e,$$

where

$$\tilde{u}^e = \mathbf{N}^e(x)\mathbf{U}^e(t).$$

Then, using the functional (2.75),

$$I = \sum_e \left\{ c^2 \int_0^{\Delta t} \mathbf{U}^{e^T} \left(\int_0^1 \boldsymbol{\alpha}^T \boldsymbol{\alpha} \, dx \right) \mathbf{U}^e \, dt - \int_0^{\Delta t} \dot{\mathbf{U}}^{e^T} \left(\int_0^1 \mathbf{N}^{e^T} \mathbf{N}^e \, dx \right) \dot{\mathbf{U}}^e \, dt \right\},$$

(5.47)

where, as usual, $\boldsymbol{\alpha} = d\mathbf{N}^e/dx$.

For the linear element of length h discussed in Section 3.5, the x integrations are easily performed to yield the two matrices

$$\frac{1}{h} \begin{bmatrix} 1 & -1 \\ -1 & 1 \end{bmatrix} \quad \text{and} \quad \frac{h}{6} \begin{bmatrix} 2 & 1 \\ 1 & 2 \end{bmatrix}.$$

These are just the usual element stiffness and mass or conductivity matrices, see Exercise 3.24 and Example 5.4.

To satisfy the essential conditions at node i at times $t = 0$ and $t = \Delta t$, set $U(0) = f_i$ and $U(\Delta t) = h_i$. Now $U(t)$ may be interpolated between $t = 0$ and $t = \Delta t$ as a quadratic function

(5.48) $U_i(t) = (1 - \tau^2)f_i + \tau^2 h_i + a_i \, \Delta t(\tau - \tau^2),$

where $\tau = t/\Delta t$ and $a_i = \dot{U}_i(0)$. Also,

$$\dot{U}_i(t) = \frac{2}{\Delta t}(h_i - f_i) + a_i(1 - 2\tau).$$

At this stage a_i is treated as unknown, so that substitution of eqn (5.48) into eqn (5.47) gives $I = I(a_1, \ldots, a_n)$.

Stationary values of I occur when $\partial I/\partial a_i = 0$, $i = 1, \ldots, n$. If $I = \sum_e I^e$, then

$$\frac{\partial I^e}{\partial a_i} = 2 \int_0^1 \left\{ \frac{c^2}{h}(U_i - U_{i-1}) \, \Delta t(\tau - \tau^2) - \frac{h}{6}(2\dot{U}_i + \dot{U}_{i-1})(1 - 2\tau) \right\} \Delta t \, d\tau$$

$$= h \left\{ -\left(\frac{1}{9} + \frac{7\mu^2}{30} \right) f_{i-1} + \left(\frac{7\mu^2}{30} - \frac{2}{9} \right) f_i \right.$$

$$+ \left(\frac{1}{9} - \frac{\mu^2}{30} \right) h_{i-1} + \left(\frac{\mu^2}{10} + \frac{2}{9} \right) h_i$$

$$+ \Delta t \left[-\left(\frac{1}{9} + \frac{\mu^2}{15} \right) a_{i-1} + \left(\frac{\mu^2}{15} - \frac{2}{9} \right) a_i \right] \right\},$$

where $\mu = c \, \Delta t/h$.

A similiar result may be obtained when element $[e]$ contains nodes i and $i + 1$. Assembling the complete equations $\sum_e \partial I^e / \partial a_i = 0$ yields

$$-\left(\frac{1}{9} + \frac{7\mu^2}{30} \right) f_{i-1} + 2 \left(\frac{7\mu^2}{30} - \frac{2}{9} \right) f_i - \left(\frac{1}{9} + \frac{7\mu^2}{30} \right) f_{i+1}$$

$$+ \left(\frac{1}{9} - \frac{\mu^2}{30} \right) h_{i-1} + 2 \left(\frac{\mu^2}{10} + \frac{2}{9} \right) h_i + \left(\frac{1}{9} - \frac{\mu^2}{30} \right) h_{i+1}$$

$$(5.49) \quad + \Delta t \left[-\left(\frac{1}{9} + \frac{\mu^2}{15} \right) a_{i-1} + 2 \left(\frac{\mu^2}{15} - \frac{2}{9} \right) a_i - \left(\frac{1}{9} + \frac{\mu^2}{15} \right) a_{i+1} \right] = 0.$$

This system of equations allows h_i to be found, since f_i and a_i are known.
To illustrate the technique, suppose that $c = 1$ and

$$f(x) = \begin{cases} x, & 0 \le x \le \frac{1}{2}, \\ 1 - x, & \frac{1}{2} \le x \le 1, \end{cases} \qquad g(x) = 0.$$

Take four equal elements along the x-axis and suppose that $\Delta t = 0.1$, so that $\mu = 0.4$. Then the difference equation for the unknown function h is

$$0.1058h_{i-1} + 0.4764h_i + 0.01058h_{i+1} = 0.1484(f_{i-1} + f_{i+1}) + 0.3698f_i$$

$$+ 0.01218(a_{i-1} + a_{i+1}) + 0.04231a_i,$$

$$(5.50) \qquad\qquad i = 2, 3, 4.$$

Now the boundary conditions give

$$f_1 = f_5 = h_1 = h_5 = 0,$$

and the initial conditions give $f_2 = 0.25$, $f_3 = 0.5$, $f_4 = 0.25$, $a_i = 0$. Substituting these values in eqn (5.49) and solving the resulting equation yields

$$h_2 = h_4 = 0.254, \quad h_3 = 0.431$$

and

$$(5.51) \qquad \dot{U}(0.1) = 20(h_i - f_i) - a_i = b_i, \quad \text{say.}$$

However, $h_i - f_i$ is the difference between two nearly equal numbers and, to avoid inaccuracies, h_i is eliminated between eqns (5.50) and (5.51) to give

$$0.00529b_{i-1} + 0.02382b_i + 0.00529b_{i+1} = 0.0426(f_{i-1} + f_{i+1}) - 0.1066f_i$$
$$+ 0.00689(a_{i-1} + a_{i+1}) + 0.01849a_i.$$

Using the initial conditions and boundary conditions previously stated, together with $b_1 = b_5 = 0$, three equations may be obtained, to yield the solutions

$$b_2 = b_4 = 0.082, \quad b_3 = -1.380.$$

The procedure may now be repeated using the calculated values at $t = \Delta t$ to step forward in time once again. Thus a step-by-step method has been developed, and the solution at any time t may be obtained by taking a sufficiently large number of steps.

The use of a time-dependent variational principle in the finite element solution of the heat equation is considered in Exercise 5.18. The principle involved is in fact related to another principle, developed by Gurtin (1964) using the Laplace transform. This will not be discussed here; instead, a procedure using the Laplace transform directly will be considered in the next section.

5.5   The Laplace transform

Until recently, this approach has not been particularly popular. However, the use of Stehfest's numerical inversion method (Stehfest 1970a,b) has developed a new interest in the Laplace transform associated with finite element methods (Moridis and Reddell 1991). A very good introduction to the use of the Laplace transform for diffusion problems has been given, in the context of boundary integral methods, by Zhu (1999).

Example 5.7 Consider the diffusion equation

$$(5.52) \qquad \frac{\partial^2 u}{\partial x^2} = \frac{1}{\alpha} \frac{\partial u}{\partial t}, \quad 0 < x < l, \quad t > 0,$$

subject to the boundary conditions

$$u(0, t) = g_1(t), \quad u(l, t) = g_2(t)$$

and the initial condition

$$u(x, 0) = f(x).$$

If

$$\bar{u}(x;\lambda) = \int_0^\infty e^{-\lambda t} u(x,t)\, dt$$

is the Laplace transform in time of $u(x,t)$, then, using the Laplace transform, eqn (5.52) becomes the ordinary differential equation

$$\frac{d^2\bar{u}}{dx^2} = \frac{1}{\alpha}(\lambda\bar{u} - f(x)),$$

which we write as

$$-\frac{d^2\bar{u}}{dx^2} + \frac{\lambda}{\alpha}\bar{u} = \frac{1}{\alpha}f(x).$$

The boundary conditions transform to

$$\bar{u}(0;\lambda) = \bar{g}_1(\lambda), \quad \bar{u}(l;\lambda) = \bar{g}_2(\lambda).$$

The effect of the Laplace transform is to remove the time dependence and the initial condition, leaving a two-point boundary-value problem to be solved. It is not difficult to see that for two-dimensional diffusion problems, the parabolic nature is removed, leaving an elliptic boundary-value problem.

To illustrate the approach, consider again the one-dimensional problem presented in Example 5.3. Then, using the Laplace transform, the equation and boundary conditions become

$$-\bar{u}'' + \lambda\bar{u} = \tfrac{1}{2}x^2,$$

$$\bar{u}(0;\lambda) = \frac{1}{\lambda^2}, \quad \bar{u}(2;\lambda) = \frac{2}{\lambda} + \frac{1}{\lambda^2}.$$

Using two equal linear elements, the overall matrices may be obtained from Example 5.4 and Exercise 3.24 as

$$\mathbf{K} = \begin{bmatrix} 1 & -1 & 0 \\ -1 & 2 & -1 \\ 0 & -1 & 1 \end{bmatrix}, \quad \mathbf{C} = \tfrac{1}{6}\begin{bmatrix} 2 & 1 & 0 \\ 1 & 4 & 1 \\ 0 & 1 & 2 \end{bmatrix}.$$

The element force vectors are

$$\mathbf{f}^1 = \int_0^1 \tfrac{1}{2}x^2 \begin{bmatrix} 1-x & x \end{bmatrix}^T dx = \tfrac{1}{24}\begin{bmatrix} 1 & 3 \end{bmatrix}^T,$$

$$\mathbf{f}^2 = \int_1^2 \tfrac{1}{2}x^2 \begin{bmatrix} 2-x & x-1 \end{bmatrix}^T dx = \tfrac{1}{24}\begin{bmatrix} 11 & 17 \end{bmatrix}^T.$$

Thus, $\mathbf{F} = \tfrac{1}{24}\begin{bmatrix} 1 & 14 & 17 \end{bmatrix}^T$. Now, after enforcing the essential boundary conditions

$$\bar{U}_1 = \frac{1}{\lambda^2}, \quad \bar{U}_3 = \frac{2}{\lambda} + \frac{1}{\lambda^2},$$

there is just one equation for $\bar{U}_2$:

$$-\frac{1}{\lambda^2} + 2\bar{U}_2 - \left(\frac{2}{\lambda} + \frac{1}{\lambda^2}\right) + \frac{\lambda}{6}\left(\frac{1}{\lambda^2} + 4\bar{U}_2 + \frac{2}{\lambda} + \frac{1}{\lambda^2}\right) = \frac{7}{12},$$

which yields

$$\bar{U}_2 = \frac{1}{\lambda} + \frac{1}{\lambda^2} - \frac{1}{8(3 + \lambda)}.$$

Inverting then gives

(5.53) $U_2(t) = \frac{1}{2} + t - \frac{1}{8}e^{-3t}.$

A drawback in this approach is the fact that the transform variable λ is carried through the analysis. This causes no difficulty in the simple hand calculation given, but it is not at all convenient for a numerical procedure. In practice, the equations could be solved for discrete set of values $\lambda_1, \lambda_2, \ldots, \lambda_M$ and the resulting transforms inverted numerically. Before we consider this approach, we shall illustrate the use of the convolution theorem.

Example 5.8 Consider the wave equation with boundary and initial conditions as in Example 5.6. Then, taking the Laplace transform,

$$c^2\frac{d^2\bar{u}}{dx^2} = \lambda^2\bar{u} - \lambda f(x) - g(x),$$

i.e.

$$\frac{-c^2}{\lambda^2}\frac{d^2\bar{u}}{dx^2} + \bar{u} = \frac{1}{\lambda}f(x) + \frac{1}{\lambda^2}g(x).$$

This equation may be inverted as

(5.54) $-c^2\frac{d^2}{dx^2}(t * u) + u = f(x) + tg(x),$

where the convolution integral is given by

$$F(t) * G(t) = \int_0^t F(t - u)G(u)\,du.$$

One way of solving the integro-differential equation (5.54) is to use an appropriate functional (Martin and Carey 1973). An alternative procedure, which will be used here, is to assume a time variation for u in the interval $0 \le t \le \Delta t$, evaluate the convolution integral and use a finite element analysis to solve the resulting two-point boundary-value problem.

Suppose that the approximation is

$$\tilde{u} = \sum_e \tilde{u}^e,$$

where

$$\tilde{u}^e = \sum_{i \in [e]} N_i^e(x) U_i(t).$$

The time variation for U_i is taken to be given by

$$U_i(t) = [(1 - \tau^2) \quad \tau^2 \quad \Delta t(\tau - \tau^2)][U_i(0) \quad U_i(\Delta t) \quad \dot{U}_i(0)]^T$$

with $\tau = t/\Delta t$. Then

$$t * \tilde{u} = \sum_e \sum_{i \in [e]} N_i^e(x) t * U_i$$

$$= \sum_e \sum_{i \in [e]} N_i^e(x) \Delta t^2 \left[\tfrac{5}{12} U_i(0) + \tfrac{1}{12} U_i(\Delta t) + \tfrac{17}{60} \Delta t \, \dot{U}_i(0) \right] \quad \text{at } t = \Delta t,$$

$$= \sum_e \sum_{i \in [e]} N_i^e(x) \frac{\Delta t^2}{60} (25 f_i + 5 h_i + 7 \Delta t \, a_i)$$

using the notation of eqn (5.48). Thus, using the two-node linear element in the space variable x (see Fig. 3.7), it follows that the element matrices are

$$\mathbf{k}^e = \frac{c^2}{h} \begin{bmatrix} 1 & -1 \\ -1 & 1 \end{bmatrix}, \quad \mathbf{m}^c = \frac{h}{6} \begin{bmatrix} 2 & 1 \\ 1 & 2 \end{bmatrix}$$

and

$$\mathbf{f}^e = \int_{-1}^{1} \frac{1}{2} \begin{bmatrix} 1 - \xi \\ 1 + \xi \end{bmatrix} [f(x) + \Delta t \, g(x)] \frac{h}{2} \, d\xi$$

$$= \mathbf{m}^e \begin{bmatrix} f_{i-1} + \Delta t \, a_{i-1} \\ f_i + \Delta t \, a_i \end{bmatrix}.$$

Thus the overall system is assembled in the usual way to give the following difference equation at node i:

$$-\left(\frac{1}{6} + \frac{5\mu^2}{12} \right) f_{i-1} + \left(\frac{5\mu^2}{6} - \frac{2}{3} \right) f_i - \left(\frac{1}{6} + \frac{5\mu^2}{12} \right) f_{i+1}$$

(5.55) $\quad + \left(\frac{1}{6} - \frac{\mu^2}{12} \right) h_{i-1} + \left(\frac{2}{3} + \frac{\mu^2}{6} \right) h_i + \left(\frac{1}{6} - \frac{\mu^2}{12} \right) h_{i+1}$

$$+ \Delta t \left[-\left(\frac{1}{6} + \frac{17\mu^2}{60} \right) a_{i-1} + \left(\frac{17\mu^2}{30} - \frac{2}{3} \right) a_i - \left(\frac{1}{6} + \frac{17\mu^2}{60} \right) a_{i+1} \right] = 0,$$

where $\mu = c \, \Delta t / h$.

Then, just as for eqn (5.49), this set of equations can be solved to find h_i, since f_i and a_i are known. The value of $U_i(\Delta t)$ is then found using

$$(5.56) \qquad\qquad U_i(\Delta t) = \frac{2}{\Delta t}(h_i - f_i) - a_i$$

and then eliminating h_i between eqns (5.55) and (5.56) as before.

In Examples 5.7 and 5.8, the inversion of the Laplace transform has been effected in closed form. In practice, this is usually not possible, and numerical inversion is required. There are many possibilities (Davies and Martin 1979), and for diffusion problems it has been shown that Stehfest's method provides a suitable approach. One advantage of the method is that the solution is found at a specific time τ without the necessity for solutions at preceding times, as is the case with the finite difference approach (see Example 5.4). Consequently, the method does not exhibit the stability problems associated with the finite difference method; see Section 7.5.

In Stehfest's method we choose a specific time value, τ, and a set of transform parameters

$$(5.57) \qquad \lambda_j = j\frac{\ln 2}{\tau}, \quad j = 1, 2, \ldots, M, \qquad \text{where } M \text{ is even.}$$

If $\bar{u}(x; \lambda)$ is the Laplace transform of $u(x, t)$, then

$$u(x, \tau) \approx \frac{\ln 2}{\tau} \sum_{j=1}^{M} w_j \bar{u}(x; \lambda_j),$$

where the weights, w_j, are given by (Stehfest 1970a,b)

$$w_j = (-1)^{M/2+j} \sum_{[(1/2)(1+j)]}^{\min(j, M/2)} \frac{k^{M/2}(2k)!}{(M/2 - k)! \, k! \, (k-1)! \, (j-k)! \, (2k-j)!},$$

and some values of w_j are shown in Appendix E (Davies and Crann 2004).

Example 5.9 Consider again the problem of Example 5.7 with two linear elements, leading to the transformed value

$$\bar{U}_2 = \frac{1}{\lambda} + \frac{1}{\lambda^2} - \frac{1}{8(3 + \lambda)}.$$

We shall evaluate the solution at $t = 1$ with $M = 2$:

$$\lambda_1 = 1\frac{\ln 2}{1} = 0.693147, \quad \lambda_2 = 2\frac{\ln 2}{1} = 1.386294,$$

$$\bar{U}_2(\lambda_1) = 3.490218, \quad \bar{U}_2(\lambda_2) = 1.213192,$$

$$U_2(1) \approx 3.16.$$

This compares rather poorly with the exact inversion, $U_2(1) \approx 1.994$, given by eqn (5.53). This is due to the fact that we have used $M = 2$. Much better results are obtained with higher values of M; see Exercise 5.20.

Now consider the two-dimensional problem

(5.58)
$$\nabla^2 u = \frac{1}{\alpha} \frac{\partial u}{\partial t} \qquad \text{in } D$$

subject to the boundary conditions

$$u = g(s, t) \qquad \text{on } C_1,$$

$$\frac{\partial u}{\partial n} + \sigma(s) u = h(s, t) \qquad \text{on } C_2$$

and the initial condition

$$u(x, y, 0) = u_0(x, y).$$

Then, taking the Laplace transform of eqn (5.58), we obtain

(5.59)
$$-\nabla^2 \bar{u} + \frac{\lambda}{\alpha} \bar{u} - \frac{1}{\alpha} u_0 = 0 \qquad \text{in } D$$

subject to the boundary conditions

$$\bar{u} = \bar{g}(s; \lambda) \qquad \text{on } C_1$$

and

$$\frac{\partial \bar{u}}{\partial n} + \sigma(s) \bar{u} = \bar{h}(s; \lambda) \qquad \text{on } C_2.$$

Just as in Sections 3.6 and 5.3, we can set up the finite element equations for eqn (5.59) in the form

$$\mathbf{K} \bar{\mathbf{U}} + \frac{\lambda}{\alpha} \mathbf{C} \bar{\mathbf{U}} = \mathbf{F}.$$

Example 5.10 Use three equal linear elements to solve the problem

$$\frac{\partial^2 u}{\partial x^2} = \frac{\partial u}{\partial t} \qquad 0 < x < 1,$$

subject to

$$u(0, t) = 0, \quad u(1, t) = 0,$$

$$u(x, 0) = 2.$$

Using the results of Exercise 3.24, the overall system of equations for the Laplace transform $\bar{U} = [\bar{U}_2 \quad \bar{U}_3]^T$ is

$$\frac{1}{h}\begin{bmatrix} 2 & -1 \\ -1 & 2 \end{bmatrix}\begin{bmatrix} \bar{U}_2 \\ \bar{U}_3 \end{bmatrix} + \frac{\lambda h}{6}\begin{bmatrix} 4 & 1 \\ 1 & 4 \end{bmatrix}\begin{bmatrix} \bar{U}_2 \\ \bar{U}_3 \end{bmatrix} = h\begin{bmatrix} 2 \\ 2 \end{bmatrix},$$

which, with $h = \frac{1}{3}$, simplifies to

$$a\bar{U}_2 - b\bar{U}_3 = \tfrac{2}{3},$$
$$-b\bar{U}_2 + a\bar{U}_3 = \tfrac{2}{3},$$

where $a = 6 + 2\lambda/9$ and $b = 3 - \lambda/18$.

Solving these equations yields

$$\bar{U}_2 = \frac{2}{3(a-b)} = \bar{U}_3.$$

U_2 and U_3 may be obtained using Stehfest's inversion. A spreadsheet solution is shown in Fig. 5.3.

We see from Fig. 5.3 that at $t = 0.1$,

$$U_2 = U_3 = 0.813.$$

The exact solution of eqn (5.58) subject to the given conditions is

$$u(x,t) = \frac{8}{\pi}\sum_{m=0}^{\infty}\frac{1}{2m+1}\sin((2m+1)\pi x)\exp(-(2m+1)^2\pi^2 t),$$

and the approximate values U_2 and U_3 compare very well with the exact values

$$u\left(\tfrac{1}{3}, 0.1\right) = u\left(\tfrac{2}{3}, 0.1\right) = 0.822.$$

	A	B	C	D	E	F	G	H	I
1	Example 5.10								
2	tau	0.1							
3	j	1	2	3	4	5	6	7	8
4	w_j	-0.33333333	48.3333333	-906	5464.667	-14376.6667	18730	-11946.7	2986.667
5	lambda_j	6.931471806	13.86294361	20.79442	27.72589	34.65735903	41.58883	48.5203	55.45177
6	a	7.540327068	9.080654136	10.62098	12.16131	13.70163534	15.24196	16.78229	18.32262
7	b	2.614918233	2.229836466	1.844755	1.459673	1.074591165	0.689509	0.304428	-0.08065
8	U2_bar_j	0.135352554	0.097311985	0.075963	0.062296	0.052796732	0.045811	0.040458	0.036225
9	w_j*U2_bar_j	-0.045117518	4.703412608	-68.8223	340.4256	-759.041017	858.0455	-483.342	108.1933
10									
11	U_2(tau)	0.813381802							

Fig. 5.3 Spreadsheet for the problem in Example 5.10.

5.6 Exercises and solutions

Exercise 5.1 A beam on an elastic foundation has a functional given by eqn (2.77) in Exercise 2.16 as

$$I[u] = \int_0^1 \left\{ ku^2 + EI(u'')^2 + 2uf \right\} dx.$$

The functional here contains second derivatives, so that it is necessary that the trial functions have continuous first derivatives. A suitable element for the problem is shown in Fig. 4.6, with the shape functions given by eqns (4.6)–(4.9). Obtain the element stiffness and force matrices.

Exercise 5.2 Repeat Exercise 3.7 using the variational approach.

Exercise 5.3 Consider the boundary-value problem

$$\mathcal{L}u = f \quad \text{in } D$$

subject to

$$\mathcal{B}u = b \quad \text{on } C,$$

to be solved by the overdetermined collocation method. Substitute the usual finite element approximation into the residual and show that in each element, the following set of equations may be obtained:

$$(5.60) \qquad\qquad \mathbf{L}^e \mathbf{U}^e = \mathbf{f}^e,$$

where the ith rows of $\mathbf{L}^e$ and $\mathbf{f}^e$ are obtained by evaluating $\mathcal{L}(\mathbf{N}^e)$ and f at collocation point i in element e.

Use the method of least squares to reduce eqn (5.60) to the form

$$\mathbf{k}^e \mathbf{U}^e = \tilde{\mathbf{f}}^e.$$

Illustrate the procedure for the case

$$\mathcal{L} \equiv -\frac{d^2}{dx^2}, \quad f(x) = x, \quad 0 < x < 1,$$

with $u(0) = u(1) = 0$.

Exercise 5.4 Poisson's equation (5.22) is to be solved using the least squares finite element method. Introduce auxiliary variables ξ and η so that the system of equations to be solved is given by eqns (5.23) and (5.24).

By taking linear variations in ξ and η and quadratic variations in u, set up the element approximation in the form $\mathbf{v}^e = \mathbf{N}^e \mathbf{U}^e$ for a triangular element with six nodes; see Fig. 4.5. Hence obtain expressions for the element stiffness and force matrices. The results of Exercise 4.6 allow an explicit form for $\mathbf{k}^e$ to be obtained in this case.

Exercise 5.5 Show how Galerkin's method may be used to solve a system of equations of the form

(5.61) $$\mathbf{Lu} = \mathbf{f},$$

where $\mathbf{L}$ is a matrix of differential operators given by

$$\mathbf{L} = [\mathcal{L}_{ij}]$$

and

$$\mathbf{u} = [u_1 \quad \cdots \quad u_p]^T.$$

By considering the one-dimensional form of Poisson's equation, $-d^2u/dx^2 = f$, which may be written as a system of equations by introducing the variable $q = -du/dx$, show that the Galerkin approach is not unique and that the symmetry of the resulting system of equations depends on the ordering of eqn (5.61).

Exercise 5.6 The diffusion equation is to be solved using triangular space–time elements. Obtain the stiffness matrices for the two triangles in the xt plane shown in Fig. 5.4. Hence obtain the difference equation for node i at time step j.

Exercise 5.7 Find the solution at time $t = 1$ to the problem in Example 5.4 using six space–time elements of the type used in Exercise 5.6.

Exercise 5.8 Obtain an expression for the stiffness matrix for the wave equation using elements in which the shape functions are functions of position and time.

Exercise 5.9 Rework Exercise 5.7 using three equal space elements and a forward difference scheme in time given by eqn (5.33).

Exercise 5.10 Write down the difference equation equivalent to eqn (5.34), valid for $(j-1)\Delta t \le t \le j\,\Delta t$. Using Galerkin's method in time, show that the

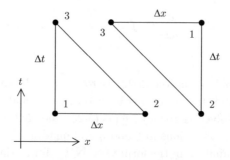

Fig. 5.4 Two triangles in the xt plane for Exercise 5.6.

following three-level difference equation may be obtained:

$$\left[\frac{1}{\Delta t}\mathbf{C}+\frac{1}{3}\mathbf{K}\right]\mathbf{U}_{j+1}+\frac{4}{3}\mathbf{K}\mathbf{U}_j+\left[-\frac{1}{\Delta t}\mathbf{C}+\frac{1}{3}\mathbf{K}\right]\mathbf{U}_{j-1}=\frac{1}{3}\left[\mathbf{F}_{j+1}+4\mathbf{F}_j+\mathbf{F}_{j-1}\right].$$

Exercise 5.11 Consider the equation of telegraphy, the wave equation with damping:

$$\nabla^2 u = \frac{1}{c^2}\frac{\partial^2 u}{\partial x^2}+\mu\frac{\partial u}{\partial t}.$$

Show that if a finite element idealization in space is used, with the nodal values considered as functions of time, then the overall system of equations is of the form

(5.62) $$\mathbf{M}\ddot{\mathbf{U}}+\mathbf{C}\dot{\mathbf{U}}+\mathbf{K}\mathbf{U}=\mathbf{F}.$$

Exercise 5.12 For the overall system of equations (5.62), suppose that $U(t)$ is interpolated between the three time levels $j-1, j, j+1$ by

$$U_i(t) = \left[\tfrac{1}{2}(\tau^2-\tau) \quad 1-\tau^2 \quad \tfrac{1}{2}(\tau^2+\tau)\right]\left[U_{ij-1} \quad U_{ij} \quad U_{ij+1}\right]^T,$$

where $t = (j+\tau)\Delta t$ with $-1 \le \tau \le 1$. Suppose also that F_i is interpolated in the same way. By using a weighting function W, set up a recurrence relation between U_{j-1}, U_j and U_{j+1} in terms of the parameters

$$\beta = \int_{-1}^{1}\tfrac{1}{2}(\tau^2+\tau)W\,d\tau \left/ \int_{-1}^{1} W\,d\tau \right.,$$

$$\gamma = \int_{-1}^{1}(\tau+\tfrac{1}{2})W\,d\tau \left/ \int_{-1}^{1} W\,d\tau \right..$$

Exercise 5.13 Solve the problem of Example 5.5 using three linear elements.

Exercise 5.14 Solve the non-linear two-point boundary-value problem

$$(e^u u')' = e^x, \quad 0 < x < 1,$$
$$u(0) = 0, \quad u(1) = 1,$$

using two linear elements.

Exercise 5.15 Repeat Exercise 5.13 making use of the Newton–Raphson scheme given by eqn (5.41); perform one iteration.

Exercise 5.16 Repeat Exercise 5.13 using an incremental method. Choose suitable starting vectors and use two equal increments for the force vector.

Exercise 5.17 The magnetic scalar potential ϕ in a magnetic material with zero current density satisfies the equation div $\{\mu(H)\,\text{grad}\,\phi\} = 0$, where the permeability μ is a function of the magnetic field strength; ϕ is related to the magnetic field by $\mathbf{H} = -\text{grad } \phi$ (see Appendix A). Show that the matrix $\mathbf{J}$ of eqn (5.40) can be assembled from the element matrices given by

$$\mathbf{j}^e = \mathbf{k}^e + \hat{\mathbf{k}}^e,$$

where $\mathbf{k}^e$ is the usual element stiffness matrix and

$$\hat{\mathbf{k}}^e = \iint_{[e]} \frac{d\mu}{dH} H^2 \boldsymbol{\alpha}^T \boldsymbol{\alpha} \, dx \, dy.$$

Exercise 5.18 Using the functional (2.76) and the method of Example 5.6, show that the difference equation for the heat equation

$$\frac{\partial^2 u}{\partial x^2} = \frac{1}{\alpha} \frac{\partial u}{\partial t}, \quad 0 < x < 1, \quad t > 0,$$

subject to the initial condition $u(x, 0) = f(x)$, is

$$4(1 + \mu)h_i + (1 - 2\mu)(h_{i-1} + h_{i+1}) = 4(1 - 2\mu)f_i + (1 + 4\mu)(f_{i-1} + f_{i+1}),$$

where $f_i = U(0)$, $h_i = U(\Delta t)$ and $\mu = k \, \Delta t / h^2$.

Exercise 5.19 Using the Laplace transform and two linear space elements, find the solution at $x = \frac{1}{2}$ to the wave equation

$$\frac{\partial^2 u}{\partial x^2} = \frac{\partial^2 u}{\partial t^2}, \quad 0 < x < 1, \quad t > 0,$$

subject to the initial conditions

$$u(x, 0) = 0, \quad \frac{\partial u}{\partial t}(x, 0) = \pi \sin \pi x$$

and the boundary conditions $u(0, t) = u(1, t) = 0$.

Exercise 5.20 Solve the problem of Example 5.9 using two linear elements together with the Stehfest Laplace transform inversion with $M = 8$.

Exercise 5.21 Solve the problem of Example 5.10 using four equal elements, and compare the results with the exact solution.

Solution 5.1 The procedure for obtaining the finite element equations follows in exactly the same way as in Chapter 3; in this case

$$\frac{\partial I^e}{\partial u_i} = \int_{-1}^{1} \left(kN_i \sum_{j=1}^{4} N_j u_j + \frac{4EI}{h^2} N_i'' \sum_{j=1}^{4} N_j'' u_j + N_i f \right) \frac{h}{2} \, d\xi.$$

Thus it follows that

$$k_{ij} = \int_{-1}^{1} \left(k N_i N_j + \frac{4EI}{h^2} N_i'' N_j'' \right) \frac{h}{2} \, d\xi,$$

$$f_i = - \int_{-1}^{1} N_i f \frac{h}{2} d\,\xi.$$

Performing the integrations,

$$\mathbf{k}^e = \frac{kh}{420} \begin{bmatrix} 156 & 22h & 54 & -13h \\ & 4h^2 & 13h & -3h^2 \\ & & 156 & -22h \\ \text{sym} & & & 4h^2 \end{bmatrix} + \frac{EI}{h^3} \begin{bmatrix} 12 & 6h & -12 & 6h \\ & 4h^2 & -6h & 2h^2 \\ & & 12 & -6h \\ \text{sym} & & & 4h^2 \end{bmatrix},$$

$$\mathbf{f}^e = \frac{fh}{12} \begin{bmatrix} -6 & -h & -6 & h \end{bmatrix}^T.$$

Solution 5.2 For the generalized Poisson equation

$$-\mathrm{div}(\kappa \,\mathrm{grad}\, u) = f$$

subject to Dirichlet boundary conditions, the corresponding functional is given by eqn (2.44) as

$$I[u] = \iint_D \{\mathrm{grad}\, u \cdot (\kappa\, \mathrm{grad}\, u) - 2uf\} \, dx\, dy + \int_{C_2} (\sigma u^2 - 2uh) \, ds.$$

Thus, dividing the region D into E elements in the usual way, the approximate solution is given by

$$\tilde{u}(x, y) = \sum_e \tilde{u}^e(x, y)$$

$$= \sum_e \mathbf{N}^e \mathbf{U}^e.$$

Thus, as in Section 5.1,

$$I[\tilde{u}] = \sum_e I^e,$$

where in this case

$$I^e = \iint_{[e]} \{\mathrm{grad}\, \tilde{u}^e \cdot (\kappa\, \mathrm{grad}\, \tilde{u}^e) - 2\tilde{u}^e f\} \, dx\, dy$$

$$= \mathbf{U}^{e^T} \left(\iint_{[e]} \boldsymbol{\alpha}^T \kappa \boldsymbol{\alpha} \, dx\, dy \right) \mathbf{U}^e - 2 \left(\iint_{[e]} f \mathbf{N}^{e^T} dx\, dy \right) \mathbf{U}^e$$

$$+ \left(\mathbf{U}^{e^T} \int_{C_2^e} \sigma \mathbf{N}^{e^T} \mathbf{N}^e \right) \mathbf{U}^e - 2 \left(\int_{C_2^e} h \mathbf{N}^{e^T} ds \right) \mathbf{U}^e,$$

where

$$\boldsymbol{\alpha} = \begin{bmatrix} \partial/\partial x \\ \partial/\partial y \end{bmatrix} \mathbf{N}^e.$$

Thus

$$\frac{\partial I^e}{\partial \mathbf{U}} = \left(\int\int_{[e]} \boldsymbol{\alpha}(\boldsymbol{\kappa} + \boldsymbol{\kappa}^T)\boldsymbol{\alpha}\, dx\, dy \right) \mathbf{U}^e - 2 \int\int_{[e]} f\mathbf{N}^{e^T} dx\, dy$$

$$+ 2 \left(\int_{C_2^e} \sigma\mathbf{N}^{e^T}\mathbf{N}^e ds \right) \mathbf{U}^e - 2 \int_{C_2^e} h\mathbf{N}^{e^T} ds.$$

Hence we see that the element matrices $\mathbf{f}^e$, $\bar{\mathbf{k}}^e$ and $\bar{\mathbf{f}}^e$ are exactly as given in Section 3.6 in eqns (3.47), (3.46) and (3.48).

The element stiffness matrix is

$$\mathbf{k}^e = \frac{1}{2} \int\int_{[e]} \boldsymbol{\alpha}^T(\boldsymbol{\kappa} + \boldsymbol{\kappa}^T)\boldsymbol{\alpha}\, dx\, dy,$$

and when $\boldsymbol{\kappa}$ is symmetric,

$$\mathbf{k}^e = \int\int_{[e]} \boldsymbol{\alpha}^T \boldsymbol{\kappa} \boldsymbol{\alpha}\, dx\, dy.$$

Solution 5.3 (Kwok *et al.* 1977) Substituting the usual finite element approximation

$$\tilde{u} = \sum_e \mathbf{N}^e \mathbf{U}^e$$

gives the residual

$$r = \sum_e \mathcal{L}(\mathbf{N}^e)\mathbf{U}^e.$$

The element residual is thus

$$r^e = \mathcal{L}(\mathbf{N}^e)\mathbf{U}^e - f.$$

This residual is now minimized by taking m^e collocation points inside element $[e]$. If there are M^e degrees of freedom in each element, then $m^e \gg M^e$. Suppose that at the collocation point P_i, $\mathcal{L}(\mathbf{N}^e) = \mathbf{L}_i^e$ and $f = f_i$; then the residual at P_i is

$$r_i^e = \mathbf{L}_i^e \mathbf{U}^e - f_i,$$

and this residual is chosen to be zero.

Thus, in each element, there is a set of equations of the form

$$\mathbf{L}^e \mathbf{U}^e = \mathbf{f}^e.$$

Since $\mathbf{L}^e$ is a rectangular $m^e \times M^e$ matrix, this constitutes an overdetermined set, which may be solved by the least squares method, setting

$$\frac{\partial}{\partial U_i} \sum_j \left(r_j^e\right)^2 = 0.$$

This then yields the symmetric set of equations

$$\mathbf{k}^e \mathbf{U}^e = \tilde{\mathbf{f}}^e,$$

where

$$\mathbf{k}^e = \mathbf{L}^{e^T} \mathbf{L}^e \quad \text{and} \quad \tilde{\mathbf{f}}^e = \mathbf{L}^{e^T} \mathbf{f}^e.$$

The overall system is then assembled in the usual way.

For the problem $-u'' = x, u(0) = u(1) = 0$, it is necessary that the shape functions have a continuous first derivative since $\mathcal{L}$ is a second-order differential operator. Thus suppose that the Hermite elements of the type shown in Fig. 4.6 are used. The shape functions are given by eqn (4.8) and (4.9), so that

$$\mathcal{L}(\mathbf{N}^e) = \frac{1}{h^2} \left[-6\xi \quad h(1 - 3\xi) \quad 6\xi \quad -h(1 + 3\xi) \right].$$

Consider the one-element solution with collocation at $x = 0, \frac{1}{4}, \frac{1}{2}, \frac{3}{4}, 1$; then

$$\mathbf{L} = \begin{bmatrix} 6 & 4 & -6 & 2 \\ 3 & 2.5 & -3 & 0.5 \\ 0 & 1 & 0 & -1 \\ -3 & -0.5 & 3 & -2.5 \\ -6 & -2 & 6 & -4 \end{bmatrix}$$

and $\mathbf{f} = \begin{bmatrix} 0 & 0.25 & 0.5 & 0.75 & 1 \end{bmatrix}^T$. Thus

$$\mathbf{L}^T \mathbf{L} = \begin{bmatrix} 90 & 45 & -90 & 45 \\ 45 & 27.5 & -45 & 17.5 \\ -90 & -45 & 90 & -45 \\ 45 & 17.5 & -45 & 27.5 \end{bmatrix}$$

and $\mathbf{L}^T \mathbf{f} = \begin{bmatrix} -17.5 & 1.25 & 7.5 & -6.25 \end{bmatrix}^T$.

Enforcing the essential boundary conditions $U_1 = U_2 = 0$ and solving the resulting equations yields $U_1' = \frac{1}{6}, U_2' = -\frac{1}{3}$.

Solution 5.4 (Lynn and Arya 1973)

$$\xi = L_1 \xi_1 + L_2 \xi_2 + L_3 \xi_3,$$

$$\eta = L_1 \eta_1 + L_2 \eta_2 + L_3 \eta_3,$$

$$u = N_1 U_1 + N_2 U_2 + \ldots + N_6 U_6,$$

where the L_i are the usual triangular coordinates and the N_i are the shape functions for the quadratic triangle (see eqns (4.4) and (4.5)), i.e. $\mathbf{v}^e = \mathbf{N}^e \mathbf{U}^e$, where

$$\mathbf{v}^e = [\xi \quad \eta \quad u]^T, \quad \mathbf{U}^e = [\xi_1, \ldots, \eta_1, \ldots, U_1, \ldots, U_6]^T$$

and

$$\mathbf{N}^e = \begin{bmatrix} L_1\ L_2\ L_3 & & \\ & L_1\ L_2\ L_3 & \\ & & N_1\ N_2\ N_3\ N_4\ N_5\ N_6 \end{bmatrix}.$$

Now eqns (5.23) and (5.24) may be expressed as $\mathbf{Lv} = \mathbf{f}$, where

$$\mathbf{L} = \begin{bmatrix} -1 & 0 & \partial/\partial x \\ 0 & -1 & \partial/\partial y \\ \partial/\partial x & \partial/\partial y & 0 \end{bmatrix} \quad \text{and} \quad \mathbf{f} = \begin{bmatrix} 0 \\ 0 \\ -f \end{bmatrix}.$$

Thus the element residual vector is

$$\mathbf{r} = \mathbf{L}\mathbf{N}^e\mathbf{U}^e - \mathbf{f}$$

$$= \boldsymbol{\alpha}\mathbf{U}^e - \mathbf{f}, \quad \text{say,}$$

where

$$\boldsymbol{\alpha} = \begin{bmatrix} -\boldsymbol{\omega} & \frac{1}{2A}\boldsymbol{\omega}\mathbf{B} \\ \frac{1}{2A}\mathbf{e} & 0 \end{bmatrix}.$$

The matrices $\boldsymbol{\omega}$ and $\mathbf{B}$ are defined in the solution to Exercise 4.6, and

$$\mathbf{e} = [b_1 \quad b_2 \quad b_3 \quad c_1 \quad c_2 \quad c_3]^T.$$

Using the least squares method to minimize the residual gives

$$\frac{\partial}{\partial v_i} \sum_e \iint_{[e]} \mathbf{r}^{e^T} \mathbf{r}^e \, dx \, dy = 0,$$

i.e.

$$\sum_e \iint_{[e]} \frac{\partial}{\partial v_i} \left([\mathbf{U}^{e^T} \boldsymbol{\alpha}^T - \mathbf{f}^T][\boldsymbol{\alpha}\mathbf{U}^e - \mathbf{f}] \right) dx \, dy = 0.$$

The element equilibrium equations follow, with the stiffness and force matrices given by

$$\mathbf{k}^e = \iint_{[e]} \boldsymbol{\alpha}^T \boldsymbol{\alpha} \, dx \, dy, \quad \mathbf{f}^e = \iint_{[e]} \boldsymbol{\alpha}^T \mathbf{f} \, dx \, dy.$$

Using the results and notation of Exercise 4.6, it follows that

$$\mathbf{k}^e = \begin{bmatrix} \mathbf{D} + \frac{1}{4A}\mathbf{e}^T\mathbf{e} & -\frac{1}{2A}\mathbf{DB} \\ -\frac{1}{2A}\mathbf{B}^T\mathbf{D} & \frac{1}{4A^2}\mathbf{B}^T\mathbf{DB} \end{bmatrix}.$$

Solution 5.5 The finite element approximation, in terms of nodal functions, is easily generalized to problems involving more than one dependent variable to give

$$\tilde{\mathbf{u}} = \sum_{i=1}^{n} \mathbf{n}_i \mathbf{U}_i^e.$$

Then the Galerkin method for the solution of eqn (5.61) yields

$$\iint_D \mathbf{n}_i^T \left[\mathbf{LU} - \mathbf{f} \right] dx\, dy = 0.$$

In the case where $\mathcal{L} = -d/dx$,

$$-u'' = f$$

is replaced by the system

$$q + u' = 0, \quad q' = f.$$

Write

$$q = \sum_i w_{qi} Q_i = \sum_e \sum_i N_{qi}^e Q_i,$$

$$u = \sum_i w_{ui} U_i = \sum_e \sum_i N_{ui}^e U_i.$$

Now

$$\mathbf{L} = \begin{bmatrix} 1 & d/dx \\ d/dx & 0 \end{bmatrix};$$

thus it follows that the element stiffness matrix is of the form

$$\mathbf{k}^e = \begin{bmatrix} \mathbf{k}^q & 0 \\ 0 & \mathbf{k}^u \end{bmatrix},$$

where

$$k_{ij}^s = \int_{-1}^{1} \begin{bmatrix} N_{si}N_{sj} & N_{si}\frac{dN_{sj}}{dx} \\ N_{si}\frac{dN_{sj}}{dx} & 0 \end{bmatrix} \frac{h}{2}\, d\xi, \quad s = q, u.$$

Notice that $\mathbf{k}^e$ is symmetric.

If, however, the equations are written in the reverse order so that

$$\mathbf{L} = \begin{bmatrix} d/dx & 0 \\ 1 & d/dx \end{bmatrix},$$

then

$$k_{ij}^{s} = \int_{-1}^{1} \begin{bmatrix} N_{si}\frac{dN_{sj}}{dx} & 0 \\ N_{si}N_{sj} & N_{si}\frac{dN_{sj}}{dx} \end{bmatrix} \frac{h}{2}\, d\xi, \quad s = q, u,$$

and a different form is obtained in which the symmetry is lost.

Solution 5.6 For element 1,

$$b_1 = -\Delta t, \quad b_2 = \Delta t, \quad b_3 = 0,$$

$$c_1 = -\Delta x, \quad c_2 = 0, \quad c_3 = \Delta x.$$

Thus

$$\mathbf{k}^1 = \frac{\Delta t}{6\,\Delta x} \begin{bmatrix} 3-r & -3 & r \\ -3-r & 3 & r \\ -r & 0 & r \end{bmatrix} \begin{matrix} 1 \\ 2 \\ 3 \end{matrix}, \quad \text{where} \quad r = \frac{\Delta x^2}{k\,\Delta t}.$$

For element 2,

$$b_1 = \Delta t, \quad b_2 = -\Delta t, \quad b_3 = 0,$$

$$c_1 = \Delta x, \quad c_2 = 0, \quad c_3 = -\Delta x.$$

Thus

$$\mathbf{k}^2 = \frac{\Delta t}{6\,\Delta x} \begin{bmatrix} 3+r & -3 & -r \\ -3+r & 3 & -r \\ r & 0 & -r \end{bmatrix} \begin{matrix} 1 \\ 2 \\ 3 \end{matrix}.$$

The difference equation for node i at time step j is thus

$$(-r-r)U_{ij-1} + (0-r)U_{i+1j-1} + (-3)U_{i-1j} + (3+r+r+3)U_{ij}$$
$$+ (-3+r)U_{i+1j} = 0,$$

i.e.

$$(6+2r)U_{ij} - (2rU_{ij-1} + rU_{i+1j-1} + 3U_{i-1j} + (3-r)U_{i+1j}) = 0.$$

Solution 5.7 In this case, $r = \frac{4}{9}$ and the only unknowns are U_{22} and U_{32}; thus the difference equations are

$$\tfrac{62}{9}U_{i2} - \left(\tfrac{8}{9}U_{i1} + \tfrac{4}{9}U_{i+11} + 3U_{i2} + \tfrac{23}{9}U_{i+12}\right) = 0, \quad i = 2, 3.$$

Enforcing the initial and boundary conditions

$$U_{21} = \tfrac{2}{9}, U_{31} = \tfrac{8}{9}, U_{41} = 2, U_{12} = 1, U_{42} = 3$$

and solving the resulting pair of equations yields

$$U_{22} = \tfrac{11}{9}, U_{32} = \tfrac{17}{9}.$$

Solution 5.8 For the wave equation

$$\nabla^2 u = \frac{1}{c^2} \frac{\partial^2 u}{\partial t^2} \quad \text{in } D$$

with Dirichlet/Robin boundary conditions and initial conditions on u and $\partial u/\partial t$, the only difference occurring in the Galerkin formulation from that for Poisson's equation in Sections 5.1 and 5.2 is that the time derivative term becomes

$$\int_0^T \iint_D \frac{1}{c^2} \frac{\partial^2 \tilde{u}}{\partial t^2} w_i \, dx \, dy \, dt = \iint_D \frac{1}{c^2} \left\{ \left. \left\{ \frac{\partial \tilde{u}}{\partial t} w_i \right\} \right|_0^T - \int_0^T \frac{\partial \tilde{u}}{\partial t} \frac{\partial w_i}{\partial t} dt \right\} dx \, dy$$

after integrating by parts. This then gives a contribution

$$\iint_{[e]} \frac{1}{c^2} \left\{ \left. N_i^e \frac{\partial N_j^e}{\partial t} \right|_0^T - \int_0^T \frac{\partial N_i^e}{\partial t} \frac{\partial N_j^e}{\partial t} dt \right\} dx \, dy$$

to the element stiffness matrix.

Solution 5.9 In this case $h = \tfrac{2}{3}$, so that

$$C = \frac{2}{18} \begin{bmatrix} 2 & 1 & 0 & 0 \\ & 4 & 1 & 0 \\ & & 4 & 1 \\ \text{sym} & & & 2 \end{bmatrix}, \quad K = \frac{3}{2} \begin{bmatrix} 1 & -1 & 0 & 0 \\ & 2 & -1 & 0 \\ & & 2 & -1 \\ \text{sym} & & & 1 \end{bmatrix}.$$

Thus the difference equation given by eqn (5.33) with $\Delta t = 1$, after the conditions

$$U_{11} = 0, U_{21} = \tfrac{2}{9}, U_{31} = \tfrac{8}{9}, U_{41} = 2, U_{12} = 1, U_{42} = 3$$

are enforced, becomes

$$\begin{bmatrix} 2 & 8 & 2 & 0 \\ 0 & 2 & 8 & 2 \end{bmatrix} \begin{bmatrix} 1 \\ U_{22} \\ U_{32} \\ 3 \end{bmatrix} + \begin{bmatrix} -29 & 46 & -29 & 0 \\ 0 & -29 & 46 & -29 \end{bmatrix} \begin{bmatrix} 0 \\ \tfrac{2}{9} \\ \tfrac{8}{9} \\ 2 \end{bmatrix} = \begin{bmatrix} 0 \\ 0 \\ 0 \\ 0 \end{bmatrix},$$

which yields $U_{22} = \tfrac{11}{9}, U_{32} = \tfrac{17}{9}$.

Solution 5.10 Equation (5.34) gives an approximation to the system (5.32) for values of t between $j\,\Delta t$ and $(j+1)\,\Delta t$. Similarly, for values of t between $(j-1)\,\Delta t$ and $j\,\Delta t$,

$$(5.63) \qquad \frac{1}{\Delta t}\mathbf{C}[\mathbf{U}_j - \mathbf{U}_{j-1}] + \mathbf{K}[\tau\mathbf{U}_j + (1-\tau)\mathbf{U}_{j-1}] = \tau\mathbf{F}_j + (1-\tau)\mathbf{F}_{j-1}.$$

Using Galerkin's method to minimize the residual for eqns (5.34) and (5.63), the weighting function is

$$W(t) = \begin{cases} \tau & (j-1)\ \Delta t \le t \le j\Delta t, \\ 1-\tau, & j\ \ \Delta t \le t \le (j+1)\Delta t, \end{cases}$$

which gives

$$\left[\frac{1}{\Delta t}\mathbf{C} + \frac{2}{3}\mathbf{K}\right]\mathbf{U}_j + \left[-\frac{1}{\Delta t}\mathbf{C} + \frac{1}{3}\mathbf{K}\right]\mathbf{U}_{j-1} - \frac{2}{3}\mathbf{F}_j - \frac{1}{3}\mathbf{F}_{j-1}$$

$$+ \left[\frac{1}{\Delta t}\mathbf{C} + \frac{1}{3}\mathbf{K}\right]\mathbf{U}_{j+1} + \left[-\frac{1}{\Delta t}\mathbf{C} + \frac{2}{3}\mathbf{K}\right]\mathbf{U}_j - \frac{1}{3}\mathbf{F}_{j+1} - \frac{2}{3}\mathbf{F}_j = 0.$$

Collecting the terms together gives the required result.

Solution 5.11 The finite element approximation is

$$\tilde{u} = \sum_e \tilde{u}^e,$$

where

$$\tilde{u}^e = \mathbf{N}^e(x,y)\mathbf{U}^e(t).$$

Using Galerkin's method, Green's theorem is used to reduce the order of the space derivatives occurring in the weighted residual equations. Then, just as in Example 5.4, the system of equations becomes

$$\mathbf{M}\ddot{\mathbf{U}} + \mathbf{C}\dot{\mathbf{U}} + \mathbf{K}\mathbf{U} = \mathbf{F},$$

where the stiffness and force matrices have their usual form. The element damping and mass matrices are given by

$$\mathbf{c}^e = \iint_{[e]} \mu \mathbf{N}^{e^T}\mathbf{N}^e \, dx\,dy$$

and

$$\mathbf{m}^e = \iint_{[e]} \frac{1}{c^2}\mathbf{N}^{e^T}\mathbf{N}^e \, dx\,dy.$$

Solution 5.12 (Zienkiewicz and Taylor 2000a)

$$\dot{U}_i = \frac{1}{\Delta t} \left[\tau - \tfrac{1}{2} \quad -2\tau \quad \tau + \tfrac{1}{2} \right] [U_{ij-1} \quad U_{ij} \quad U_{ij+1}]^T$$

and

$$\ddot{U}_i = \frac{1}{\Delta t^2} [1 \quad -2 \quad 1][U_{ij-1} \quad U_{ij} \quad U_{ij+1}]^T.$$

Thus the weighted residual form of eqn (5.62) becomes

$$\int_{-1}^{1} \left\{ \frac{1}{\Delta t^2} \mathbf{M}[\mathbf{U}_{j-1} - 2\mathbf{U}_j + \mathbf{U}_{j+1}] \right.$$
$$+ \frac{1}{\Delta t} \mathbf{C} \left[\left(\tau - \tfrac{1}{2} \right) \mathbf{U}_{j-1} - 2\tau \mathbf{U}_j + \left(\tau + \tfrac{1}{2} \right) \mathbf{U}_{j+1} \right]$$
$$+ \mathbf{K} \left[\tfrac{1}{2}(\tau^2 - \tau)\mathbf{U}_{j-1} + (1 - \tau^2)\mathbf{U}_j + \tfrac{1}{2}(\tau^2 + \tau)\mathbf{U}_{j+1} \right]$$
$$\left. - \left[\tfrac{1}{2}(\tau^2 - \tau)\mathbf{F}_{j-1} + (1 - \tau^2)\mathbf{F}_j + \tfrac{1}{2}(\tau^2 + \tau)\mathbf{F}_{j+1} \right] \right\} \mathbf{W} \, d\tau = 0.$$

Thus it follows that

(5.64)
$$[\mathbf{M} + \gamma \, \Delta t \, \mathbf{C} + \beta \, \Delta t^2 \, \mathbf{K}]\mathbf{U}_{j+1}$$
$$+ \left[-2\mathbf{M} + (1 - 2\gamma) \, \Delta t \, \mathbf{C} + \left(\tfrac{1}{2} - 2\beta + \gamma \right) \, \Delta t^2 \, \mathbf{K} \right] \mathbf{U}_j$$
$$+ \left[\mathbf{M} + (\gamma - 1) \, \Delta t \, \mathbf{C} + \left(\tfrac{1}{2} + \beta - \gamma \right) \, \Delta t^2 \, \mathbf{K} \right] \mathbf{U}_{j-1}$$
$$= \beta \, \Delta t^2 \, \mathbf{F}_{j+1} + \left(\tfrac{1}{2} - 2\beta + \gamma \right) \, \Delta t^2 \, \mathbf{F}_j + \left(\tfrac{1}{2} + \beta - \gamma \right) \, \Delta t^2 \, \mathbf{F}_{j-1}.$$

Solution 5.13 Using the results of the problem in Example 5.5, the overall stiffness and force matrices are

$$\mathbf{K} = \frac{3}{2} \begin{bmatrix} U_1 + U_2 & -U_1 - U_2 & 0 & 0 \\ & U_1 + 2U_2 + U_3 & -U_2 - U_3 & 0 \\ & & U_2 + 2U_3 + U_4 & -U_3 - U_4 \\ \text{sym} & & & U_3 + U_4 \end{bmatrix},$$

$$\mathbf{F} = \tfrac{1}{6}[1 \quad 2 \quad 2 \quad 1]^T.$$

Since $U_1 = 0$ and $U_4 = 1$, the finite element equations are thus

$$-\tfrac{3}{2}(1 + U_2) + \tfrac{3}{2}(1 + 2U_2 + U_3)U_2 - \tfrac{3}{2}(U_2 + U_3)U_3 = -\frac{1}{3},$$

$$-\tfrac{3}{2}(U_2 + U_3) + \tfrac{3}{2}(U_2 + 2U_3)U_3 = -\frac{1}{3}.$$

Solving these yields $U_2^2 = \tfrac{4}{9}, U_3^2 = \tfrac{1}{9}.$

Just as in Example 5.5, the positive roots are the admissible ones, so that $U_2 = \frac{2}{3}, U_3 = \frac{1}{3}$.

Solution 5.14 Following the method of the problem in Example 5.5 and using the notation of Fig. 3.7,

$$\mathbf{k}^e(u) = \frac{1}{2h}\begin{bmatrix} 1 & -1 \\ -1 & 1 \end{bmatrix}\int_{-1}^{1} e^u\, d\xi,$$

so that

$$\mathbf{k}^e(U_A, U_B) = \frac{1}{2h}\begin{bmatrix} 1 & -1 \\ -1 & 1 \end{bmatrix}\int_{-1}^{1}\exp\left\{\frac{1}{2}(1-\xi)U_A + \frac{1}{2}(1+\xi)U_B\right\}d\xi$$

$$= \frac{e^{U_B} - e^{U_A}}{h(U_B - U_A)}\begin{bmatrix} 1 & -1 \\ -1 & 1 \end{bmatrix}.$$

$\mathbf{f}^e$ may be obtained from the integral formula given in Solution 3.5. Putting $h = \frac{1}{2}$ and remembering to change the sign by virtue of the definition of the differential operator involved, it follows that

$$\mathbf{f}^e = -e^{xm}\begin{bmatrix} 0.23165 & 0.27358 \end{bmatrix}^T.$$

Thus, assembling the equation for the one unknown U_2 and enforcing the essential boundary conditions $U_1 = 0, U_3 = 1$,

$$\left[\frac{e^{U_2} - 1}{U_2} + \frac{e - e^{U_2}}{1 - U_2}\right]U_2 - \frac{e - e^{U_2}}{1 - U_2} = -0.42084.$$

This non-linear equation may be solved by the Newton–Raphson method to give the solution $U_2 = 0.5$.

Solution 5.15 Using the results of Exercise 5.13, the non-linear algebraic equations are

$$6U_2^2 - 3U_3^2 - \frac{7}{3} = 0,$$

$$-3U_2^2 + 6U_3^2 + \frac{2}{3} = 0.$$

Thus

$$\mathbf{G} = \begin{bmatrix} 12U_2 & -6U_3 \\ -6U_2 & 12U_3 \end{bmatrix}.$$

Taking a first approximation $U_0 = [0.7 \quad 0.3]^T$, it follows from eqn (5.41) that the second approximation is

$$\mathbf{U}_1 = \begin{bmatrix} 0.7 \\ 0.3 \end{bmatrix} - \begin{bmatrix} 0.1587 & 0.0794 \\ 0.1852 & 0.3704 \end{bmatrix} \begin{bmatrix} 0.3367 \\ -0.2633 \end{bmatrix}$$

$$= \begin{bmatrix} 0.667 \\ 0.335 \end{bmatrix}.$$

Similarly,

$$\mathbf{U}_2 = \begin{bmatrix} 0.667 \\ 0.335 \end{bmatrix} - \begin{bmatrix} 0.1666 & 0.0833 \\ 0.1658 & 0.3317 \end{bmatrix} \begin{bmatrix} -0.674 \\ 5.350 \end{bmatrix} \times 10^{-3}$$

$$= \begin{bmatrix} 0.667 \\ 0.333 \end{bmatrix}.$$

Solution 5.16 Since in this case $\mathbf{F}$ is independent of $\mathbf{U}$, the matrix $\mathbf{J}$ of eqn (5.40) is identical with the matrix $\mathbf{G}$ of eqn (5.39). Thus, with starting values

$$\mathbf{U}_0 = [0.7 \quad 0.3]^T, \quad \mathbf{F}_0 = [2.67 \quad -0.93]^T,$$

it follows that

$$\Delta \mathbf{F}_0 = [-0.1683 \quad 0.1367]^T,$$

since two equal increments are to be used. Equation (5.42) then gives

$$\mathbf{U}_1 = \begin{bmatrix} 0.7 \\ 0.3 \end{bmatrix} - \begin{bmatrix} 0.1587 & 0.0794 \\ 0.1852 & 0.3704 \end{bmatrix} \begin{bmatrix} -0.1683 \\ 0.1367 \end{bmatrix}$$

$$= \begin{bmatrix} 0.6837 \\ 0.3195 \end{bmatrix}.$$

Similarly,

$$\mathbf{U}_2 = \begin{bmatrix} 0.6837 \\ 0.3195 \end{bmatrix} - \begin{bmatrix} 0.1625 & 0.0813 \\ 0.1739 & 0.3478 \end{bmatrix} \begin{bmatrix} -0.1683 \\ 0.1367 \end{bmatrix}$$

$$= \begin{bmatrix} 0.667 \\ 0.338 \end{bmatrix}.$$

Solution 5.17 (Zienkiewicz, Lyness *et al.* 1977) The overall stiffness matrix is given by eqn (5.36) as

$$\mathbf{K}(\mathbf{U}) = \sum_e \iint_{[e]} \mu \boldsymbol{\alpha}^T \boldsymbol{\alpha} \, dx \, dy,$$

with $\mu = \mu(H)$ and $\mathbf{H} = -\mathrm{grad}\ u$. In this case the matrix $\mathbf{J}$ is given by

$$J_{ij} = \frac{\partial}{\partial U_j} \left(\sum_{p=1}^{n} K_{ip} U_p \right)$$

$$= K_{ij} + \sum_{p=1}^{n} \frac{\partial K_{ip}}{\partial U_j} U_p.$$

Thus $\mathbf{J}$ may be assembled from the following element matrices:

$$\mathbf{j}^e = \mathbf{k}^e + \sum_{j \in [e]} \left[\frac{\partial}{\partial U_j} (\mathbf{k}^e) \right] \mathbf{U}^e$$

$$= \mathbf{k}^e + \hat{\mathbf{k}}^e.$$

Now

$$\frac{\partial}{\partial U_j} (\mathbf{k}^e) = \iint_{[e]} \frac{\partial \mu}{\partial U_j} \boldsymbol{\alpha}^T \boldsymbol{\alpha}\, dx\, dy$$

and

$$\frac{\partial \mu}{\partial U_j} = \frac{d\mu}{dH} \frac{\partial H}{\partial U_j}$$

$$= \frac{d\mu}{dH} \frac{1}{H} \left(H_x \frac{\partial H_x}{\partial U_j} + H_y \frac{\partial H_y}{\partial U_j} \right)$$

$$= \frac{d\mu}{dH} \frac{1}{H} \mathbf{H}^T \left[\frac{\partial}{\partial U_j} (\mathbf{H}) \right]$$

$$= \frac{d\mu}{dH} \frac{1}{H} \mathbf{H}^T [-\alpha_{1j} \quad - \alpha_{2j}]^T.$$

Thus

$$\hat{\mathbf{k}}^e = \iint_{[e]} \frac{d\mu}{dH} \mathbf{H}^T \sum_{j \in [e]} [-\alpha_{1j} \quad - \alpha_{2j}]^T [U_j] \boldsymbol{\alpha}^T \boldsymbol{\alpha}\, dx\, dy$$

$$= \iint_{[e]} \frac{d\mu}{dH} \mathbf{H}^T \mathbf{H} \boldsymbol{\alpha}^T \boldsymbol{\alpha}\, dx\, dy$$

$$= \iint_{[e]} \frac{d\mu}{dH} H^2 \boldsymbol{\alpha}^T \boldsymbol{\alpha}\, dx\, dy.$$

Solution 5.18 Interpolating $U_i(t)$ in $0 \le t \le \Delta t$ by

$$U_i(t) = f_i + a_i\, \Delta t\, \tau, \quad 0 \le \tau \le 1,$$

substituting in the functional (2.76) and performing the space integrations as in Example 5.6,

$$
\begin{aligned}
I^e = \int_0^1 & \left\{ [U_{i-1}(1-\tau) \quad U_i(1-\tau)] \frac{k}{h} \begin{bmatrix} 1 & -1 \\ -1 & 1 \end{bmatrix} \begin{bmatrix} U_{i-1}(\tau) \\ U_i(\tau) \end{bmatrix} \right. \\
& + [U_{i-1}(1-\tau) \quad U_i(1-\tau)] \frac{h}{6} \begin{bmatrix} 2 & 1 \\ 1 & 2 \end{bmatrix} \begin{bmatrix} a_{i-1} \\ a_i \end{bmatrix} \right\} \Delta t \, d\tau \\
& - [f_{i-1} \quad f_i] \frac{h}{6} \begin{bmatrix} 2 & 1 \\ 1 & 2 \end{bmatrix} \begin{bmatrix} U_{i-1}(1) \\ U_i(1) \end{bmatrix}.
\end{aligned}
$$

Thus

$$
\begin{aligned}
\frac{\partial I^e}{\partial a_i} = & \frac{h \, \Delta t}{6} (-1 - 4\mu) f_{i-1} + \frac{h \, \Delta t}{3} (2\mu - 1) f_i \\
& + \frac{h \, \Delta t}{6} (-2\mu + 1) h_{i-1} + \frac{h \, \Delta t}{3} (1 + \mu) h_i.
\end{aligned}
$$

Assembling the overall equations just as in Example 5.6 gives the stated difference formula.

Solution 5.19 Taking the Laplace transform gives the two-point boundary-value problem

$$
-\frac{d^2 \bar{u}}{dx^2} + \lambda^2 \bar{u} = \pi \sin \pi x,
$$

$$
\bar{u}(0; \lambda) = \bar{u}(1; \lambda) = 0.
$$

Using two equal linear elements, the overall stiffness and mass matrices are

$$
\mathbf{K} = 2 \begin{bmatrix} 1 & -1 & 0 \\ -1 & 2 & -1 \\ 0 & -1 & 1 \end{bmatrix}, \quad \mathbf{C} = \frac{1}{12} \begin{bmatrix} 2 & 1 & 0 \\ 1 & 4 & 1 \\ 0 & 1 & 2 \end{bmatrix}.
$$

The element force matrices are

$$
\mathbf{f}^1 = \int_0^{1/2} \pi \sin \pi x \begin{bmatrix} \frac{1}{2} - x & x \end{bmatrix}^T dx = \begin{bmatrix} \frac{1}{2} - \frac{1}{\pi} & \frac{1}{\pi} \end{bmatrix}^T,
$$

$$
\mathbf{f}^2 = \int_{1/2}^1 \pi \sin \pi x \begin{bmatrix} 1 - x & x - \frac{1}{2} \end{bmatrix}^T dx = \begin{bmatrix} \frac{1}{\pi} & \frac{1}{2} - \frac{1}{\pi} \end{bmatrix}^T.
$$

Thus the overall equation for the only unknown $\bar{U}_2$ is, enforcing the boundary conditions $\bar{U}_1 = \bar{U}_3 = 0$,

$$
4\bar{U}_2 + \frac{\lambda^2}{12} (4\bar{U}_2) = \frac{2}{\pi}.
$$

Thus

$$\bar{U}_2 = \frac{6}{\pi(12 + \lambda^2)};$$

therefore $U_2(t) = 0.551 \sin(3.46t)$.

Solution 5.20 Just as in Example 5.7, the Laplace transform of $U_2(t)$ is given by

$$\bar{U}_2 = \frac{1}{\lambda} + \frac{1}{\lambda^2} - \frac{1}{8(3 + \lambda)}.$$

A spreadsheet inversion of $\bar{U}_2$ is shown in Figure 5.5.

From Figure 5.5, we see that Stehfest's method yields the value $U_2(1) = 1.994$. This agrees to three decimal places with the exact inversion given by eqn (5.53).

Solution 5.21 The overall system of equations for the Laplace transform is

$$\frac{1}{h} \begin{bmatrix} 2 & -1 & 0 \\ -1 & 2 & -1 \\ 0 & -1 & 2 \end{bmatrix} \begin{bmatrix} \bar{U}_2 \\ \bar{U}_3 \\ \bar{U}_4 \end{bmatrix} + \frac{\lambda h}{6} \begin{bmatrix} 4 & 1 & 0 \\ 1 & 4 & 1 \\ 0 & 1 & 4 \end{bmatrix} \begin{bmatrix} \bar{U}_2 \\ \bar{U}_3 \\ \bar{U}_4 \end{bmatrix} = \frac{1}{4} \begin{bmatrix} 2 \\ 2 \\ 2 \end{bmatrix},$$

which yields, with $h = \frac{1}{4}$,

$$\left(8 + \frac{\lambda}{6}\right)\bar{U}_2 - \left(4 - \frac{\lambda}{24}\right)\bar{U}_3 = \frac{1}{2},$$

$$-\left(4 - \frac{\lambda}{24}\right)\bar{U}_2 + \left(8 + \frac{\lambda}{6}\right)\bar{U}_3 - \left(4 - \frac{\lambda}{24}\right)\bar{U}_4 = \frac{1}{2},$$

$$-\left(4 - \frac{\lambda}{24}\right)\bar{U}_3 + \left(8 + \frac{\lambda}{6}\right)\bar{U}_4 = \frac{1}{2}.$$

	A	B	C	D	E	F	G	H	I
1	Exercise 5.20								
2	tau	1							
3	j	1	2	3	4	5	6	7	8
4	w_j	-0.33333	48.33333	-906	5464.667	-14376.6667	18730	-11946.7	2986.667
5	lambda_j	0.693147	1.386294	2.079442	2.772589	3.465735903	4.158883	4.85203	5.545177
6	U2_bar_j	3.490218	1.213192	0.687553	0.469105	0.352461088	0.280804	0.232657	0.19823
7	w_j*U2_bar_j	-1.16341	58.63761	-622.923	2563.504	-5067.21558	5259.462	-2779.47	592.0473
8									
9	U_2(tau)	1.993675							

Fig. 5.5 Spreadsheet for Solution 5.20.

	A	B	C	D	E	F	G	H	I
1	Exercise 5.21								
2	tau	0.1							
3	j	1	2	3	4	5	6	7	8
4	w_j	-0.33333333	48.3333333	-906	5464.667	-14376.6667	18730	-11946.7	2986.667
5	lambda_j	6.931471806	13.86294361	20.79442	27.72589	34.65735903	41.58883	48.5203	55.45177
6	a	9.155245301	10.3104906	11.46574	12.62098	13.7762265	14.93147	16.08672	17.24196
7	b	3.711188675	3.42237735	3.133566	2.844755	2.555943374	2.267132	1.978321	1.689509
8	U_2_bar_j	0.11432222	0.082847018	0.065278	0.054037	0.046209506	0.040435	0.035993	0.032464
9	U_3_bar_j	0.147297272	0.103493379	0.079289	0.063976	0.053441177	0.045765	0.039934	0.035361
10	U_4_bar_j	0.11432222	0.082847018	0.065278	0.054037	0.046209506	0.040435	0.035993	0.032464
11									
12	w_j*U2_bar_j	-0.038107406	4.004272516	-59.1416	295.2927	-664.338666	757.35	-429.991	96.95915
13	w_j*U3_bar_j	-0.04909909	5.00217999	-71.8356	349.6086	-768.305991	857.1844	-477.08	105.6121
14	w_j*U4_bar_j	-0.038107406	4.004272516	-59.1416	295.2927	-664.338666	757.35	-429.991	96.95915
15									
16	U_2(tau)	0.667923788		U_3(tau)	0.945424		U_4(tau)	0.667924	

Fig. 5.6 Spreadsheet solution for the problem in Exercise 5.21.

Since this system is diagonally dominant, we expect a Jacobi iteration to converge (Jennings and McKeown 1992):

$$\bar{U}_2^{(n+1)} = \frac{1}{8 + \lambda/6} \left(\frac{1}{2} + \left(4 - \frac{\lambda}{24} \right) \bar{U}_3^{(n)} \right),$$

$$\bar{U}_3^{(n+1)} = \frac{1}{8 + \lambda/6} \left(\frac{1}{2} + \left(4 - \frac{\lambda}{24} \right) \bar{U}_2^{(n)} - \left(4 - \frac{\lambda}{24} \right) \bar{U}_4^{(n)} \right),$$

$$\bar{U}_4^{(n+1)} = \frac{1}{8 + \lambda/6} \left(\frac{1}{2} + \left(4 - \frac{\lambda}{24} \right) \bar{U}_3^{(n)} \right).$$

A spreadsheet solution at $t = 0.1$ is shown in Fig. 5.6. Here we have used the Microsoft Excel spreadsheet and employed the Solver facility to implement the iterative solution to obtain $\bar{U}_2(\lambda_j), \bar{U}_3(\lambda_j), \bar{U}_4(\lambda_j), j = 1, 2, \ldots, M$. Stehfest inversion yields

$$U_2 = 0.671 = U_4, \quad U_3 = 0.949.$$

These results agree with the exact values to three decimal places; see Example 5.10.

6 Convergence of the finite element method

In this chapter, an introduction to the mathematical basis of the finite element method is presented. It is by no means a detailed discussion, since the concepts required are beyond the scope of this text. However, the chapter gives a flavour of the way in which the ideas may be developed, and the interested reader may take the subject matter further by consulting the references. In particular, the text by Brenner and Scott (1994) gives a good picture of the mathematical theory underpinning the finite element method.

6.1 A one-dimensional example

Consider the following positive definite, self-adjoint, two-point boundary-value problem:

$$(6.1) \qquad -\frac{d}{dx}\left(p(x)\frac{du}{dx}\right) + q(x)u = f(x), \quad a < x < b,$$

subject to the Dirichlet boundary conditions

$$(6.2) \qquad u(a) = \alpha, \quad u(b) = \beta.$$

In eqn (6.1), the functions p and q satisfy

$$(6.3) \qquad p(x) > 0 \quad \text{and} \quad q(x) \geq 0 \quad \text{for} \quad a < x < b.$$

It follows from the one-dimensional form of the functional (2.44) of Section 2.6 that the solution u_0 of eqn (6.1) subject to eqn (6.2) minimizes

$$(6.4) \qquad I[u] = \int_a^b \left\{ p(u')^2 + qu^2 - 2uf \right\} dx.$$

It is convenient at this stage to introduce the following *inner product* notations:

$$(6.5) \qquad (u, v) = \int_a^b uv \, dx,$$

$$(6.6) \qquad A(u, v) = \int_a^b (pu'v' + quv) \, dx.$$

By analogy with the functional of eqn (2.27) in Section 2.4 and the physical interpretation of the terms, $\frac{1}{2}A(u, u)$ is often called the *energy* of the function u. Consequently, $A(u, u)$ is called the *energy norm* of the function u. If the error e in a sequence of approximations is such that $A(e, e) \to 0$, then the method used to obtain the approximations is said to converge in energy, or in the energy norm. The norm of the function u is defined by

$$(6.7) \qquad \|u\|_p = \left\{ (u, u) + (u', u') + \ldots + (u^{(p)}, u^{(p)}) \right\}^{1/2}.$$

An important property of a *properly posed* problem is that the solution depends continuously on the data. This implies that there exist positive constants C and M such that

$$(6.8) \qquad \|u_0\|_2 \le C \, \|f\|_0 < M.$$

In Chapter 2, nothing was said about the class of functions which are admissible in the functional (6.4), except that they must satisfy the essential boundary conditions. In fact, the required class is known as the Hilbert space of functions defined on $[a, b]$ which have a finite norm $\|u\|_1$ and which satisfy the essential boundary conditions. Using the notation of Strang and Fix (1973), we denote this space by H_E^1; the superscript shows the order of derivatives which have finite norm and the subscript shows that admissible functions must satisfy the essential boundary conditions.

In the finite element method, an approximate solution is sought amongst functions which belong to a closed subspace, S^h, of H_E^1; for example, the approximating function considered in Section 3.5 is a piecewise linear function. The questions that then arise are 'does the method converge as the mesh size decreases, i.e. as $h \to 0$?' and 'can error bounds be obtained in terms of h?' Theorem 6.4 answers these questions; but first, three lemmas are required.

Lemma 6.1 *Suppose that the minimum of $I[v]$ as v varies over S^h occurs at the function u^h; then $A(u^h, v) = (f, v)$ for any $v \in S^h$. In particular, if $S^h = H_E^1$, then $A(u_0, v) = (f, v)$ for any $v \in H_E^1$.*

Proof Since u^h minimizes $I[v]$, it follows that for any ϵ and $v \in S^h$,

$$I[u^h] \le I\left[u^h + \epsilon\right]$$
$$= I[u^h] + 2\epsilon \left\{ A(u^h, v) - (f, v) \right\}$$
$$+ \epsilon^2 A(v, v) \quad \text{(see Exercise 6.1)}.$$

Thus

$$2\epsilon \left\{ A(u^h, v) - (f, v) \right\} + \epsilon^2 A(v, v) \ge 0.$$

Since this result must hold for arbitrary ϵ, it follows that

(6.9) $$A(u^h, v) = (f, v), \quad v \in S^h.$$

In the special case that S^h is the whole space H_E^1, $u^h = u_0$, and thus

(6.10) $$A(u_0, v) = (f, v), \quad v \in H_E^1.$$

$\square$

Lemma 6.2 $A(u_0 - u^h, v) = 0$ *for all* $v \in S^h$.

Proof Since eqn (6.10) holds for all $v \in H_E^1$, it certainly holds for $v \in S^h$, i.e.

(6.11) $$A(u_0, v) = (f, v), \quad v \in S^h.$$

Subtracting eqn (6.9) from eqn (6.11) yields the result. $\square$

Lemma 6.3 *The minimum of $I[v]$ and the minimum of $A(u_0 - v, u_0 - v)$ for $v \in S^h$ are achieved by the same function u^h.*

Proof

$$A(u_0 - u^h - v, u_0 - u^h - v) = A(u_0 - u^h, u_0 - u^h) - 2A(u_0 - u^h, v) + A(v, v);$$

thus, by virtue of Lemma 6.2, it follows that

$$A(u_0 - u^h - v, u_0 - u^h - v) \geq A(u_0 - u^h, u_0 - u^h),$$

equality occurring only if $A(v, v) = 0$, i.e. if $v = 0$. Thus u^h is the unique function which minimizes $A(u_0 - v, u_0 - v)$, and the result is established. $\square$

Theorem 6.4 *The error $e = u_0 - u^h$ in the finite element method using linear elements satisfies*

(6.12) $$A(e, e) \leq Kh^2 \, \|f\|_0^2$$

and

(6.13) $$\|e\|_0 \leq K_1 h^2 \, \|f\|_0^2.$$

Proof Consider the function $\tilde{u}_I$ which agrees exactly with u_0 at the nodes and interpolates it linearly between them. This function $\tilde{u}_I$ is of course unknown, since u_0 itself is unknown; however, it is a more convenient function to deal with. Notice first that any results established for $\tilde{u}_I$ will also hold for u^h, since, by virtue of Lemma 6.3, u^h is at least as good an approximation as $\tilde{u}_I$, in the sense that u^h minimizes the energy norm.

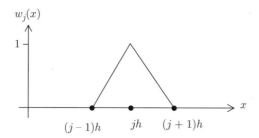

Fig. 6.1 The nodal function $n_j(x)$ associated with node j for linear elements.

Consider the case of linear interpolation for u_0, so that

$$\tilde{u}_I = \sum_{j=1}^{n} w_j U_j,$$

where w_j is the nodal function associated with node j (*cf.* eqn (3.35)); see Fig. 6.1. It may be shown (see Exercise 6.2) that

$$A(u_0 - \tilde{u}_I, u_0 - \tilde{u}_I) \leq \left(\frac{h^2}{\pi^2}P + \frac{h^4}{\pi^4}Q\right)||u_0''||_0^2,$$

where $P = \max p(x)$ and $Q = \max q(x)$, $a \leq x \leq b$. Thus there exists a constant C_1 such that

$$A(u_0 - \tilde{u}_I, u_0 - \tilde{u}_I) \leq C_1 h^2 ||u_0''||_0^2.$$

Now

$$
\begin{aligned}
A(e, e) &\leq A(u_0 - \tilde{u}_I, u_0 - \tilde{u}_I) \quad \text{using Lemma 6.3} \\
&\leq C_1 h^2 ||u_0''||_0^2 \\
&\leq C_1 h^2 C ||f||_0^2 \quad \text{using eqn (6.8)} \\
&= K h^2 ||f||_0^2.
\end{aligned}
$$

(6.14)

Finally, then, as $h \to 0$, $A(e, e) \to 0$, and the finite element method converges in the energy norm.

To obtain the result (6.13), suppose that in eqn (6.1), $f = u_0 - u^h = e$, and that w is the corresponding solution.

Then, using eqn (6.10) with $v = e$,

$$A(w, e) = (e, e) = ||e||_0^2.$$

From Lemma 6.2, $A(v, e) = 0$ for all $v \in S^h$; thus it follows that $A(w - v, e) = ||e||_0^2$. Now eqn (6.14) gives

$$\{A(e, e)\}^{1/2} \leq C_2 h ||u_0''||_0.$$

and

$$\{A(w - v, w - v)\}^{1/2} \le C_2 h \, ||w''||_0.$$

The Schwartz inequality states that

$$|A(w - v, e)| \le \{A(w - v, w - v)A(e, e)\}^{1/2} \, ;$$

thus

$$||e||_0^2 \le C_2^2 h^2 \, ||u_0''||_0 \, ||w''||_0$$

$$\le C_2^2 h^2 \, ||u_0''||_0 \, C_3 \, ||e||_0,$$

using an inequality of the form (6.8) for w.

Finally, then, it follows, using eqn (6.8), that

$$||e||_0 \le K_1 h^2 \, ||f||_0. \qquad \qquad \square$$

So far, the convergence proofs have assumed that the actual function f is used in the finite element calculations. Often an interpolate f_I is used, and thus the corresponding finite element solution $\tilde{u}^h$ differs from u^h.

Theorem 6.5 *The error $E = u^h - \tilde{u}^h$ due to the replacement of f by its inter-polate f_I satisfies*

$$A(E, E) \le K_2 h^4 \, ||f''||_0^2.$$

Proof The exact solution $u_0 - \tilde{u}$ corresponding to the data $f - f_I$ is bounded by

$$||u_0 - \tilde{u}||_2 \le C \, ||f - f_I||_0.$$

Now, we can show in a manner similar to that in Solution 6.2 that

$$||f - f_I||_0 \le \frac{h^2}{\pi^2} \, ||f''||_0.$$

Thus

$$||u_0 - \tilde{u}||_2 \le \frac{Ch^2}{\pi^2} \, ||f''||_0^2.$$

Also,

$$A(u_0 - \tilde{u}, u_0 - \tilde{u}) \le C_3 \, ||u_0 - \tilde{u}||_0^2$$

$$\le K_2 h^4 ||f''||_0^2.$$

Finally,

$$\begin{aligned} A(E, E) &= A(u_0 - \tilde{u}, u_0 - \tilde{u}) \\ &\quad - A\left([u_0 - \tilde{u}] - [u^h - \tilde{u}^h], [u_0 - \tilde{u}] - [u^h - \tilde{u}^h]\right) \\ &\le A(u_0 - \tilde{u}, u_0 - \tilde{u}) \\ &\le K_2 h^4 \, \|f''\|_0^2. \end{aligned}$$

Thus it follows that the error due to the linear interpolation of f is smaller, in the energy sense, than the inherent error in the method based on linear elements. □

The results in this section have been deduced using arguments based on variational methods; the reason for this is that the same ideas may be used for problems in two or three dimensions. A finite difference approach may also be used for the one-dimensional case, but it is not easy to extend it to higher dimensions. However, it is interesting to use this approach here because it brings out a remarkable property of the finite element solution to the two-point boundary-value problem.

Consider, for simplicity, the equation

(6.15) $$-u'' = f, \quad a < x < b,$$

subject to the boundary conditions

(6.16) $$u(a) = \alpha, \quad u(b) = \beta.$$

The corresponding functional is

(6.17) $$I[u] = \int_a^b \{(u')^2 - 2uf\} \, dx,$$

and all trial functions must satisfy the essential boundary conditions (6.16). Only Dirichlet conditions are considered here; similar reasoning follows if mixed conditions are given.

Using the usual finite element approximation in terms of nodal functions

$$\tilde{u} = \sum_{j=1}^n w_j(x) U_j,$$

where $w_j(x)$ is as shown in Fig. 6.1, the error is given by $e = u_0 - \tilde{u}$. From Lemma 6.2, since $\tilde{u}$ is chosen to minimize eqn (6.17),

$$\int_a^b \frac{d}{dx}(u_0 - \tilde{u})\frac{dw_j}{dx} \, dx = 0, \quad j = 2, \dots, n-1,$$

i.e.

$$\int_a^b \frac{de}{dx} \frac{dw_j}{dx} \, dx = 0, \quad j = 2, \ldots, n-1.$$

Since $w_j \equiv 0$ for $x \le (j-1)h$ or $x \ge (j+1)h$, it follows that

$$\int_{(j-1)h}^{jh} \frac{de}{dx} \left(+\frac{1}{h}\right) dx + \int_{jh}^{(j+1)h} \frac{de}{dx} \left(-\frac{1}{h}\right) dx = 0, \quad j = 2, \ldots, n-1.$$

Therefore

$$\frac{1}{h}(-e_{j-1} + 2e_j - e_{j+1}) = 0, \quad j = 2, \ldots, n-1.$$

Now $e_0 = e_n = 0$; thus it follows that $e_1 = e_2 = \ldots = e_{n-1} = 0$, i.e. the finite element solution $\tilde{u}$ coincides with the exact solution u_0 at the nodes. This phenomenon is easily seen in Example 3.1 (see Fig. 3.9), and is referred to as *superconvergence* in the literature (Douglas and Dupont 1974).

6.2   Two-dimensional problems involving Poisson's equation

For two-dimensional elements, error bounds are found in terms of the element diameters; for example, the diameter of a triangle is the length of the longest side and the diameter of a quadrilateral is the length of the longer diagonal.

Consider a discretization of some two-dimensional region D by means of triangles and suppose that h is the maximum diameter for these triangles. Error analysis in this case is far more complicated than for the two-point boundary-value problem of Section 6.1, and only a statement of the error bounds is given here; the interested reader is referred to the books by Strang and Fix (1973) and Wait and Mitchell (1985) for more details and further references.

The norm used here is given by

$$||u||_p = \left[\int\!\!\int_D \left\{ u^2 + \left(\frac{\partial u}{\partial x}\right)^2 + \left(\frac{\partial u}{\partial y}\right)^2 + \left(\frac{\partial^2 u}{\partial x^2}\right)^2 + \ldots + \left(\frac{\partial^p u}{\partial y^p}\right)^2 \right\} dx \, dy \right]^{1/2}.$$

(6.18)

The error $e = u_0 - \tilde{u}$ may be shown to satisfy an inequality of the form

(6.19) $$||e||_1 \le Ch^2 \max\left(|u_{xx}|, |u_{xy}|, |u_{yy}|\right),$$

i.e. just as in the one-dimensional case, the norm of the error behaves like h^2 as $h \to 0$.

Table 6.1 The solution, as the mesh is subdivided, of Poisson's equation $-\nabla^2 u = -2e^{x+y}$ in the region of the problem of Exercise 3.15; the exact solution is e^{x+y}

No. of elements	h	Finite element solution at $\left(\frac{1}{2}, \frac{1}{2}\right)$	Error $\times 10^2$
4	0.7071	2.6802	-3.81
16	0.3536	2.7079	-1.04
64	0.1768	2.7156	-0.27

Although the bounds on the error show that the method converges as $h \to 0$, the manner in which convergence occurs is not apparent. Melosh (1963) gave the following sufficient condition under which the method gives monotonic convergence:

If each subdivision of the finite element mesh contains the previous one as a subset then the convergence will be monotonic.

Table 6.1 shows such convergence for Poisson's equation in a square, the mesh being obtained by successively halving the dimensions of the triangles in Fig. 3.39.

In eqn (6.19) it is assumed that the smallest angle α of the triangle is bounded away from zero, i.e. the aspect ratio remains finite. In practice, it is usually desirable that the aspect ratio be chosen to be as near to unity as possible, unless it is known a priori that the solution possesses high gradients in some direction.

A similar result holds for bilinear rectangular elements, i.e. the norm of the error behaves like h^2 as $h \to 0$.

The remarks concerning monotonicity of convergence and the desirability of using elements with unit aspect ratio also hold for higher-order elements. Table 6.2 shows the monotonic convergence and Table 6.3 shows the effect of aspect ratio for the solution of Poisson's equation using eight-node rectangular elements.

A further point of interest here concerns the approximation of an infinite boundary. The convergence proofs referred to in this section are valid only for finite regions. In practice, the mathematical model of a physical problem may well include an infinite region, and the finite element method has often been used very successfully in such cases. The boundary at infinity may be replaced by a boundary at a relatively large finite distance from the region of interest. This technique has been very popular; however, the development of infinite elements (Bettess 1977, 1992) allows the complete region to be modelled. Finally, an alternative procedure involves a coupling of the finite element method for the

Table 6.2 The solution at $(0.5, 1)$ of Laplace's equation in the square with vertices at $(0,\ 0)$, $(2,\ 0)$, $(2,\ 2)$, $(0,\ 2)$. The boundary conditions are $u(x, 2) = \sin(\pi x/2)$, $u = 0$ on the remaining boundaries. The exact solution is $\sin(\pi x/2)\sinh(\pi y/2)/\sinh\pi$

No. of elements	h	Finite element solution	Error $\times 10^2$
8	0.7071	0.14120	2.99
32	0.3536	0.14093	0.24
128	0.1768	0.14091	0.02

Table 6.3 Effect of aspect ratio on the solution at $(0.5, 1)$ to the problem in Table 6.2

Aspect ratio	16	4	1
Finite element solution	0.1584	0.1462	0.1435
Error	0.0174	0.0053	0.0026

local region of interest with the boundary integral equation method for the far field (Zienkiewicz, Kelly *et al.* 1977). In fact, the boundary integral equation method has been couched in finite element terms for potential problems by Brebbia and Dominguez (1977), who used a weighted residual approach and referred to it as the boundary element method. We shall give an introduction to the boundary element method in Chapter 8.

6.3 Isoparametric elements: numerical integration

The isoparametric concept was introduced in Section 4.6. There, it was noted that the forms of the integrands obtained are usually too complicated to be evaluated analytically and are invariably obtained numerically using Gauss quadrature. Consequently, another source of error is introduced. The effect of numerical integration is considered in this section. Typically, for Poisson's equation, the integrals involved in computing k_{ij}^e are of the form of eqn (3.40),

$$\int\int_{[e]} k(x, y) \left(\frac{\partial N_i^e}{\partial x}\frac{\partial N_j^e}{\partial x} + \frac{\partial N_i^e}{\partial y}\frac{\partial N_j^e}{\partial y} \right) dx\, dy.$$

This expression was obtained using Galerkin's method and can be considered to come from an expression of the form

$$\int \int_{[e]} k(x,y) \operatorname{grad} \tilde{u}^e \cdot \operatorname{grad} W \, dx \, dy,$$

where $\tilde{u}^e$ is a trial function and W a weighting function. In the case where W is a linear polynomial, the integral is of the form

(6.20) $$\int \int_{[e]} k(x,y) \left(c_1 \frac{\partial \tilde{u}^e}{\partial x} + c_2 \frac{\partial \tilde{u}^e}{\partial y} \right) dx \, dy.$$

Using the isoparametric transformation given by eqns (4.15) and (4.16), the first term in eqn (6.20) becomes

(6.21) $$c_1 \int \int_{[e]} k(\xi, \eta) \left(\frac{\partial \tilde{u}^e}{\partial \xi} \frac{\partial \xi}{\partial x} + \frac{\partial \tilde{u}^e}{\partial \eta} \frac{\partial \eta}{\partial x} \right) |\det \mathbf{J}| \, d\xi \, d\eta.$$

Now the Jacobian matrix $\mathbf{J}$ is given by eqn (4.14), so that

(6.22) $$\begin{bmatrix} \partial \xi / \partial x & \partial \eta / \partial x \\ \partial \xi / \partial y & \partial \eta / \partial y \end{bmatrix} = \mathbf{J}^{-1} = \frac{1}{\det \mathbf{J}} \begin{bmatrix} \partial y / \partial \eta & -\partial y / \partial \xi \\ -\partial x / \partial \eta & \partial x / \partial \xi \end{bmatrix}.$$

Thus expression (6.21) is proportional to

(6.23) $$\int \int_{[e]} k(\xi, \eta) \left(\frac{\partial \tilde{u}^e}{\partial \xi} \frac{\partial y}{\partial \eta} - \frac{\partial \tilde{u}^e}{\partial \eta} \frac{\partial y}{\partial \xi} \right) d\xi \, d\eta,$$

i.e. the rational functions in the integrand in eqn (6.21) have been replaced by polynomials, and the method will certainly converge if this integral is computed exactly.

It is interesting to note that this condition may be interpreted by saying that it is necessary that the area of the element must be computed exactly by the quadrature rule. This follows since the expression $(\partial \tilde{u}^e / \partial \xi)(\partial y / \partial \eta) - (\partial \tilde{u}^e / \partial \eta)(\partial y / \partial \xi)$ in the integrand has the same form as $\det \mathbf{J}$ and the area of the element is given by $\int \int_{[e]} |\det \mathbf{J}| \, d\xi \, d\eta$. In the Gauss quadrature method, integrals of the form $\int \int_{[e]} F(\xi, \eta) \, d\xi \, d\eta$ are evaluated as a sum $\sum_g w_g F(\xi_g, \eta_g)$, where (ξ_g, η_g) is a sampling point inside the element and w_g is an associated weight. It is not the intention of this section to give a detailed derivation of the Gauss quadrature method. However, for the interested reader who is not familiar with the ideas, an introduction is provided in Exercises 6.3–6.7.

The question to be answered is 'under what conditions can eqn (6.23) be evaluated exactly?' Strang and Fix (1973) showed that there are two parts to the answer. Firstly, to ensure positive definiteness of the approximate functional, there is a minimum number N of sampling points in the region. If the trial functions are polynomials of degree p, then $N \geq p - 1$ ensures that the quadrature errors are of the same order as those due to the piecewise polynomial approximation. Secondly, for convergence as the mesh size decreases, it is necessary that $N \geq m$, where m is the order of the highest derivative occurring in the energy functional. Thus, provided that a quadrature scheme is chosen for which $N \geq \max(p-1, m)$, such a scheme will be suitable.

An important point regarding the use of isoparametric elements has not so far been considered. This concerns the possible vanishing of det $\mathbf{J}$. It is clear from eqn (6.22) that det $\mathbf{J}$ must not vanish inside the element; this imposes restrictions on the geometry of the chosen element.

Example 6.1 Consider the linear isoparametric quadrilateral introduced in Example 4.4; see Fig. 4.10. The parent element is a conforming element in terms of local coordinates (ξ, η) (see Section 3.7); consequently, it follows from the remarks made in Section 4.6 that boundaries of adjacent distorted elements coincide. Thus the only way in which the transformation may not be invertible is if det $\mathbf{J} = 0$ somewhere in the parent element.

Using the results of Example 4.4,

$$\det \mathbf{J} = \frac{1}{16} \begin{vmatrix} -x_1 + x_2 + x_3 - x_4 + A\eta & -y_1 + y_2 + y_3 - y_4 + B\eta \\ -x_1 - x_2 + x_3 + x_4 + A\xi & -y_1 - y_2 + y_3 + y_4 + B\xi \end{vmatrix},$$

where $A = x_1 - x_2 + x_3 - x_4$, $B = y_1 - y_2 + y_3 - y_4$. Thus det $\mathbf{J}$ is a linear function of ξ and η, and thus it follows that if det $\mathbf{J}$ has the same sign at all four nodes of the parent element then it cannot vanish inside it.

Now, at node 1,

$$\det \mathbf{J} = \tfrac{1}{4}\left\{(x_2 - x_1)(y_4 - y_1) - (y_2 - y_1)(x_4 - x_1)\right\}$$
$$= \tfrac{1}{4} ab \sin \theta,$$

where a and b are the lengths of sides 1,4 and 1,2, respectively, and θ is the interior angle at node 1. The notation here is that shown in Fig. 4.10(b).

Similar results hold for nodes 2, 3 and 4. Thus it follows that det $\mathbf{J}$ will be of the same sign at all nodes if and only if all interior angles are less than π, i.e. the quadrilateral is convex. For higher-order isoparametric elements, the condition that det $\mathbf{J}$ must not vanish inside the element places restrictions on the possible position of the nodes on the curved side, see Exercise 6.9.

6.4   Non-conforming elements: the patch test

Throughout the text, it has been assumed that the trial functions are such that continuity across interelement boundaries is preserved, i.e. the elements considered have been conforming elements. In Section 3.7, a rectangular element was developed; this element is non-conforming in terms of global Cartesian coordinates if the element sides are not parallel to the coordinate axes. In practice, certain non-conforming elements are used, and sometimes these work very well indeed, although the convergence proofs outlined in this chapter do not hold for such elements. The following test, called the *patch test*, may be used to check that a certain non-conforming element will yield convergence.

Suppose that u_0 is a known solution to $\mathcal{L}u = f$ and that round the perimeter of any arbitrary patch of elements, values of u are chosen to be equal to u_0. If the approximate solution U to this problem, inside the patch, is identical with u_0 there, then the test is satisfied and the element will yield convergence.

The patch test for the finite element method may be considered to be analogous to the test of consistency for the finite difference method (Smith 1985).

Example 6.2 Consider the solution of Laplace's equation in the region shown in Fig. 4.15, using four equal bilinear rectangular elements. Suppose that a test solution $x - y + 1$ is considered within the patch. Then the corresponding nodal values are

$$U_1 = U_2 = U_3 = 1, \quad U_4 = U_6 = 2, \quad U_7 = U_8 = U_9 = 3.$$

Thus the overall equation for the internal node 5 is, using the results of Exercise 3.13,

$$-2U_1 - 2U_2 - 2U_3 - 2U_4 + 16U_5 - 2U_6 - 2U_7 - 2U_8 - 2U_9 = 0,$$

which yields $U_5 = 2$.

Substituting these values into the element approximation for $\tilde{u}^e$ yields

$$\tilde{u}^1 = \frac{1}{4}\left[(1-\xi)(1-\eta) \quad (1+\xi)(1-\eta) \quad (1+\xi)(1+\eta) \quad (1-\xi)(1+\eta)\right][1\ 2\ 2\ 1]^T,$$

where (ξ, η) are the usual local coordinates. Thus

$$\tilde{u}^1 = \tfrac{1}{2}(3 + \xi)$$
$$= x - y + 1.$$

Similarly, $\tilde{u}^2 = \tilde{u}^3 = \tilde{u}^4 = x - y + 1$; hence the element passes the patch test.

6.5   Comparison with the finite difference method: stability

Part of the procedure in the finite element method is to reduce a partial differential equation to a set of algebraic equations for unknown nodal variables. The reader with a knowledge of finite differences will have recognized already some of the usual finite difference replacements for particular equations; for example, in Exercise 3.18, the use of linear triangles led to the usual five-point formula for the torsion equation,

(6.24) $$\frac{1}{h^2}(U_2 + U_3 + U_4 + U_5 - 4U_1) = -2;$$

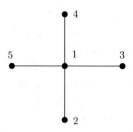

Fig. 6.2 Nodal arrangement on a square mesh of side h.

the notation is defined in Fig. 6.2. However, a different orientation of the elements yielded the same overall stiffness matrix but with a different load vector, giving

(6.25)
$$\frac{1}{h^2}(U_2 + U_3 + U_4 + U_5 - 4U_1) = -\tfrac{8}{3}.$$

Consider now Poisson's equation with a general non-homogeneous term $f(x, y)$. The finite difference replacement of $f(x, y)$ at (x_i, y_i) is simply $f_i = f(x_i, y_i)$, which gives

(6.26)
$$\frac{1}{h^2}(U_2 + U_3 + U_4 + U_5 - 4U_1) = -f_1,$$

i.e. the forces are lumped at the nodes. In the finite element method, however, consistent forces are used, so that for the triangular element of Fig. 3.24, with f interpolated linearly throughout the element, the result of Exercise 3.11 gives the element force vector as

$$\frac{h^2}{24}\,[2f_1 + f_2 + f_3 \quad f_1 + 2f_2 + f_3 \quad f_1 + f_2 + 2f_3]^T.$$

Thus, with the orientation shown in Fig. 3.31(a), the difference equation given by the finite element method is

$$\frac{1}{h^2}(U_2 + U_3 + U_4 + U_5 - 4U_1) = -\tfrac{1}{12}(6f_1 + f_2 + f_3 + f_4 + f_5 + f_6 + f_7 + f_8).$$

It is interesting here to illustrate the relationship between the two methods, showing why they yield the same overall stiffness matrix for Poisson's equation. Equation (2.38) gives the functional for $-\nabla^2 u = f$ as

(6.27)
$$I[u] = \iint_D \left\{ \left(\frac{\partial u}{\partial x}\right)^2 + \left(\frac{\partial u}{\partial y}\right)^2 - 2uf \right\} dx\, dy.$$

For a triangulation based on a square mesh parallel to the coordinate axes, each triangle has one or two positions relative to the axes; see Fig. 6.3. The finite element approximation yields values for $\partial \tilde{u}^e/\partial x$ and $\partial \tilde{u}^e/\partial y$, which may be substituted in eqn (6.27); these values are given in Exercise 6.10. Substituting and performing the integration yields

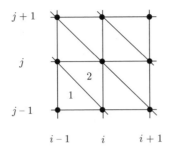

Fig. 6.3 Nodal numbering for a square mesh, showing triangular elements of types 1 and 2.

$$I^e = \tfrac{1}{2}h \left\{ \left(\frac{U_{i+1j} - U_{ij}}{h} \right)^2 + \left(\frac{U_{ij+1} - U_{ij}}{h} \right)^2 \right\} + \text{linear terms in } U_{ij}.$$

Now the only terms in I containing U_{ij} are those due to the sum over the six elements surrounding node i, j. Performing the sum and setting $\partial I / \partial U_{ij} = 0$ yields, after a little algebra,

$$(U_{i+1j} - 2U_{ij} + U_{i-1j})/h^2 + (U_{ij+1} - 2U_{ij} + U_{ij-1})/h^2 = \text{constant}.$$

The left-hand side is readily seen to be the central difference approximation to the Laplacian operator (Smith 1985), and the finite element method reproduces the well-known finite difference approach. Why then use finite elements at all? So far, only the replacement of the governing partial differential equation has been considered; no attention has been paid to the boundary conditions, and it is here where the methods differ and where the finite element method is particularly advantageous. A finite element mesh can be made to fit a given curved region arbitrarily closely, with its nodes actually on the boundary. However, in general, it requires many more nodes to fit a finite difference mesh with the same accuracy, and for this reason the finite difference mesh is often chosen in such a way that the nodes do not actually lie on the boundary at all. Difference operators with unequal arms are then used to approximate the given boundary conditions. Unfortunately, the accuracy is reduced and the procedure can be very clumsy from a programming point of view. Thus, in the finite difference method, the application of the boundary conditions may be a very complicated process. In the finite element method, however, the boundary conditions cause no problems. Dirichlet conditions are enforced without difficulty, and natural conditions are accommodated by using a suitable variational or weighted residual approach.

Another difficulty associated with the finite difference approach concerns the effect of a mixed boundary condition on the system of equations. Suppose that nodes 2, 5, 8 in Fig. 3.31 are on the boundary and that, for simplicity, a homogeneous Neumann boundary condition holds there. Then the left-hand side of the finite difference expression about node 5, $U_2 + U_4 + U_6 + U_8 - 4U_5$, becomes, approximating $\partial u / \partial y = 0$ by $(U_4 - U_6)/2h = 0$, $U_2 + 2U_6 + U_8 - 4U_5$. Thus, in the overall stiffness matrix, $K_{56} = 2$. Now, from the difference equation for node 6, it is seen that $K_{65} = 1$. Consequently, $K_{56} \neq K_{65}$, i.e. in the finite difference method the overall system of equations is not necessarily symmetric. This of course is not the case in the finite element method, where, for elliptic problems, the stiffness matrix is always symmetric.

The equivalence at interior points of the overall equations is well known and it is often said that one method is a special case of the other. The decision to use finite elements or finite differences is in some ways a matter of individual preference and depends on the problem under consideration. Each method has advantages and disadvantages; it is probably best to have both methods available and to utilize the best of each whenever possible. Finite element methods are particularly useful for describing boundary-value problems, while finite differences are often used for time-dependent terms, see Section 5.3.

One of the difficulties associated with time-dependent problems is that of the stability of the numerical scheme. A scheme is said to be *stable* if local errors remain bounded as the method proceeds from one time step to the next. There is a great deal of literature on the stability of finite difference schemes (Smith 1985, Lambert 2000), and the ideas presented there may be utilized for problems set up by the finite element method. To illustrate these ideas, one example will be considered here.

Example 6.3 Equation (5.32),

$$(6.28) \qquad C\dot{U} + KU = F,$$

gives a system of ordinary differential equations which occurs when a finite element procedure in space is used to solve the heat equation. A weighted residual method in time then yields eqn (5.35),

$$(6.29) \qquad \left[\frac{1}{\Delta t}C + rK\right]U_{j+1} + \left[-\frac{1}{\Delta t}C + (1 - r)K\right]U_j = rF_{j+1} + (1 - r)F_j,$$

where the well-known finite difference replacements are obtained by using suitable values of r.

As it stands, eqn (6.29) is not particulary susceptible to stability analysis. A change of variable is performed so that the equations (6.28) are uncoupled. Suppose that the time variation is given by

$$U = Ve^{-\omega t};$$

then the homogeneous form of eqn (6.28) yields the eigenvalue problem

$$(6.30) \qquad [K - \omega C]V = 0,$$

with eigenvalues ω_i and corresponding eigenvectors $\mathbf{V}_i$. Since $\mathbf{K}$ and $\mathbf{C}$ are positive definite, the eigenvalues are real and positive. Moreover, since $\mathbf{K}$ and $\mathbf{C}$ are symmetric, the eigenfunctions corresponding to distinct eigenvalues are orthogonal (Wilkinson 1999) in the sense that

$$\mathbf{V}_i^T \mathbf{K} \mathbf{V}_j = \mathbf{V}_i^T \mathbf{C} \mathbf{V}_j = 0, \quad i \neq j.$$

Also,

$$\mathbf{V}_i^T \mathbf{K} \mathbf{V}_i = k_i, \quad \mathbf{V}_i^T \mathbf{C} \mathbf{V}_i = c_i$$

with

(6.31)
$$\omega_i = k_i / c_i.$$

Suppose now that the solution is written as a linear combination of these modes,

$$\mathbf{U} = \sum_j \alpha_j(t) \mathbf{V}_j;$$

then substituting in eqn (6.28) and pre-multiplying by $\mathbf{V}_i^T$ gives

(6.32)
$$c_i \dot{\alpha}_i = k_i \alpha_i = f_i,$$

where

$$f_i = \mathbf{v}_i^T \mathbf{F}.$$

Equation (6.32) represents a set of uncoupled differential equations for the unknowns α_i, and it is this set which will be used rather than eqn (6.28). Starting with eqn (6.32) and applying the weighted residual process in time, just as in Section 5.3, the following difference equation is obtained (*cf.* eqn (6.29)):

$$(c_i/\Delta t + r k_i)(\alpha_i)_{j+1} + (-c_i/\Delta t + (1-r)k_i)(\alpha_i)_j = r(f_i)_{j+1} + (1-r)(f_i)_j.$$

(6.33)

For the purposes of stability analysis, consider the homogeneous form of eqn (6.33), i.e. $f_i \equiv 0$, and suppose that

(6.34)
$$(\alpha_i)_{j+1} = \lambda(\alpha_i)_j;$$

then

(6.35)
$$\lambda(c_i/\Delta t + r k_i) + (-c_i/\Delta t + (1-r)k_i) = 0.$$

From eqn (6.34), it may be seen that if $|\lambda| < 1$, the scheme is stable. Thus eqns (6.35) and (6.31) give the following condition for stability,

$$\left| \frac{1 - (1 - r)\omega_i \, \Delta t}{1 + r\omega_i \, \Delta t} \right| < 1,$$

which yields the inequality

(6.36) $$\omega_i \, \Delta t \, (1 - 2r) < 2.$$

Thus the scheme is *unconditionally stable* for $r \geq \frac{1}{2}$, since eqn (6.36) will always be satisfied no matter how large is the time step Δt. However, if $r \leq \frac{1}{2}$, the stability is conditional, since the time step Δt must be such that

$$\Delta t < \frac{2}{(1 - 2r)\omega_i}.$$

A similar procedure may be adopted for the stability analysis of the time-stepping method for the wave equation, see Exercise 6.12.

6.6 Exercises and solutions

Exercise 6.1 Show that, with the notation of Lemma 6.1,

$$I[u^h + \epsilon v] = I[u^h] + 2\epsilon[A(u^h, v) - (f, v)] + \epsilon^2 A(v, v).$$

Exercise 6.2 With the notation of Section 6.1, consider an element $[e]$ of length h. Let $\Delta(x) = u_0(x) - \tilde{u}(x)$, and suppose that in element $[e]$ it is given as a Fourier sine series by

(6.37) $$\Delta(x) = \sum_1^\infty b_n \sin(n\pi x/h).$$

Use Parseval's identity to obtain expressions for

$$\int_{(i-1)h}^{ih} (\Delta')^2 \, dx \quad \text{and} \quad \int_{(i-1)h}^{ih} \Delta^2 \, dx,$$

where $[e]$ contains nodes $i - 1$ and i. Hence show that

$$||u_0' - \tilde{u}_I'||_0 \leq \frac{h}{\pi} \, ||u_0''||_0$$

and

$$||u_0 - \tilde{u}_I||_0 \leq \frac{h}{\pi} \, ||u_0''||_0.$$

Deduce that

$$A(u_0 - \tilde{u}_I, u_0 - \tilde{u}_I) \leq \left(\frac{h^2}{\pi^2} P + \frac{h^4}{\pi^4} Q \right) ||u_0''||_0^2,$$

where $P = \max p(x)$ and $Q = \max q(x)$, $a \leq x \leq b$.

Exercise 6.3 Consider $I = \int_{-1}^{1} f(x)\,dx$. Integrals over the general range $[a, b]$ may be converted to $[-1, 1]$ by the change of variable $t = (2x - a - b)/(b - a)$. Suppose that the integral is to be approximated by

$$I \approx w_1 f(x_1) + w_2 f(x_2).$$

Find the four constants w_1, w_2, x_1, x_2 so that the formula is exact for an arbitrary cubic.

Exercise 6.4 Repeat Exercise 3.5 using the Gauss two-point formula, obtained in Exercise 6.3, to evaluate the integrals.

Exercise 6.5 Suppose that the integral I of Exercise 6.3 is to be approximated using p sampling points by

$$I \approx w_1 f(x_1) + w_2 f(x_2) + \ldots + w_p f(x_p),$$

in such a manner that an arbitrary polynomial $S_{2p-1}(x)$, of degree $2p - 1$, may be integrated exactly.

By writing $S_{2p-1}(x) = P_p(x)q(x) + r(x)$, where $P_p(x)$ is the Legendre polynomial of degree p, $q(x)$ is a polynomial of degree $p - 1$ and $r(x)$ is a polynomial of degree less than p, show that

$$\int_{-1}^{1} S_{2p-1}(x)\,dx = \int_{-1}^{1} r(x)\,dx.$$

Deduce that

$$\int_{-1}^{1} S_{2p-1}(x)\,dx = \sum_{g=1}^{p} w_g S_{2p-1}(x_g),$$

where x_j are the zeros of $P_p(x)$; these zeros are all real. How may the weights be found? A table of sample points and weights is given in Appendix D.

Exercise 6.6 (i) Using eqn (4.1), find $\mathbf{k}^e$ for the quadratic element using both one-point and two-point Gauss quadrature. (ii) Using eqn (4.10), find $\mathbf{k}^e$ for the Hermite element using one-point, two-point and three-point Gauss quadrature.

Exercise 6.7 Show how the one-dimensional form of Gauss quadrature over $[-1, 1]$ may be used to obtain numerical quadrature formulae for

$$I = \int_{-1}^{1} \int_{-1}^{1} f(x, y)\,dx\,dy.$$

Exercise 6.8 It is required to find a quadrature rule for

$$\int_{0}^{1} \int_{0}^{1-L_3} f(L_1, L_2, L_3)\,dL_2\,dL_3,$$

where L_1, L_2, L_3 are the usual area coordinates for a triangle. Consider an integration formula which shows no bias for any one coordinate. If three sampling points $(a, b, c), (b, c, a), (c, a, b)$ are taken, then there are only six independent parameters (a, b, c, w_1, w_2, w_3). Thus an arbitrary quadratic can be integrated by this rule. Find the integration points and the corresponding weights. A table of quadrature formulae for triangles is given in Appendix D.

Exercise 6.9 A typical boundary triangular element will have one curved side. Consider the case of such a quadratic triangle with the straight sides parallel to the coordinate axes, see Fig. 6.4(b). If $\det \mathbf{J} \neq 0$ in the element, show that node 5 must lie inside the shaded region in Fig. 6.4(b).

Exercise 6.10 Consider the linear triangular element with three mid-side nodes shown in Fig. 6.5. Find a suitable shape function matrix and hence find the element stiffness matrix. Show that, in general, this is not a conforming element. By considering the solution of Laplace's equation in a unit square comprising two such elements, show that it passes the patch test.

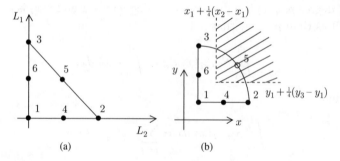

(a)   (b)

Fig. 6.4 A quadratic triangle with one curved side: (a) parent element; (b) distorted element.

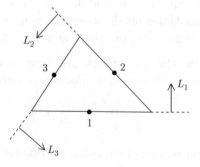

Fig. 6.5 A linear triangle with three mid-side nodes.

Exercise 6.11 Using the results of Exercise 5.6 and eqns (3.65) and (3.66), find expressions for $\partial u^e / \partial x$ and $\partial u^e / \partial y$ inside the elements of type 1 and 2 in Fig. 6.3. Hence show that for each type,

$$\frac{\partial \tilde{u}^e}{\partial x} = \frac{U_{i+1j} - U_{ij}}{h}, \qquad \frac{\partial \tilde{u}^e}{\partial y} = \frac{U_{i+1j} - U_{ij}}{h}.$$

Exercise 6.12 In Exercises 5.11 and 5.12, it is shown that the system of equations (5.60) arises in the finite element idealization of the equation of telegraphy, and a weighted residual approach in time yields the difference equation (5.64). The results for the unforced wave equation may be obtained by putting $\mu = 0$, which gives $\mathbf{C} = \mathbf{0}$, so that

$$(6.38) \qquad\qquad\qquad \mathbf{M\ddot{U} + KU = 0}.$$

Obtain the uncoupled form of eqn (5.64) in this case, and hence show that the scheme is unconditionally stable if

$$\beta \geq \tfrac{1}{4}\left(\gamma + \tfrac{1}{2}\right)^2, \qquad \gamma \geq \tfrac{1}{2}, \qquad \tfrac{1}{2} + \gamma + \beta \geq 0.$$

Solution 6.1

$$I[u] = \int_a^b \left\{ p(u')^2 + qu - 2fu \right\} dx$$

$$= A(u, u) - 2(f, u).$$

Thus

$$I[u^h + \epsilon v] = A(u^h + \epsilon v, u^h + \epsilon v) - 2(f, u^h + \epsilon v)$$

$$= A(u^h, u^h) + \epsilon A(v, u^h) + \epsilon A(u^h, v) + \epsilon^2 A(v, v)$$

$$- 2(f, u^h) - 2\epsilon(f, v)$$

$$= I[u^h] + 2\epsilon[A(u^h, v) - (f, v)] + \epsilon^2 A(v, v).$$

Solution 6.2 With the notation of Fig. 3.7, $\Delta(x) = 0$ at nodes A and B and is given in the element by eqn (6.37). Differentiating this series and using Parseval's identity, it follows that

$$\int_{(i-1)h}^{ih} (\Delta')^2 \, dx = \frac{h}{2} \sum \frac{n^2 \pi^2}{h^2} a_n^2,$$

$$\int_{(i-1)h}^{ih} (\Delta'')^2 \, dx = \frac{h}{2} \sum \frac{n^4 \pi^4}{h^4} a_n^2.$$

Since $n \geq 1$,

$$\frac{n^2 \pi^2 a_n^2}{h^2} \leq \frac{h^2}{\pi^2} \frac{n^4 \pi^4 a_n^2}{h^4}.$$

Therefore

$$\int_{(i-1)h}^{ih} (\Delta')^2 \, dx \leq \frac{h^2}{\pi^2} \int_{(i-1)h}^{ih} (\Delta'')^2 \, dx$$

$$= \frac{h^2}{\pi^2} \int_{(i-1)h}^{ih} (u_0'')^2 \, dx,$$

since $\Delta'' = u_0'' - \tilde{u}'' = u_0''$ because $\tilde{u}$ is linear. Thus, summing over all elements,

$$\|u_0' - \tilde{u}'\|_0^2 = \int_a^b (\Delta')^2 \, dx \leq \frac{h^2}{\pi^2} \|u_0''\|_0^2.$$

Similarly,

$$\|u_0 - \tilde{u}\|_0^2 = \int_a^b \Delta^2 \, dx \leq \frac{h^4}{\pi^4} \|u_0''\|_0^2.$$

Thus, since $A(u_0 - \tilde{u}, u_0 - \tilde{u}) = \int_a^b \{p(\Delta')^2 + q\Delta^2\} \, dx$, the final result follows.

Solution 6.3 The formula is to be exact for the arbitrary cubic $a_0 + a_1 x + a_2 x^2 + a_3 x^3$; thus it must be exact for each of the four linearly independent terms $1, x, x^2, x^3$.

Now,

$$\int_{-1}^1 1 \, dx = 2, \quad \int_{-1}^1 x \, dx = 0, \quad \int_{-1}^1 x^2 \, dx = \tfrac{2}{3} \quad \int_{-1}^1 x^3 \, dx = 0.$$

Thus $w_1 + w_2 = 2$, $w_1 x_1 + w_2 x_2 = 0$, $w_1 x_1^2 + w_2 x_2^2 = \tfrac{2}{3}$, $w_1 x_1^3 + w_2 x_2^3 = 0$.

Solving these equations gives $x_1 = -1/\sqrt{3}$, $x_2 = 1/\sqrt{3}$, $w_1 = 1$, $w_2 = 1$. Thus these sampling points and weights give the Gauss two-point formula

$$\int_{-1}^1 f(x) \, dx \approx f\left(-\tfrac{1}{\sqrt{3}}\right) + f\left(\tfrac{1}{\sqrt{3}}\right).$$

Solution 6.4 In the evaluation of the element stiffnesses, constants only are integrated. These, of course, are integrated exactly using the Gauss two-point formula; thus $\mathbf{k}^e$ is as given in Example 3.1. Using the result of Exercise 3.5,

$$\mathbf{f}^e \approx \frac{0.25}{4} e^{x_m} \left[\left(1 + \tfrac{1}{\sqrt{3}}\right) e^{-\alpha} + \left(1 - \tfrac{1}{\sqrt{3}}\right) e^{\alpha} \quad \left(1 - \tfrac{1}{\sqrt{3}}\right) e^{-\alpha} + \left(1 + \tfrac{1}{\sqrt{3}}\right) e^{\alpha}\right]^T,$$

$$\text{where } \alpha = 0.125/\sqrt{3},$$

$$= e^{x_m} [0.12011 \quad 0.13054]^T.$$

This agrees to five decimal places with the result in Exercise 3.5, and hence the finite element solution will be the same as that given there.

Solution 6.5 $S_{2p-1}(x) = P_p(x)q(x) = r(x)$. Since $q(x)$ is a polynomial of degree $p-1$, it follows that it may be written as a linear combination of $P_0, P_1, \ldots, P_{p-1}$. Now, the Legendre polynomials are orthogonal over $[-1, 1]$; thus

$$\int_{-1}^{1} P_p(x)q(x) \, dx = 0$$

and it follows that

$$\int_{-1}^{1} S_{2p-1}(x) \, dx = \int_{-1}^{1} r(x) \, dx;$$

$r(x)$ is a polynomial of degree less than p, and consequently it is integrated exactly by the quadrature rule:

$$\int_{-1}^{1} r(x) \, dx = \sum_{g=1}^{p} w_g r(x_g).$$

If the x_g are chosen such that $P_p(x_g) = 0$, then $S_{2p-1}(x_g) = r(x_g)$ and hence

$$\int_{-1}^{1} S_{2p-1}(x) \, dx = \sum_{g=1}^{p} w_g S_{2p-1}(x_g).$$

The weights w_g may be found from the system of p linear equations

$$\int_{-1}^{1} x^{j-1} \, dx = \sum_{g=1}^{p} w_g x_g^{j-1}, \quad j = 1, \ldots, p.$$

The $(2p-1)$-point Gauss quadrature rule integrates exactly polynomials of degree p.

Solution 6.6 (i) Using eqn (4.1), we obtain the following values for $\mathbf{k}^e$ for the quadratic element.

One-point quadrature :
$$\frac{1}{h} \begin{bmatrix} 1 & 0 & -1 \\ 0 & 0 & 0 \\ -1 & 0 & 1 \end{bmatrix}.$$

Two-point quadrature :
$$\frac{1}{h} \begin{bmatrix} 2.33333 & -2.66667 & 0.33333 \\ & 5.33333 & -2.66667 \\ \text{sym} & & 2.33333 \end{bmatrix}.$$

(ii) Using eqn (4.10), we obtain the following values for $\mathbf{k}^e$ for the Hermite element.

One-point quadrature :
$$\frac{1}{h}\begin{bmatrix} 4.5 & 0.75h & -0.45 & 0.75h \\ & 0.125h^2 & -0.75h & -0.125h^2 \\ & & 4.5 & -0.75h \\ \text{sym} & & & 0.125h^2 \end{bmatrix}.$$

Two-point quadrature :
$$\frac{1}{h}\begin{bmatrix} 1 & 0 & -1 & 0 \\ & 0.08333h^2 & 0 & -0.08333h^2 \\ & & 1 & 0 \\ \text{sym} & & & 0.08333h^2 \end{bmatrix}.$$

Three-point quadrature :
$$\frac{1}{h}\begin{bmatrix} 1.2 & 0.1h & -1.2 & 0.1h \\ & 0.133333h^2 & -0.1h & -0.03333h^2 \\ & & 1.2 & -0.1h \\ \text{sym} & & & 0.13333h^2 \end{bmatrix}.$$

We notice that for the quadratic element, the one-point formula is very poor but we obtain the exact values with the two-point formula. For the Hermite element, neither the one-point nor the two-point formula is sufficient. We require to use the three-point formula to obtain the exact value. The reason is that the integrals for the quadratic element are themselves quadratic. Those for the Hermite element are quartic. These results are consistent with the result of Exercise 6.5: a polynomial of degree $2p - 1$ is integrated exactly by a Gauss formula of order p.

Solution 6.7 $I = \int_{-1}^{1} \int_{-1}^{1} f(x, y) \, dx \, dy$. If the x integral is performed, keeping y constant and using the results of Exercise 6.5, then

$$I \approx \int_{-1}^{1} \left\{ \sum_{g=1}^{p_1} w_g f(x_g, y) \right\} dy.$$

Using the results of Exercise 6.5 again for the y integral,

$$I \approx \sum_{k=1}^{p_2} w_k \left\{ \sum_{g=1}^{p_1} w_g f(x_g, y_k) \right\}$$

$$= \sum_{k=1}^{p_2} \sum_{g=1}^{p_1} w_k w_g f(x_g, y_k).$$

Polynomials of order $2p_1 - 1$ in the x-direction and $2p_2 - 1$ in the y-direction are evaluated exactly by this formula.

Solution 6.8 $I \approx w_1 f(a, b, c) + w_2 f(b, c, a) + w_3 f(c, a, b)$ is to be exact for an arbitrary quadratic function. Consider the six linearly independent functions $1, L_2, L_3, L_2^2, L_3^2, L_2 L_3$; integrating these in turn using eqn (3.67) gives

$$\tfrac{1}{2} = w_1 + w_2 + w_3,$$

$$\tfrac{1}{6} = w_1 b + w_2 c + w_3 a = w_1 c + w_2 a + w_3 b,$$

$$\tfrac{1}{12} = w_1 b^2 + w_2 c^2 + w_3 a^2 = w_1 c^2 + w_2 a^2 + w_3 b^2,$$

$$\tfrac{1}{12} = w_1 bc + w_2 ca + w_3 ab.$$

This set of non-linear algebraic equations has a solution

$$w_1 = w_2 = w_3 = \tfrac{1}{6}, \quad a = b = \tfrac{1}{2}, \quad c = 0.$$

Solution 6.9 The transformation from the xy plane to the $L_1 L_2$ plane is given by

$$x = x_1 + (x_2 - x_1)L_1 + (4x_5 - 2x_1 - 2x_2)L_1 L_2,$$

$$y = y_1 + (y_3 - y_1)L_2 + (4y_5 - 2y_1 - 2y_3)L_1 L_2.$$

Thus

$$\mathbf{J} = \begin{bmatrix} (x_2 - x_1) + (4x_5 - 2x_1 - 2x_2)L_2 & (4y_5 - 2y_1 - 2y_3)L_2 \\ (4x_5 - 2x_1 - 2x_2)L_1 & (y_3 - y_1) + (4y_5 - 2y_1 - 2y_3)L_1 \end{bmatrix}.$$

Hence

$$\det \mathbf{J} = (x_2 - x_1)(y_3 - y_1) + (y_3 - y_1)(4x_5 - 2x_1 - 2x_2)L_2$$

$$+ (x_2 - x_1)(4y_5 - 2y_1 - 2y_3)L_1.$$

At node 1, $\det \mathbf{J} > 0$. Thus, since it is linear, a necessary and sufficient condition for the non-vanishing of $\det \mathbf{J}$ inside the element is that it is positive at nodes 2 and 3.

Therefore

$$(y_3 - y_1) + (4y_5 - 2y_1 - 2y_3) > 0 \text{ and } (x_2 - x_1) + (4x_5 - 2x_1 - 2x_2) > 0,$$

and thus

$$y_5 > y_1 + (y_3 - y_1)/4 \text{ and } x_5 > x_1 + (x_2 - x_1)/4;$$

i.e. node 5 must lie inside the shaded region of Fig. 6.4(b).

Solution 6.10 A suitable shape function matrix is

$$\mathbf{N}^e = [1 - 2L_1 \quad 1 - 2L_2 \quad 1 - 2L_3].$$

Now,

$$K_{ij} = \int \int_{[e]} \left(\frac{\partial N_i}{\partial x} \frac{\partial N_j}{\partial x} + \frac{\partial N_i}{\partial y} \frac{\partial N_j}{\partial y} \right) dx\, dy$$

$$= \frac{1}{A}(b_i b_j + c_i c_j)$$

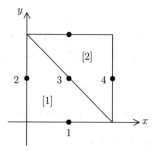

Fig. 6.6 Unit square divided into two triangular elements.

using the notation and results of Section 3.8.

Consider Laplace's equation in the unit square shown in Fig. 6.6. A solution is $u = x - y$, and the boundary conditions consistent with this solution are $U_1 = \frac{1}{2}, U_2 = -\frac{1}{2}, U_4 = \frac{1}{2}, U_5 = -\frac{1}{2}$. Thus, using the results of Example 3.3,

$$
\mathbf{k}^1 = \begin{array}{c} 1 \\ 2 \\ 3 \end{array}
\begin{array}{ccc} 1 & 2 & 3 \end{array}
\begin{bmatrix}
2 & 0 & -2 \\
 & 2 & -2 \\
\text{sym} & & 4
\end{bmatrix}
\begin{array}{c} 5 \\ 4 \\ 3 \end{array}
= \mathbf{k}^2.
$$

Thus the overall equation for the unknown U_3 is

$$2(-U_1 - U_2 + 4U_3 - U_4 - U_5) = 0,$$

which gives $U_3 = 0$. Interpolating the solution in each element yields

$$
\begin{aligned}
\tilde{u}^1 &= (1 - 2L_1)\left(\tfrac{1}{2}\right) + (1 - 2L_2)\left(-\tfrac{1}{2}\right) \\
&= L_2 - L_1 \\
&= x - y, \\
\tilde{u}^2 &= (1 - 2L_2)\left(\tfrac{1}{2}\right) + (1 - 2L_3)\left(-\tfrac{1}{2}\right) \\
&= L_3 - L_2 \\
&= x - y,
\end{aligned}
$$

i.e. the finite element solution coincides with the exact solution and the element passes the patch test.

Solution 6.11 With the local node numbering of Fig. 5.4,

$$
\frac{\partial \tilde{u}^e}{\partial x} = U_1 \frac{b_1}{2A} + U_2 \frac{b_2}{2A} + U_3 \frac{b_3}{2A},
$$

$$
\frac{\partial \tilde{u}^e}{\partial y} = U_1 \frac{c_1}{2A} + U_2 \frac{c_2}{2A} + U_3 \frac{c_3}{2A}.
$$

Thus for elements of type 1,

$$\frac{\partial \tilde{u}^e}{\partial x} = \frac{U_2 - U_1}{h}, \quad \frac{\partial \tilde{u}^e}{\partial y} = \frac{U_3 - U_1}{h},$$

and for elements of type 2,

$$\frac{\partial \tilde{u}^e}{\partial x} = \frac{U_1 - U_2}{h}, \quad \frac{\partial \tilde{u}^e}{\partial y} = \frac{U_1 - U_3}{h}.$$

Thus in each case, with the node numbering of Fig. 6.3, the result follows.

Solution 6.12 (Zienkiewicz *et al.* 2005). Consider the following oscillatory time variation:

$$\mathbf{U} = \mathbf{V}e^{i\omega t}.$$

Equation (6.38) then yields the eigenvalue problem

$$[\mathbf{K} - \omega^2 \mathbf{M}]\mathbf{V} = 0,$$

which, since $\mathbf{M}$ is symmetric and positive definite, is of the same form as eqn (6.30), and analogous results hold for the eigenvalues and eigenvectors. Thus, just as in Example 6.3, eqn (6.38) may be uncoupled, and hence eqn (5.64) becomes

$$[1 + \beta(\omega_i \,\Delta t)^2](\alpha_i)_{j+1} + \left[-2 + \left(\tfrac{1}{2} - 2\beta + \gamma\right)(\omega_i \,\Delta t)^2\right](\alpha_i)_j$$
$$+ \left[1 + \left(\tfrac{1}{2} + \beta - \gamma\right)(\omega_i \,\Delta t)^2\right](\alpha_i)_{j-1} = 0.$$

Suppose that $(\alpha_i)_{j+1} = \lambda(\alpha_i)_j$; then

$$\lambda^2(1 + \beta\Omega_i) + \lambda \left\{-2 + \left(\tfrac{1}{2} - 2\beta + \gamma\right)\Omega_i\right\} + \left\{1 + \left(\frac{1}{2} + \beta - \gamma\right)\Omega_i\right\} = 0,$$

where $\Omega_i = (\omega_i \,\Delta t)^2$. In the case when the roots are complex,

$$\Omega_i \left\{4\beta - \left(\tfrac{1}{2} + \gamma\right)^2\right\} > -4,$$

$$|\lambda|^2 = 1 + \left(\tfrac{1}{2} - \gamma\right)\Omega_i/(1 + \beta\Omega_i).$$

Now the scheme will be stable if $|\lambda| \leq 1$, and, after a little algebra, this condition yields the given inequalities.

7   The boundary element method

7.1   Integral formulation of boundary-value problems

In principle, the boundary element method is just another aspect of the finite element method. However, there is sufficient difference to warrant the new name, which was first coined by Brebbia and Dominguez (1977).

The underlying idea is that a boundary-value problem such as that in Section 3.4, involving a partial differential equation of the form

$$(7.1) \qquad\qquad\qquad \mathcal{L}u = f \quad \text{in } D$$

subject to the boundary condition

$$(7.2) \qquad\qquad\qquad \mathcal{B}u = g \quad \text{on } C,$$

can be transformed to an integral equation

$$(7.3) \qquad\qquad \int\int_D v\mathcal{L}u \, dA = \int\int_D vf \, dA$$

using any weighting function v. However, there are circumstance in which the integral $\int\int_{vD} \mathcal{L}u \, dA$ may be reduced to an integral over the boundary C by use of a *reciprocal theorem*, for example Green's theorem for potential-type problems (Green 1828) or Somigliana's 1886 identity (cited by Becker 1992) for elasticity problems.

In the case of a homogeneous equation, $f \equiv 0$, so that the integral equation is taken only on the boundary, thus reducing the dimensions of the problem. If the equation is non-homogeneous, then there is a domain integral in eqn (7.3) and it would appear that the boundary nature is lost. There are techniques to overcome this problem, for example the dual reciprocity method (Partridge *et al.* 1992, Wrobel 2002), but we shall not discuss these here.

The weighting function used in the boundary element method approach is the fundamental solution (Kythe 1996) associated with $\mathcal{L}$. Suppose that $\mathcal{L} = \nabla^2$.

N.B. Throughout this text, we have used the positive definite operator $-\nabla^2$. However, since the positive definite nature of the operator is unimportant from a boundary element perspective, we use the standard operator ∇^2 here, in common with other boundary element authors.

Also, for convenience, we shall suppose that the boundary condition on C is of the form

(7.4) $$u = g(s) \qquad \text{on } C_1,$$

(7.5) $$q \equiv \frac{\partial u}{\partial n} = h(s) \qquad \text{on } C_2,$$

where $C = C_1 + C_2$.

The fundamental solution, sometimes called the free-space Green's function, associated with ∇^2 in two dimensions is (see Exercise 7.1)

(7.6) $$u^* = -\frac{1}{2\pi} \ln R,$$

where $\mathbf{R}$ is the position vector of the *field point* Q (i.e. a general point in D) relative to the source point P (i.e. a point at which the solution is sought); see Fig. 7.1. We notice that the fundamental solution u^* has a singularity at P.

Define a disc D_ϵ, of radius ϵ, circumference C_ϵ and centre P (see Fig. 7.1), and consider what happens as $\epsilon \to 0$.

We apply the integral equation (7.3) to the region $D - D_\epsilon$ with $v = u^*$, since in $D - D_\epsilon$ both u and u^* are well defined. We shall develop the form for Poisson's equation in the first instance:

$$\iint_{D-D_\epsilon} u^* \, \nabla^2 u \, dA = \iint_{D-D_\epsilon} u^* f \, dA.$$

Then, using the first form of Green's theorem, we obtain

$$\iint_{D-D_\epsilon} u \, \nabla^2 u^* \, dA + \oint_{C+C_\epsilon} \left(u^* \frac{\partial u}{\partial n} - u \frac{\partial u^*}{\partial n} \right) ds = \iint_{D-D_\epsilon} u^* f \, dA$$

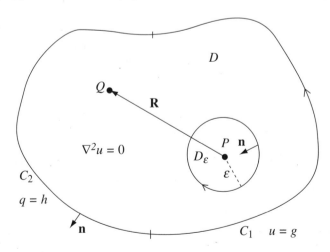

Fig. 7.1 Small disc D_ϵ in the region D.

or, using the notation $q \equiv \partial u/\partial n$ and $q^* \equiv \partial u^*/\partial n$ and noting that $\nabla^2 u^* = 0$ in $D - D_\epsilon$,

$$(7.7) \qquad \oint_{C+C_\epsilon} (u^* q - u q^*)\, ds = \int\int_{D-D_\epsilon} u^* f \, dA.$$

Now, on C_ϵ, $\partial/\partial n \equiv -\partial/\partial R$ and $R = \epsilon$. Also, on C_ϵ, $u(s) = u_P + \eta_1(s)$ and $q(s) = q_P + \eta_2(s)$, where $\max(|\eta_1|, |\eta_2|) \to 0$ as $\epsilon \to 0$.

We write

$$I_1 = \oint_{C_\epsilon} u^* q \, ds$$

and

$$\oint_{C_\epsilon} u q^* ds = \oint_{C_\epsilon} u_P q^* ds + I_2,$$

where

$$I_2 = \oint_{C_\epsilon} \eta_2 q^* ds.$$

Then, on C_ϵ,

$$|I_1| = \left| \oint_{C_\epsilon} \left(-\frac{1}{2\pi} \ln R \right) (q_P + \eta_2)\, ds \right|$$

$$\leq \frac{1}{2\pi} \ln \epsilon \left(|q_P| + |\eta_2| \right) 2\pi\epsilon$$

$$\to 0 \text{ as } \epsilon \to 0.$$

Also,

$$|I_2| = \left| \oint_{C_\epsilon} \eta_2 \left(-\frac{\partial}{\partial R} \left(-\frac{1}{2\pi} \ln R \right) \right) ds \right|$$

$$\leq |\eta_2| \frac{1}{2\pi} \frac{1}{\epsilon} 2\pi\epsilon$$

$$\to 0 \text{ as } \epsilon \to 0.$$

Finally,

$$\oint_{C_\epsilon} u_P q^* ds = u_P \frac{1}{2\pi} \frac{1}{\epsilon} 2\pi\epsilon$$

$$= u_P.$$

Hence, as $\epsilon \to 0$,

$$\oint_{C_\epsilon} (u^* q - u q^*)\, ds \to u_P$$

and we have the integral equation

$$(7.8) \qquad u_P = \oint_C (u^*q - q^*u) + \int\int_D u^*f \, dA$$

for values of u inside D in terms of values of u and q on the boundary.

In Exercise 7.3, we obtain the corresponding integral equations for points on the boundary and for points outside the boundary in the form

$$(7.9) \qquad c_P u_P = \oint_C (u^*q - q^*u) + \int\int_D u^*f \, dA,$$

where

$$(7.10) \qquad c_P = \begin{cases} 1, & P \in D, \\ \alpha_P/2\pi, & P \in C, \\ 0, & P \in D \cup C, \end{cases}$$

and where α_P is the internal angle between the left- and right-hand tangents at P. If the boundary is smooth at P, then $c_P = \frac{1}{2}$. In the terminology of integral equations, the functions u^* and q^* in eqn (7.9) are called *kernels* of the integral equation (Manzhirov and Polyanin 2008). A result which provides a useful check in the boundary element method is obtained by choosing $v = 1$ in the first form of Green's theorem to obtain

$$\int\int_D \nabla^2 u \, dA = \oint_C \frac{\partial u}{\partial n} \, ds,$$

so that if u satisfies Laplace's equation, then

$$(7.11) \qquad \oint_C q \, ds = 0.$$

7.2 Boundary element idealization for Laplace's equation

We proceed in a manner analogous to that for the finite element method in Chapter 3. We shall approximate the boundary C by a polygon, C_n, and it is the polygon edges which are the boundary elements. We also choose a set of nodes, at which we seek approximations U and Q to u and q, respectively. We shall consider in the first instance the so-called constant element. In such an element, the geometry is a straight line segment with just one node at the centre; see Fig. 7.2.

N.B. (i) In this element, the approximation of the function is of a lower order than that for the geometry, and the element is known as a *superparametric* element. Similarly, we can define *subparametric* elements (*cf.* the isoparametric elements of Chapter 4).

(ii) There is no requirement for interelement continuity in the boundary element, non-conforming elements are frequently used.

Fig. 7.2 Constant element.

We choose a set of basis functions $\{w_j(s) : j = 1, 2, \ldots, n\}$ to approximate u and q as follows:

$$(7.12) \qquad \tilde{u} = \sum_{j=1}^{n} w_j(s) U_j \quad \text{and} \quad \tilde{q} = \sum_{j=1}^{n} w_j(s) Q_j,$$

where s is a measure of the arc length around C_n (*cf.* eqn (3.16) for a finite element approximation). The weighting functions for the constant element are given by

$$(7.13) \qquad w_j(s) = \begin{cases} 1, & j \in [e], \\ 0, & j \notin [e]. \end{cases}$$

We now use point collocation with the approximations (7.12) in eqn (7.9), remembering that for Laplace's equation $f \equiv 0$, the curve C is replaced by C_n and the boundary point P is chosen to be, successively, the nodes $1, 2, \ldots, n$:

$$c_i U_i = \frac{1}{2\pi} \oint_{C_n} \left(\sum_{j=1}^{n} w_j(s) Q_i \right) (-\ln R_i) \, ds - \frac{1}{2\pi} \oint_{C_n} \sum_{j=1}^{n} (w_j(s) U_j) \left(-\frac{\partial}{\partial n} (\ln R_i) \right) ds$$

or

$$\alpha_i U_i = \sum_{j=1}^{n} \left(\int_{[j]} \frac{\partial}{\partial n} (\ln R_i) \, ds \right) U_j - \sum_{j=1}^{n} \left(\int_{[j]} \ln R_i \, ds \right) Q_j,$$

$$j = 1, 2, \ldots, n,$$

where $R = |\mathbf{R}_i|$, $\mathbf{R}_i(s)$ is the position vector of a boundary point s relative to node i, and $\alpha_i = 2\pi c_i$.

We can rewrite these equations as

$$(7.14) \qquad \sum_{j=1}^{n} H_{ij} U_j + \sum_{j=1}^{n} G_{ij} Q_j = 0, \qquad j = 1, 2, \ldots, n,$$

where

$$(7.15) \qquad H_{ij} = \int_{[j]} \frac{\partial}{\partial n} (\ln R_{ij}) \, ds - \alpha_i \delta_{ij}$$

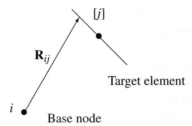

Fig. 7.3 Target element $[j]$ relative to base node i.

and

$$(7.16) \qquad G_{ij} = -\int_{[j]} \ln R_{ij} \; ds,$$

with $R_{ij} = |\mathbf{R}_{ij}|$ and where $\mathbf{R}_{ij}(s)$ is the position vector of a point in the *target element* $[j]$ relative to the *base node* i; see Fig. 7.3.

Finally, then, the system of equations (7.14) may be written in matrix form as

$$(7.17) \qquad \mathbf{HU} + \mathbf{GQ} = \mathbf{0},$$

where $\mathbf{U}$ and $\mathbf{Q}$ are vectors of the approximations to the boundary values of u and q, respectively.

For properly posed problems, we have boundary conditions of the form of eqns (7.4) and (7.5) so that at any point we know only the value of either u or q, and we partition the matrices in eqn (7.13) appropriately. Suppose that $\mathbf{U}_1$ and $\mathbf{Q}_2$ are vectors of the known values of u and q and that $\mathbf{U}_2$ and $\mathbf{Q}_1$ are vectors of unknown values. Then we may write eqn (7.17) as

$$(7.18) \qquad \begin{bmatrix} \mathbf{H}^1 \mathbf{H}^2 \end{bmatrix} \begin{bmatrix} \mathbf{U}_1 \\ \mathbf{U}_2 \end{bmatrix} + \begin{bmatrix} \mathbf{G}^1 \mathbf{G}^2 \end{bmatrix} \begin{bmatrix} \mathbf{Q}_1 \\ \mathbf{Q}_2 \end{bmatrix} = \mathbf{0},$$

and hence we can rearrange it in the form

$$(7.19) \qquad \mathbf{Ax} = \mathbf{b},$$

where

$$(7.20) \qquad \mathbf{A} = \begin{bmatrix} \mathbf{H}^2 \mathbf{G}^1 \end{bmatrix}, \quad \mathbf{b} = -\begin{bmatrix} \mathbf{H}^1 \mathbf{G}^2 \end{bmatrix} \begin{bmatrix} \mathbf{U}_1 \\ \mathbf{Q}_2 \end{bmatrix}$$

and the unknowns are given by the vector

$$(7.21) \qquad \mathbf{x} = \begin{bmatrix} \mathbf{U}_2 \\ \mathbf{Q}_1 \end{bmatrix}.$$

Equation (7.19) is an $n \times n$ system of algebraic equations for the n unknowns x_i. We note here that the system matrix $\mathbf{A}$ is densely populated, non-symmetric and non-positive definite. This is in stark contrast to the stiffness matrix for the finite element method in Section 3.6.

The solution of the system (7.19) yields the nodal vectors $\mathbf{U}$ and $\mathbf{Q}$, and values at k internal points, $k = 1, 2, \ldots, m$, may be obtained using the discrete form of eqn (7.9) for k inside the boundary:

$$U_k = \frac{1}{2\pi} \oint_{C_n} \left[\left(\sum_{j=1}^{n} w_j(s) U_j \right) \frac{\partial}{\partial n} \ln R_k - \left(\sum_{j=1}^{n} w_j(s) Q_j \right) \ln R_k \right] ds,$$

$$k = 1, 2, \ldots, m.$$

We write this in matrix form as

(7.22) $$\mathbf{U}_I = \mathbf{\check{H}U} + \mathbf{\check{G}Q},$$

where

(7.23) $$\check{H}_{kj} = \frac{1}{2\pi} \int_{[j]} \frac{\partial}{\partial n} (\ln R_{kj}) \, ds \quad \text{and} \quad \check{G}_{kj} = -\frac{1}{2\pi} \int_{[j]} \ln R_{kj} \, ds.$$

The integrals required to evaluate H_{ij} and G_{ij}, eqns (7.15) and (7.16), involve the kernels $(\partial/\partial n)(\ln R)$ and $\ln R$ and may be considered to be of two types:

1. *Base node not in target element.* In this case, all the integrals are non-singular and standard Gauss quadrature can be used. It is also possible to obtain analytic values; see Exercise 7.5.

2. *Base node in target element, $i = j$ singular integral.* In this case, the integrals are singular and we cannot use a standard Gauss quadrature. We proceed as follows.

To evaluate the integrals for G_{ij}, which have a logarithmic singularity (eqn (7.16)), we can use a special logarithmic Gauss quadrature; see Appendix D. However, in this case we can obtain analytic values for the singular integrals, and we shall see how to do that in Section 7.3.

The singularity in the integral for H_{ij}, eqn (7.15), is $\mathcal{O}(1/R)$, and such integrals usually require special treatment. The term H_{ij} also requires computation of the parameter α_i. Both these difficulties may be overcome in the case of the Laplace operator. We notice that we can apply our approach to the problem whose unique solution is $u \equiv 1$. The equivalent boundary-value problem is

$$\nabla^2 u = 0 \quad \text{in } D, \qquad u = 1 \quad \text{on } C,$$

and clearly $q = 0$ on C.

Hence $\mathbf{U} = [1\,1\ldots 1]^T$ and $\mathbf{Q} = [0\,0\ldots 0]^T$, satisfying eqn (7.15) so that

$$\mathbf{HU} = \mathbf{0}.$$

Consequently, it follows that

$$(7.24) \qquad\qquad H_{ii} = -\sum_{j=1}^{n}{}' H_{ij}, \qquad i = 1, 2, \ldots, n,$$

and the diagonal terms are found from the sum of the off-diagonal terms, a very convenient result.

Finally, all integrals in the computation of $\breve{H}_{jk}$ and $\breve{G}_{jk}$ (eqn (7.21)) are non-singular and a standard Gauss quadrature can be used.

Once all elements of the matrices $\mathbf{H}$ and $\mathbf{G}$ are evaluated, the matrices $\mathbf{A}$ and $\mathbf{b}$ are assembled (eqn (7.20)), and the system (7.19) is solved by a suitable routine. This leads to values for $\mathbf{U}$ and $\mathbf{Q}$ on the boundary, and internal values are obtained using eqn (7.22). We note here that even though the evaluation of the internal values uses the approximate boundary values, the internal values are often more accurate than the computed values on the boundary. However, care is needed for internal points close to the boundary, since the values of R_{kj} in eqns (7.23) can become very small compared with the element length. A useful rule of thumb is that internal points should be no closer to the boundary than one quarter of the minimum length of a boundary element.

7.3 A constant boundary element for Laplace's equation

The element is shown in Fig. 7.3, and in Fig. 7.4 we define some of the geometry of element $[j]$, whose length is l_j.

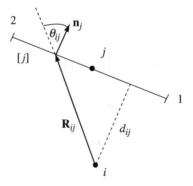

Fig. 7.4 Geometry for the constant element $[j]$.

We calculate the coefficients H_{ij} and G_{ij}. Suppose that the base node i is not in the target element $[j]$. Then

$$\frac{\partial}{\partial n}(\ln R_{ij}) = (\text{grad}R \cdot \mathbf{n})_{ij}$$

$$= \frac{1}{R_{ij}}\mathbf{R}_{ij} \cdot \mathbf{n}_j$$

$$= \frac{\cos\theta_{ij}}{R_{ij}}$$

$$= \frac{d_{ij}}{R_{ij}^2}$$

so that, using Gauss quadrature, we obtain

$$H_{ij} = \int_{[j]} \frac{d_{ij}}{R_{ij}^2}\, ds$$

$$= \frac{l_j d_{ij}}{2}\int_{-1}^{1}\frac{1}{R_{ij}^2(\xi)}\, d\xi$$

$$(7.25) \qquad \approx \frac{l_j d_{ij}}{2}\sum_{g=1}^{G}\frac{1}{R_{ij}^2(\xi_g)}w_g.$$

We note here that if node i is in element $[j]$, then $\mathbf{n}_j$ is perpendicular to $\mathbf{R}_{ij}$ and $\mathbf{R}_{ij} \cdot \mathbf{n}_j = 0$, so that $H_{ii} = -\alpha_i$. We shall use eqn (7.24) to compute H_{ii} and, where appropriate, use this result as a check:

$$G_{ij} = -\int_{[j]} \ln R_{ij}\, ds$$

$$= -\frac{l_j}{2}\int_{-1}^{1} \ln R_{ij}(\xi)\, d\xi$$

$$(7.26) \qquad \approx -\frac{l_j}{2}\sum_{g=1}^{G}\ln R_{ij}(\xi_g)w_g.$$

Now, to obtain G_{ii} we use the notation of Fig. 7.5:

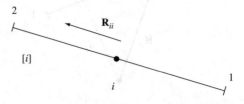

Fig. 7.5 Base node in target element.

$$G_{ii} = -\int_{[j]} \ln R_{ii}\, ds$$

$$= -2\int_0^1 \ln\left(\frac{l_i}{2}\eta\right)\frac{l_i}{2}\, d\eta$$

$$= -l_i \int_0^1 \left(\ln\left(\tfrac{1}{2}l_i\right) + \ln\eta\right)\, d\eta$$

(7.27)
$$= -l_i \ln\left(\tfrac{1}{2}l_i\right) - l_i \int_0^1 \ln\eta\, d\eta.$$

In Exercise 7.4, we show that $G_{ii} = l_i\left(1 - \ln\left(\tfrac{1}{2}l_i\right)\right)$. The coefficients H_{kj} and G_{kj} are obtained using Gauss quadrature. In Exercise 7.5, we obtain analytic expressions for H_{ij} and G_{ij}.

Example 7.1 Solve Laplace's equation in the unit square subject to the boundary conditions

$$u(x,0) = u(x,1) = 1 + x,$$

$$q(0,y) = -1, \qquad q(1,y) = 1,$$

with four constant boundary elements as shown in Fig. 7.6.

We shall use four-point Gauss quadrature for the integrals, and with the notation of Fig. 7.6 we find

$$R_{12}(\eta_g): 0.504798 \quad 0.599088 \quad 0.835995 \quad 1.056389$$

so that

$$\sum_{g=1}^4 \frac{1}{R_{12}^2}(\xi_g)w_g = 4.426960.$$

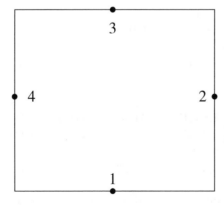

Fig. 7.6 Boundary nodes for the problem in Example 7.1.

Hence, using eqn (7.25),

$$H_{12} = \frac{\left(1 \times \frac{1}{2}\right)}{2} \times 4.426960 = 1.106740.$$

By symmetry, $H_{14} = H_{12}$. Similarly,

$$H_{13} = 0.927812.$$

Now,

$$H_{11} = -(H_{12} + H_{13} + H_{14})$$
$$= -3.140762.$$

Now $\alpha_1 = \pi$, so we see that $H_{11} \approx -\alpha_1$ as expected.

The values H_{ij} $(i = 1, 2, 3, 4; j = 1, 2, 3, 4)$ may be obtained by symmetry. Also,

$$\sum_{g=1}^{4} \ln R_{12}(\eta_g) w_g = -0.669656.$$

Hence, using eqn (7.26),

$$G_{ij} = \left(-\frac{1}{2}\right) \times (-0.669656) = 0.334828.$$

By symmetry, $G_{14} = G_{12}$. Similarly,

$$G_{13} = 0.038869.$$

Using the result of Exercise 7.3, we have $G_{11} = 1.693147$.

Hence our overall system matrices are

$$\mathbf{H} = \begin{array}{c} \\ \\ \begin{bmatrix} -3.14076 & 1.10674 & 0.92781 & 1.10674 \\ 1.10674 & -3.14076 & 1.10674 & 0.92781 \\ 0.92781 & 1.10674 & -3.14076 & 1.10674 \\ 1.10674 & 0.92781 & 1.10674 & -3.14076 \end{bmatrix} \begin{array}{c} 1 \\ 2 \\ 3 \\ 4 \end{array} \end{array},$$

$$\mathbf{G} = \begin{bmatrix} 1.69315 & 0.33483 & 0.03887 & 0.33483 \\ 0.33483 & 1.69315 & 0.33483 & 0.03887 \\ 0.03887 & 0.33483 & 1.69315 & 0.33483 \\ 0.33483 & 0.03887 & 0.33483 & 1.69315 \end{bmatrix} \begin{array}{c} 1 \\ 2 \\ 3 \\ 4 \end{array}.$$

Now we apply the boundary conditions and partition $\mathbf{H}$ and $\mathbf{G}$ accordingly. At nodes 2 and 4 we specify u, and at nodes 1 and 3 we specify q; hence we obtain

$$\mathbf{A} = \begin{bmatrix} \overset{2}{1.10674} & \overset{4}{1.10674} & \overset{1}{1.69315} & \overset{3}{0.03887} \\ -3.14076 & 0.92781 & 0.33483 & 0.33483 \\ 1.10674 & 1.10674 & 0.03887 & 1.69315 \\ 0.92781 & -3.14076 & 0.33483 & 0.33483 \end{bmatrix},$$

$$\mathbf{b} = \begin{bmatrix} \overset{2}{3.31943} & \overset{4}{-4.97450} & \overset{1}{3.31943} & \overset{3}{-1.66594} \end{bmatrix}^T,$$

and solving $\mathbf{Ax} = \mathbf{b}$ yields

$$\mathbf{x} = \begin{bmatrix} \overset{2}{1.907} & \overset{4}{1.094} & \overset{1}{-0.001} & \overset{3}{-0.001} \end{bmatrix}^T.$$

Hence we have the nodal values

$$\mathbf{U} = \begin{bmatrix} \overset{1}{1.5} & \overset{2}{1.907} & \overset{3}{1.5} & \overset{4}{1.094} \end{bmatrix}^T,$$

$$\mathbf{Q} = \begin{bmatrix} \overset{1}{-0.001} & \overset{2}{1} & \overset{3}{-0.001} & \overset{4}{-1} \end{bmatrix}^T.$$

The exact solution is $u = 1 + x$. We see that the approximate values given above compare very well, given the very coarse approximation. The approximate value of $\oint_C q\,ds$ is -0.002, which again is close the the exact value, zero (eqn (7.11)).

Equation (7.23) yields the following coefficients for the computation of internal values (*cf.* eqns (7.25) and (7.26)):

$$\breve{H}_{kj} \approx \frac{l_j d_{kj}}{4\pi} \sum_{g=1}^{G} \frac{1}{R_{kj}^2(\xi_g)} w_g, \qquad \breve{G}_{kj} \approx -\frac{l_j}{4\pi} \sum_{g=1}^{G} \ln R_{kj}(\xi_g) w_g.$$

We shall obtain the solution at $\left(\frac{1}{2}, \frac{1}{2}\right)$.

At $\left(\frac{1}{2}, \frac{1}{2}\right)$,

$$R_{11}(\xi_g): 0.659840 \quad 0.528107 \quad 0.528107 \quad 0.659840.$$

Hence

$$\breve{H}_{11} = 0.24965 = \breve{H}_{12} = \breve{H}_{13} = \breve{H}_{14}$$

and

$$\sum \breve{H}_{ij} = 0.999,$$

very close to 1; see Exercise 7.7.

Also,

$$\breve{G}_{11} = 0.08928 = \breve{G}_{12} = \breve{G}_{13} = \breve{G}_{14}.$$

Now,

$$u\left(\frac{1}{2}, \frac{1}{2}\right) = \sum_{j=1}^{4} \breve{H}_{1j}U_j + \sum_{j=1}^{4} \breve{G}_{1j}Q_j$$

$$= 1.498.$$

N.B. As mentioned earlier, this internal potential value is more accurate than the potential values on the boundary.

The solution at $\left(\frac{1}{2}, \frac{3}{4}\right)$ is given in Exercise 7.6.

7.4 A linear element for Laplace's equation

We shall consider an element with two nodes, just as we did for the linear finite element in Section 3.5, illustrated in Fig. 3.7. The geometry and parameters for the element are the same as those for the constant element, see Fig. 7.4.

In this case we find that the integrals for H_{ij}, G_{ij}, $\breve{H}_{kj}$ and $\breve{G}_{kj}$ are taken over the two elements containing node j. The details are obtained in Exercise 7.9.

Example 7.2 Consider the problem of Example 7.1 using the linear element of Exercise 7.9. We shall assume that we have a Dirichlet problem with u known at all four nodes.

$$\mathbf{H} = \begin{bmatrix} \overset{1}{-1.57081} & \overset{2}{0.43883} & \overset{3}{0.69312} & \overset{4}{0.43883} \\ 0.43883 & -1.57081 & 0.43883 & 0.69312 \\ 0.69312 & 0.43883 & -1.57081 & 0.43883 \\ 0.43883 & 0.69312 & 0.43883 & -1.57081 \end{bmatrix},$$

$$\mathbf{G} = \begin{bmatrix} \overset{1}{1.50000} & \overset{2}{0.21460} & \overset{3}{-0.19315} & \overset{4}{0.21460} \\ 0.21460 & 1.50000 & 0.21460 & -0.19315 \\ -0.19315 & 0.21460 & 1.50000 & 0.21460 \\ 0.21460 & -0.19315 & 0.21460 & 1.50000 \end{bmatrix}.$$

Since this is a Dirichlet problem, $\mathbf{A} = \mathbf{G}$ and, using the known values of u, we find

$$\mathbf{b} = \begin{bmatrix} \overset{1}{-1.13192} & \overset{2}{1.13201} & \overset{3}{1.13201} & \overset{4}{-1.13192} \end{bmatrix}^T;$$

the solution is

$$\mathbf{Q} = \begin{bmatrix} \overset{1}{-0.6684} & \overset{2}{0.6684} & \overset{3}{0.6684} & \overset{4}{-0.6684} \end{bmatrix}^T.$$

For the internal potential value, we have

$$\breve{H}_{11} = 0.24965 = \breve{H}_{12} = \breve{H}_{13} = \breve{H}_{14},$$

$$\breve{G}_{11} = 0.08928 = \breve{G}_{12} = \breve{G}_{13} = \breve{G}_{14},$$

so that $\left(\frac{1}{2}, \frac{1}{2}\right) \approx 1.498$.

The boundary values of q are significantly different from the exact values for q. The exact value for the potential is $u = 1 + x$, which gives

$$q(x, 0) = 0, \quad q(1, y) = 1, \quad q(x, 1) = 0, \quad q(0, y) = -1.$$

The problem arises because, at each node, the direction of q depends on the element under consideration. This is an example of the so-called 'corner-node problem', and there have been a variety of ways to overcome it. We shall describe just one of these ways.

The approximation for u and q as given by eqn (7.12) has the same set of basis functions for both variables; this does not have to be the case. Indeed, we may even choose different sets of nodes for the two variables. Suppose we consider the linear element and approximate u in the same manner as in Exercise 2.8. This will yield the $n \times n$ matrix $\mathbf{H}$ associated with the $n \times 1$ column vector $\mathbf{U}$. Now, for the q approximation, suppose we consider the set-up shown in Fig. 7.7 so that we have discontinuous elements for q. Using the notation of Exercise 7.9, the contributions to the boundary element equations in this case would be

$$g_{ij-1}q_j^- + g_{ij}q_j^+.$$

N.B. In Exercise 7.9 we have just one value, q_j, and the contribution is $(g_{ij-1} + g_{ij})q_j$.

So, now we have an $n \times 2n$ matrix $\mathbf{G}$ associated with a $2n \times 1$ column vector $\mathbf{Q}$. Our overall system is again of the form

$$\mathbf{HU} + \mathbf{GQ} = \mathbf{0}.$$

However, care is needed in applying the boundary conditions; we now have $2n$ unknowns but only n conditions, either Dirichlet or Neumann. We shall not give the details here as to how we handle this; an excellent description has been given

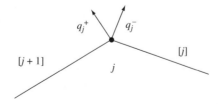

Fig. 7.7 Nodal fluxes in adjacent elements.

by París and Cañas (1997), who also explain some of the other techniques for handling the corner problem. These authors also give very good descriptions of a variety of boundary elements, including isoparametric quadratics and circular elements.

We have considered only Dirichlet or Neumann conditions. In principle, a Robin condition is not difficult to incorporate; see Exercise 7.10.

Finally, let us consider briefly Poisson's equation $\nabla^2 u = f$, which leads to the integral equation (7.9), comprising the domain integral $\int \int_D u^* f \, dA$. If f can be written as $\nabla^2 F$ for some function F, then Poisson's equation can be written as Laplace's equation

$$\nabla^2 w = 0,$$

where $w = u - F$.

The boundary conditions for u yield boundary conditions for w, and hence we can use the boundary element method for the solution.

It is clear from this chapter that there are many similarities between the development of the system matrices for the finite element method and for the boundary element method. The major difference is that the boundary element method requires the evaluation of singular integrals. In this section, we have seen how to evaluate them for the constant element. There are other approaches, and we shall mention just two.

The logarithmic integrals involved in the evaluation of G_{ii} (see eqn (7.16) and the solution to Exercise 7.4) have been evaluated analytically. However, there is an alternative approach using a form of logarithmic Gauss quadrature given by

$$\int_0^1 f(\xi) \ln \xi \, d\xi \approx -\sum_{g=1}^{n_g} w_g f(\xi_g),$$

where the Gauss points and associated weights are given in Appendix D.

An alternative approach was developed by Telles (1987), in which a coordinate transformation is developed in such a way that the Jacobian vanishes at the singularity. The effect is to render the integral amenable to standard Gauss quadrature. The major advantage of this technique is that it is also applicable to the non-singular integrals, and so there is no need to distinguish between the cases in which the base node is or is not in the target element. It is particularly useful for situations where the integrands involve functions other than logarithms, for example for those integrals involving the modified Bessel function in the following section.

7.5 Time-dependent problems

In this section, we give a brief introduction to the solution of the diffusion problem

$$(7.28) \qquad \nabla^2 u = \frac{1}{\alpha} \frac{\partial u}{\partial t} \quad \text{in } D$$

subject to the boundary conditions

$$(7.29) \qquad u = g(s) \quad \text{on } C_1,$$

$$(7.30) \qquad q = h(s) \quad \text{on } C_2$$

and the initial condition

$$(7.31) \qquad u(x, y, 0) = u_0(x, y).$$

There are three possible approaches to dealing with such problems.

Finite difference method

An explicit finite difference approach in t (Smith 1985) will yield an elliptic problem of the form

$$\nabla^2 u^{(k+1)} = \alpha \frac{u^{(k+1)} - u^{(k)}}{\Delta t}, \qquad k = 1, 2, \dots.$$

This requires the treatment of a domain integral at each time step.
    We shall not take this further, but refer the reader to Wrobel (2002).

Time-dependent fundamental solution

It can be shown (Carslaw and Jaeger 1986) that the fundamental solution of the differential operator $\nabla^2 - (1/\alpha)(\partial/\partial t)$ is given by

$$u^*(R, \tau) = \frac{1}{4\pi\alpha\tau} \exp\left(\frac{-R^2}{4\alpha\tau}\right) \quad \text{with } \tau = T - t,$$

where the solution is sought at time T.
    The corresponding boundary integral equation is of the form (Kythe 1995)

$$u(s, t) = \alpha \oint_C \int_0^t (u^* q - q^* u) \, d\tau \, ds + \int\int_D u_0^* u_0 \, dA,$$

where $u_0^* = u^*(R, T)$. A time-stepping scheme may now be used to develop the solution (Becker 1992).

The Laplace transform

We shall define the Laplace transform as in Section 5.5 and, applying it to the problem defined by eqns (7.28)–(7.31), we obtain

$$\nabla^2 \bar{u} - \frac{\lambda}{\alpha} \bar{u} = u_0 \quad \text{in } D$$

subject to

$$\bar{u} = \bar{g} \quad \text{on } C_1,$$

$$\bar{q} = \bar{h} \quad \text{on } C_2.$$

We shall consider the case in which $u_0 \equiv 0$; if $u_0 = \nabla^2 F$, we can follow the approach outlined in Section 7.4. We also write $p^2 = \lambda/\alpha$

We seek the solution of

(7.32) $$\nabla^2 \bar{u} - p^2 \bar{u} = 0 \quad \text{in } D$$

subject to boundary conditions on C. Equation (7.32) is the *modified Helmholtz equation*. In Exercise 7.2, we show that the fundamental solution is

$$\bar{u}^* = \frac{1}{2\pi} K_0(pR),$$

where K_0 is the modified Bessel function of the second kind.

Since $(d/dx)(K_0(x)) = -K_1(x)$ (Abramowitz and Stegun 1972), it follows that

$$\bar{q}^* = -\frac{1}{2\pi} p K_1(pR) \frac{1}{R} \mathbf{R} \cdot \mathbf{n},$$

and the boundary integral equation takes the form

$$\bar{u}_P = \oint_C \left(K_0(pR)\bar{q} + pK_1(pR)\frac{1}{R}\mathbf{R} \cdot \mathbf{n}\bar{u} \right) ds.$$

We can set up a boundary element solution, leading to a system of equations of the form

$$\mathbf{H}\bar{\mathbf{U}} + \mathbf{G}\bar{\mathbf{Q}} = 0$$

for the boundary values of the transformed variables. Internal values $\bar{\mathbf{U}}_I$ may also be computed. Finally, the inverse transform using Stehfest's method (see Section 5.3) will yield the solution vectors $\mathbf{U}$, $\mathbf{Q}$ and $\mathbf{U}_I$.

7.6 Exercises and solutions

Exercise 7.1 The fundamental solution for the operator $\mathcal{L}$ is defined as that function u^* which satisfies

$$\mathcal{L}u^*(\mathbf{r}, \boldsymbol{\rho}) = -\delta(\mathbf{r}, \boldsymbol{\rho})$$

over the whole space D. The point $\mathbf{r}$ is fixed, the equation is referred to this position and $\delta(\mathbf{r}, \boldsymbol{\rho})$ is the Dirac delta function, with the property

$$\delta(\mathbf{r}, \boldsymbol{\rho}) = 0, \quad \mathbf{r} \neq \boldsymbol{\rho}, \quad \int_D f(\boldsymbol{\rho}) \, \delta(\mathbf{r}, \boldsymbol{\rho}) \, dV(\boldsymbol{\rho}) = f(\mathbf{r}).$$

Obtain the fundamental solution for ∇^2 in two and three dimensions.

Exercise 7.2 Obtain the fundamental solution for the modified Helmholtz equation in two dimensions,

$$\nabla^2 u - p^2 u = 0.$$

Exercise 7.3 Show that the integral equation (7.7) becomes

$$c_P u_P = \oint_C (uq^* - uq*) \, ds,$$

where

$$c_P = \begin{cases} 1, & \text{if } P \text{ is inside the boundary,} \\ \alpha_P/2\pi, & \text{if } P \text{ is on the boundary,} \\ 0, & \text{if } P \text{ is outside the boundary.} \end{cases}$$

Exercise 7.4 Obtain an analytical expression for the coefficient G_{ii} for the constant element.

Exercise 7.5 (París and Cañas (1997)) In this exercise, we obtain analytic expressions for H_{ij} and G_{ij} $(i \neq j)$ for the constant element.

With reference to Fig 7.4, the angle θ_{ij} varies from the value θ_1 to θ_2 at ends 1 and 2, respectively, of the element. Obtain analytic expressions for H_{ij} and G_{ij}.

Using these expressions, calculate the values of H_{ij} and G_{ij} for the problem in Example 7.1.

Exercise 7.6 Obtain the solution at $\left(\frac{1}{2}, \frac{3}{4}\right)$ for the problem of Example 7.1.

Exercise 7.7 By considering the constant solution $u \equiv 1$ to Laplace's equation in a region D, show that the coefficients $\check{H}_{kj}$ for a constant element satisfy

$$\sum_{j=1}^{n} \check{H}_{kj} \approx 1,$$

and use this result to check the solution to Exercise 7.6.

Exercise 7.8 Use the results of Example 7.1 to solve Laplace's equation in the unit square subject to the boundary conditions

$$u(0, y) = u(x, 0) = 0,$$

$$q(1, y) = y \quad q(x, 1) = x.$$

Find $u\left(\frac{1}{2}, \frac{1}{2}\right)$.

Compare the results with the exact solution $u(x, y) = 1 + xy$.

Exercise 7.9 Obtain the coefficient matrices for a linear element.

Exercise 7.10 Show how a Robin boundary condition of the form given in Section 2.3,

$$\frac{\partial u}{\partial n} + \sigma(s)u = h(s),$$

may be incorporated into the boundary element method.

Solution 7.1 We seek a solution which is dependent only on the distance from our source point. Without loss of generality, suppose that this is the origin, i.e. we seek a solution $u^*(r)$ which satisfies

$$\left(r^2 \frac{d^2}{dr^2} + ar \frac{d}{dr}\right) u^* = -\delta(r),$$

where

$$a = \begin{cases} 2 & \text{in three dimensions,} \\ 1 & \text{in two dimensions.} \end{cases}$$

First we seek a solution to the homogeneous equation which has a singularity at $r = 0$:

$$\frac{d^2 u^*}{dr^2} + \frac{a}{r} \frac{du^*}{dr} = 0$$

and

$$u^*(r) = \begin{cases} k_3 r^{-1} & \text{in three dimensions,} \\ k_2 \ln r & \text{in two dimensions,} \end{cases}$$

where we have ignored the additive constant in each case. Now we consider the equation for u^*,

$$\nabla^2 u^* = -\delta(\mathbf{r}, \boldsymbol{\rho})$$

and consider first the three-dimensional case.

Suppose we construct a small sphere V_ϵ, with a surface S_ϵ of radius ϵ and centre $\mathbf{r}$; then we may write

$$\int_{V_\epsilon} \nabla^2 u^* \, dV = \int_{V_\epsilon} -\delta(\mathbf{r}, \boldsymbol{\rho}) \, dV$$
$$= -1.$$

We now apply the divergence theorem to the integral on the left-hand side to obtain

$$\int_{S_\epsilon} \operatorname{grad} u^* \cdot \hat{\mathbf{n}} \, dS,$$

so that

$$\int_{S_\epsilon} \frac{\partial u^*}{\partial n} (\mathbf{r}, \boldsymbol{\rho}) \, dS = -1 \quad \text{for } \mathbf{r} \in V_\epsilon.$$

Now we can write

$$\boldsymbol{\rho} = \mathbf{r} + R\hat{\mathbf{R}},$$

where $\hat{\mathbf{R}}$ is the usual spherical polar unit vector in the radial direction relative to an origin at $\mathbf{r}$.

Hence, on S_ϵ, $R = \epsilon$ and $\partial/\partial n \equiv \partial/\partial R$, so that

$$-1 = \int_{S_\epsilon} \frac{\partial u^*}{\partial n} dS = \int_{S_\epsilon} \frac{\partial u^*}{\partial R}\bigg|_{R=\epsilon} dS$$
$$= -\frac{k_3}{\epsilon^2} 4\pi\epsilon^2$$

and

$$k_3 = \frac{1}{4\pi}.$$

Consequently, in three dimensions,

$$u^*(r) = \frac{1}{4\pi r}.$$

In a similar manner, for the two-dimensional case we find

$$k_2 = -\frac{1}{2\pi}$$

and

$$u^*(R) = -\frac{1}{2\pi} \ln R.$$

Solution 7.2 Following in the same manner as in Exercise 7.1, we write the modified Helmholtz equation in two dimensions as

$$\frac{d^2u}{dR^2} + \frac{1}{R}\frac{du}{dR} - p^2u = 0.$$

If we let $x = pR$, then this equation becomes

$$x^2u'' + xu' - x^2u = 0,$$

the modified Bessel equation of order zero.

Two linearly independent solutions are $I_0(x)$ and $K_0(x)$, the modified Bessel functions of the first and second kind, respectively (Abramowitz and Stegun 1972). Now, $I_0(x)$ is well behaved as $x \to 0$, and $K_0(x) \sim -\ln x$. Hence, following the solution to Exercise 7.1, we have the fundamental solution for the modified Helmholtz equation in two dimensions as

$$u^*(R) = \frac{1}{2\pi}K_0(pR).$$

Solution 7.3 Suppose that P is on the boundary and that there is a discontinuity α_P in the tangent at P; see Fig. 7.8.

We then proceed in just the same manner as in Section 7.1, but in this case C_ϵ is an arc of a circle and

$$I_2 = \frac{1}{2\pi}\int_0^{\alpha_P}(u_P + \eta_2)\left(-\frac{\partial}{\partial R}\ln R\right)\Bigg|_{R=\epsilon}\epsilon\,d\theta$$

$$\to -\frac{\alpha_P}{2\pi}u_P \qquad \text{as } \epsilon \to 0,$$

and $c_P = \alpha_P/2\pi$.

If P is outside the boundary, then there is no need to introduce the limiting procedure at all, and it follows from eqn (7.7) that $c_P = 0$.

Solution 7.4

$$\int_0^1 \ln\eta\,d\eta = [\eta\ln\eta - \eta]_0^1$$

$$= -1.$$

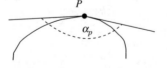

Fig. 7.8 Point P on the boundary C.

Hence, using eqn (7.27), we obtain

$$G_{ii} = l_i \left(1 - \ln \left(\tfrac{1}{2} l_i \right) \right).$$

Solution 7.5 We shall use the results of Section 7.3, dropping the subscripts ij inside the integrals

$$H_{ij} = \int_{[j]} \frac{d}{R^2} \, ds.$$

Now,

$$ds = \frac{R \, d\theta}{\cos \theta}$$

$$= \frac{R^2 \, d\theta}{d}.$$

Hence

$$H_{ij} = \int_{\theta_1}^{\theta_2} d\theta$$

$$= \theta_2 - \theta_1.$$

Also,

$$G_{ij} = - \int_{[j]} \ln R \, ds$$

$$= - \int_{\theta_1}^{\theta_2} \ln \left| \frac{d}{\cos \theta} \right| \frac{d}{\cos^2 \theta} \, d\theta$$

$$= d \left[\tan \theta \left(1 + \ln \left| \frac{\cos \theta}{d} \right| \right) - \theta \right]_{\theta_1}^{\theta_2}.$$

To find H_{12} and G_{12} in Example 7.1, we have

$$\theta_1 = 0, \quad \theta_2 = \tan^{-1} 2, \quad d = 0.5.$$

Hence $H_{12} = 1.107149$ and $G_{12} = 0.334854$.

For H_{13} and G_{13}, we have

$$\theta_1 = - \tan^{-1} \left(\tfrac{1}{2} \right), \quad \theta_2 = \tan^{-1} \left(\tfrac{1}{2} \right), \quad d = 1.$$

Hence $H_{13} = 0.927295$ and $G_{13} = 0.038867$.

Similarly,

$$\check{H}_{kj} = \frac{1}{2\pi} (\theta_2 - \theta_2) \text{ and } \check{G}_{kj} = \frac{d}{2\pi} \left[\tan \theta \left(1 + \ln \left| \frac{\cos \theta}{d} \right| \right) - \theta \right]_{\theta_1}^{\theta_2},$$

giving $\check{H}_{11} = 0.25$ and $\check{G}_{11} = 0.08931$.

Compare $H_{12} = 1.106740$, $G_{12} = 0.334828$, $H_{13} = 0.927812$, $G_{13} = 0.038869$, $\check{H}_{11} = 0.24965$ and $\check{G}_{11} = 0.08928$,   the   values   obtained   using Gauss quadrature in Example 7.1.

Solution 7.6

$$R_{21}(\xi_g) : 0.844496 \quad 0.652987 \quad 0.506361 \quad 0.531606.$$

Hence

$$\check{H}_{21} = 0.23044 = \check{H}_{23}.$$

Similarly,

$$\check{H}_{22} = 0.33974 \text{ and } \check{H}_{24} = 0.18715$$

and

$$\check{G}_{21} = 0.07960 = \check{G}_{23}, \quad \check{G}_{22} = 0.16277, \quad \check{G}_{24} = 0.03530.$$

Hence

$$u\left(\frac{1}{2}, \frac{3}{4}\right) = \sum_{j=1}^{4} \check{H}_{2j} U_j + \sum_{j=1}^{4} \check{G}_{2j} Q_j$$

$$= 1.671$$

using the values of U_j and Q_j from Example 7.1.

N.B. This result is not as good as that for $u\left(\frac{1}{2}, \frac{1}{2}\right)$ in Example 7.1, and this is because $\left(\frac{1}{2}, \frac{3}{4}\right)$ is relatively close to the boundary approximation one quarter of an element's length away, contrary to our 'rule of thumb'; see Section 7.4.

Solution 7.7 The constant solution $u \equiv 1$ has boundary values $u = 1$ and $q = 0$. Hence, using eqns (7.9) and (7.10) for an internal point, we have

$$1 = \oint_C (-q^*) \, dx$$

$$= \frac{1}{2\pi} \oint_C \frac{\partial}{\partial n} (\ln R) \, ds,$$

so for any internal point k we have

$$\frac{1}{2\pi} \oint_C \frac{\partial}{\partial n} (\ln R_k) \, ds = 1$$

and hence

$$\sum_{j=1}^{n} \check{H}_{kj} \approx 1.$$

Using the results of Exercise 7.6, we find

$$\sum_{j=1}^{4} \check{H}_{2j} = 0.988,$$

close to 1 as expected.

Solution 7.8 In this case we find the matrices

$$
\mathbf{A} =
\begin{array}{cccc}
2 & 3 & 1 & 4 \\
\end{array}
\left[
\begin{array}{cccc}
1.10674 & 0.92781 & 1.69315 & 0.33483 \\
-3.14076 & 1.10674 & 0.33483 & 0.03887 \\
1.10674 & -3.14076 & 0.03887 & 0.33483 \\
0.92781 & 1.10674 & 0.33483 & 1.69315 \\
\end{array}
\right],
$$

$$
\mathbf{b} =
\begin{array}{cccc}
2 & 3 & 1 & 4 \\
\end{array}
\left[1.84717 \quad -3.04854 \quad -3.04854 \quad 1.84717 \right]^T,
$$

and the solution is given by

$$
\mathbf{x} =
\begin{array}{cccc}
2 & 3 & 1 & 4 \\
\end{array}
\left[1.407 \quad 1.407 \quad -0.500 \quad -0.500 \right]^T.
$$

Hence we have the nodal values

$$
\mathbf{U} =
\begin{array}{cccc}
1 & 2 & 3 & 4 \\
\end{array}
\left[1 \quad 1.407 \quad 1.407 \quad 1 \right]^T,
$$

$$
\mathbf{Q} =
\begin{array}{cccc}
1 & 2 & 3 & 4 \\
\end{array}
\left[-0.500 \quad 0.5 \quad 0.5 \quad -0.500 \right]^T.
$$

The exact values are $u_2 = u_3 = 1.5$ and $q_2 = q_3 = 0.5$. Using the values of $\check{H}_{ij}$ and $\check{G}_{ij}$ from Example 7.1, we find

$$u \left(\tfrac{1}{2}, \tfrac{1}{2} \right) \approx 1.157.$$

In this case the error in the internal value is similar to the errors in the boundary values.

Solution 7.9 The basis functions are the nodal functions of Section 3.5 and we have, using the notation of Fig. 7.4,

$$H_{ij} = h_{ij} + h_{ij-1} - \alpha_i \delta_{ij}$$

with

$$h_{iJ} = \frac{d_{iJ} l_J}{4} \int_{-1}^{1} (1 \pm \xi) \frac{1}{R_{iJ}^2(\xi)} \, d\xi,$$

and

$$G_{ij} = g_{ij} + g_{ij-1}$$

with

$$g_{iJ} = -\frac{l_J}{4} \int_{-1}^{1} (1 \pm \xi) \ln R_{iJ}(\xi) \, d\xi,$$

where we take the positive or negative sign when $J = j$ or $(j - 1)$, respectively. There are three possibilities:

(1) *Base node not in target element.* The integrals are non-singular, and we can use a standard Gauss quadrature.

(2) *Base node in target element $i = j - 1$ or $i = j + 1$.* Just one of the integrals is singular; the other may be computed using a standard Gauss quadrature. As in Section 7.3, it follows that $h_{i\,i-1} = h_{i\,i+1} = 0$. H_{ii} is obtained using the 'row sum' technique of eqn (7.24):

$$g_{j\,j-1} = -\frac{l_{j-1}}{4} \int_{-1}^{1} (1 - \xi) \ln \left[\tfrac{1}{2}(1 - \xi)l_{j-1} \right] d\xi$$

$$= \frac{l_{j-1}}{2} \left(\ln l_{j-1} - \tfrac{1}{2} \right)$$

and

$$g_{j\,j+1} = \frac{l_j}{2} \left(\ln l_j - \tfrac{1}{2} \right).$$

(3) *Base node in target element $i = j$.*

$$g_{jj} = -\frac{l_j}{4} \int_{-1}^{1} (1 + \xi) \ln \left[\tfrac{1}{2}(1 - \xi)l_j \right] d\xi$$

$$= \frac{l_j}{2} \left(\ln l_j - \tfrac{3}{2} \right).$$

The singular integrals for $g_{i\,j-1}$ and g_{jj} are also obtained using logarithmic Gauss quadrature as follows. We use the change of variable $\eta = \tfrac{1}{2}(1 - \xi)$ to obtain

$$g_{j\,j-1} = l_{j-1} \int_0^1 (\eta \ln l_{j-1} - \eta \ln \eta) \, d\eta$$

and

$$g_{ij} = l_j \int_0^1 ((1 - \eta) \ln l_j - (1 - \eta) \ln \eta) \, d\eta.$$

Using any order of logarithmic Gauss quadrature, we find that

$$\int_0^1 (1-\eta) \ln \eta \, d\eta = -0.75$$

and

$$\int_0^1 \eta \ln \eta \, d\eta = -0.25$$

so that we recover the exact values for $g_{j\,j-1}$ and g_{jj}.

The expressions for $\breve{H}_{kj}$ and $\breve{G}_{kj}$ are given by

$$\breve{H}_{kj} = \breve{h}_{kj} + \breve{h}_{k\,j-1}, \quad \breve{G}_{kj} = \breve{g}_{kj} + \breve{g}_{k\,j-1},$$

where

$$\breve{h}_{kJ} = \frac{d_{kJ} l_J}{8\pi} \int_{-1}^1 (1 \pm \xi) \frac{1}{R_{kJ}^2(\xi)} \, d\xi$$

and

$$\breve{g}_{kJ} = -\frac{l_J}{8\pi} \int_{-1}^1 (1 \pm \xi) \ln(R_{kJ}(\xi)) \, d\xi.$$

Solution 7.10 We partition the matrices $\mathbf{H}$ and $\mathbf{G}$ in eqn (7.17) according to the boundary conditions

$$\begin{bmatrix} \mathbf{H}^1 & \mathbf{H}^2 & \mathbf{H}^3 \end{bmatrix} \begin{bmatrix} \mathbf{U}^1 \\ \mathbf{U}^2 \\ \mathbf{U}^3 \end{bmatrix} + \begin{bmatrix} \mathbf{G}^1 & \mathbf{G}^2 & \mathbf{G}^3 \end{bmatrix} \begin{bmatrix} \mathbf{Q}^1 \\ \mathbf{Q}^2 \\ \mathbf{Q}^3 \end{bmatrix} = 0,$$

where the superscripts 1 and 2 refer to Dirichlet and Neumann conditions, respectively, as before. Superscript 3 refers to a Robin condition of the form

$$q + \sigma u = h.$$

We can use this form to eliminate one of the variables at the appropriate node to obtain a system of equations of the form

$$\begin{bmatrix} \mathbf{H}^2 & \mathbf{G}^1 & \mathbf{H}^3 - \mathbf{G}^3 \operatorname{diag}(\sigma_j) \end{bmatrix} \begin{bmatrix} \mathbf{U}^2 \\ \mathbf{Q}^1 \\ \mathbf{U}^3 \end{bmatrix} = - \begin{bmatrix} \mathbf{H}^1 & \mathbf{G}^2 & \mathbf{G}^3 \end{bmatrix} \begin{bmatrix} \mathbf{U}^1 \\ \mathbf{Q}^2 \\ \mathbf{h} \end{bmatrix},$$

and proceed in the usual manner.

8 Computational aspects

At the time of publication of the first edition of this text, we did not consider it necessary to include details of the computational aspects of the finite element method. However, computing power and availability have moved on, almost beyond recognition, in the intervening thirty years.

The most striking progress is in the power and versatility of personal computing environments together with a variety of ready-to-use but sophisticated software. All this at very modest, previously unimaginable prices. We have suggested throughout the text how a spreadsheet may be used to perform some simple finite element analysis. We could equally well have chosen to use a computer algebra package such as MATLAB® (Kaltan 2007).

It is not the purpose of this section to provide a detailed description of the computational processes, nor is it intended to provide a finite element suite of programs. There are many texts to which the interested reader may refer; see, for example, Smith and Griffiths (2004). Also, there is a large amount of software available to users without the necessity to write finite element code. A simple Internet search will reveal a wide variety of proprietary software, ranging from general-purpose programs to code for very specific applications. Indeed, software such as MATLAB is developing very rapidly, and there is now a finite element toolbox available with the software (Fausett 2007).

In this chapter we shall discuss some of the aspects that would be considered when writing finite element code. There are, essentially, three stages: (i) the pre-processor, (ii) the solution phase and (iii) the post-processor. The development of these three phases follows very closely the seven steps outlined for the model problem in Section 3.5. For further details, see Zienkiewicz *et al.* (2005) and the references cited therein.

8.1 Pre-processor

During the pre-processing phase, the finite element environment is developed from suitable input data. This includes mesh generation, calculation of the element matrices and assembly of the overall system of equations. By far the most complicated aspect of this phase is that of mesh generation. Ideally, we would like to be able to produce a mesh which enabled automatic refinement in the region of any point to obtain an improved solution. Also, we would like the mesh to be generated in such a way that the bandwidth in the stiffness matrix is minimized. The development of such automatic mesh generators is the subject of much current research. We shall limit ourselves here to a very brief overview of mesh generation for two-dimensional problems. Three-dimensional problems provide a much stiffer challenge.

The first consideration is the type of element required. In general, if the boundary is curved, geometric errors in the mismatch between the boundary and the finite element mesh will be significantly reduced if elements with curved sides are used. A popular generation process is the Delaunay triangulation method, in which a region is covered by an array of triangles. Since a quadrilateral may be obtained by combining two triangles, the Delaunay approach can be used with either of the two elements described in Chapters 3 and 4. The Delaunay approach develops triangles in such a way that the circumcircle of each triangle contains no other node in the mesh. A consequence of this propery is that the ratio of the area of the circumcircle to the area of its triangle is not large, ensuring that the triangle does not have a large aspect ratio, a desirable property of the mesh; see Section 6.2. For further details on mesh generation methods, see Zienkiewicz *et al.* (2005) and the references therein. Once the mesh has been generated, a suitable node numbering must be imposed, and it is important that this numbering yields a satisfactory bandwidth.

The simple problem considered in Exercise 3.12 shows that the bandwidth depends on the maximum difference between the node numbers in each elemenent. This leads to the strategy that nodes should be numbered along paths that contain the fewest nodes as we move from an element to an adjacent element across the domain from one boundary to another.

Calculation of the element matrices is a straightforward process, usually involving Gauss quadrature of a suitable order to effect the integrations. The overall stiffness matrix is then assembled as the individual element matrices are developed.

We note here that the boundary element method, as described in Chapter 7, requires significantly less effort to generate a suitable mesh, since the method reduces the geometric dimensions by one: partial differential equations in two or three dimensions become integral equations in one or two dimensions, respectively.

8.2   Solution phase

The solution phase is in principle well defined; we have a system of equations, so let's solve them. Here, the important properties of the stiffness matrix described in Section 3.6 may be exploited. Probably the most important book on linear algebra is Wilkinson's classic 1965 treatise, reprinted in 1999, and most of the methods considered here, together with many more, are considered there. Where appropriate, we shall indicate more recent texts. We shall restrict ourselves to the case of linear equations of the form of eqn (3.44), which we shall write as

$$(8.1) \qquad\qquad\qquad \mathbf{Ax} = \mathbf{b}.$$

The problem is to find the solution vector $\mathbf{x}$, given the matrix of coefficients $\mathbf{A}$ and the vector of known quantities $\mathbf{b}$. We shall restrict ourselves to systems of linear equations; readers interested in the solution of non-linear systems should consult the text by Rheinboldt (1987).

There are two approaches: a *direct* approach, in which the exact solution will be obtained in a finite number of steps, and an *indirect* approach, in which an iterative process is used to obtain an approximate solution of sufficient accuracy. The choice of method is usually dependent on the form of the equations. There are many texts available which provide details of the solutions of systems of linear equations; for example, Jennings and McKeown (1992) describe matrix techniques for both systems of equations and eigenvalue problems, as well as considering sparsity in large-order systems.

The most commonly used direct solver is the *Gauss elimination* process. The system of equations (8.1) is reduced to upper triangular form

$$\mathbf{Ux} = \mathbf{b}'$$

by a sequence of row operations. Improved accuracy and avoidance of induced instability are obtained by using a partial pivoting strategy. This reduced system is now easily solved by a back-substitution process. The Gauss elimination process can be considered as a factorization process in which $\mathbf{A} = \mathbf{LU}$, where $\mathbf{L}$ is a lower triangular matrix and $\mathbf{U}$ an upper triangular matrix.

The finite element equations (3.44) are sparse, and such a property can be exploited in variants of the Gauss elimination process. The equations for the boundary element method are fully populated and are not necessarily symmetric, and as such are usually solved with a standard Gauss elimination process. There is a process, the *multipole method* (Popov and Power 2001), in which parts of the boundary element regions are lumped together, giving a sparse system matrix and enabling other solvers to be used.

In Section 3.6 we saw that the stiffness matrix is symmetric and positive definite, which allows a *Cholesky decomposition* of the form $\mathbf{A} = \mathbf{LL}^T$, where $\mathbf{L}$ is a lower triangular matrix. For large systems, the Cholesky decomposition requires approximately one half the number of arithmetic operations for an $\mathbf{LU}$ decomposition, the number for the Cholesky decomposition being $\mathcal{O}(n^3/3)$ and for the $\mathbf{LU}$ decomposition being $\mathcal{O}(2n^3/3)$ for an $n \times n$ system of equations.

Indirect approaches involve iteration, and a typical iterative method is *Gauss–Seidel* iteration. If the matrix $\mathbf{A}$ is written as

$$\mathbf{A} = \mathbf{L} + \mathbf{D} + \mathbf{U},$$

where $\mathbf{D}$ is a diagonal matrix, then the algorithm is of the form

$$[\mathbf{L} + \mathbf{D}]\mathbf{x}^{(k+1)} = \mathbf{b} - \mathbf{U}\mathbf{x}^{(k)},$$

where iteration on k is performed until some suitable accuracy is achieved. Such methods are particularly suitable for sparse systems. Unfortunately, convergence is not guaranteed unless $\max|\lambda_i| < 1$, where $\{\lambda_i\}$ is the set of eigenvalues of the iteration matrix $[\mathbf{L} + \mathbf{D}]^{-1}\mathbf{U}$. A stricter sufficient condition is that convergence will occur if the matrix $\mathbf{A}$ is strictly *diagonally dominant*, i.e.

$$|A_{ii}| > \sum_{j=1}^{n}{}'|A_{ij}|.$$

More recently, *conjugate gradient* methods have been used, since these have a particular relevance to the finite element method. The attraction is that it is possible to develop the solution method without the necessity to assemble the whole stiffness matrix. Smith and Griffiths (2004) discussed the importance of this in the context of large three-dimensional problems, for which the computer storage requirement can be reduced by an order of magnitude. The important aspect is that in the iterative cycle a matrix product of the form $\mathbf{u}^{(k)} = \mathbf{A}\mathbf{p}^{(k)}$ is developed, where $\mathbf{p}^{(k)}$ is a vector of length n. Clearly, this product can be obtained on an element-by-element basis to develop the iterate $\mathbf{u}^{(k)}$. Details of the conjugate gradient method can be found in the text by Broyden and Vespucci (2004).

An innovative idea due to Irons (1970) also enables equation solution without assembling the complete stiffness matrix. The equation solution commences while the overall stiffness matrix $\mathbf{K}$ is being developed. $\mathbf{K}$ is banded, and as each element stiffness is incorporated, a *front* progresses through the band, behind which $\mathbf{K}$ is fully assembled. As soon as a section of $\mathbf{K}$ is assembled, then the solution process can begin and follows the movement of the front down through the band.

Finally, we have said nothing about parallel computing, in which independent processes may be computed simultaneously. A very attractive environment is one in which processors are connected together and each one performs independently on a set of data. In this way, there is the potential for speed-up of calculation. However, interprocessor communication can have an adverse effect on performance, and it is more usual to consider solving larger problems in the same time rather than seeking speed-up of small problems. Much work is currently in progress using different parallel processing paradigms; in particular, the nature of the boundary element method makes it particulary attractive for implementation in a parallel environment.

This section has been deliberately short; we wish only to give a brief overview of the techniques which may be used to solve the system of equations obtained during a finite element analysis. Specific details may be found in the suggested texts.

8.3   Post-processor

The post-processing phase has received a great deal of attention recently. Specifically, modern graphics facilities allow a wide variety of output visualization, and users of finite element packages will choose those output processes best suited to their needs. In particular, finite elements are at the heart of many computer-aided design packages, and typical output pictures can be found in the references. Post-processing techniques are advancing very quickly, and it is now possible to see simulations such as vibrations of structures ranging from bridges to musical instruments and of fluid flow around aircraft and racing bicycles. In the world of animation, film producers can use such software to produce realistic fluid motions as a backdrop to scenes filmed indoors. An idea of the variety of applications can be found in the texts by Burnett (1987), Aliabadi (2002), Zienkiewicz *et al.* (2005) and Fish and Belytschko (2007).

As far as the boundary element method is concerned, similar comments to those for the finite element method apply. The major difference is that the system matrices for boundary elements are full and there is, in general, no symmetry. However, for problems of the same geometric size, the boundary element matrices are smaller. For more details, the interested reader may refer to Beer (2001).

8.4   Finite element method (FEM) or boundary element method (BEM)?

When compared one with the other, each method has advantages and disadvantages. The major advantage of the FEM is that it is significantly further advanced than the BEM and that there is a wide variety of easily available codes. Also, the mathematics associated with the FEM is much more familiar: partial differential equations are more widely known than integral equations. For non-linear material problems, the interior of regions must be modelled, which removes one of the major features of the BEM, i.e. a boundary-only geometry. This is mitigated somewhat by being able to choose only those interior points which may be particularly interesting. In the FEM, the whole interior is modelled. In general, the BEM requires less effort in pre-processing, since only the boundary is modelled, and remeshing is also easier. Clearly there is a difference in computer storage required, although we must remember that the FEM has large sparse matrices, whereas the BEM has smaller but fully populated matrices. The FEM is very good for the analysis of problems involving thin shells. In these cases the BEM performs poorly, since elements on opposite sides of the shell are very close together, leading to inaccuracies in the numerical integration.

So, neither method is 'better' than the other. It is important to have both available and to use the one which is more suitable in any particular circumstance. As a general rule, BEM for linear problems, for problems such as fracture mechanics where variables may change rapidly over small distances and for infinite regions; FEM for non-linear material problems. There are often problems which comprise finite non-linear regions and infinite linear regions. In these cases, a hybrid FEM/BEM approach is suitable.

Appendix A   Partial differential equation models in the physical sciences

We list the models according to their classification; see Chapter 2. In the following examples, we introduce the most commonly used symbols for the corresponding physical quantities.

A.1   Parabolic problems

We write our generic field equation for the unknown dependent variable u in the form

$$
\text{(A.1)} \qquad\qquad \text{div}(k \, \text{grad} \, u) = C\frac{\partial u}{\partial t} + f,
$$

where the physical properties k, C and f may be functions of position $\mathbf{r}$, time t or, in the non-linear case, u. For anisotropic problems, we replace the scalar k by a tensor quantity $\boldsymbol{\kappa}$.

Equation (A.1) is usually a representation of a conservation law coupled with a suitable constitutive law.

Heat conduction *(Carslaw and Jaeger 1986)*

     u: temperature (T);
     k: thermal conductivity;
     ρ: density; c: specific heat; $C = \rho c$;
     f: heat source (Q);
     $\mathbf{q}$: heat flux.

     Constitutive law: Fourier's law, $\mathbf{q} = -k \, \text{grad} \, T$.
     Heat conservation law: $-\text{div} \, \mathbf{q} = \rho c (\partial T / \partial t) + Q$.
     Heat conduction equation: $\text{div}(k \, \text{grad} \, T) = \rho c (\partial T / \partial t) + Q$.
     For constant parameters and no heat sources, $\nabla^2 T = (1/\alpha)(\partial T / \partial t)$, where $\alpha = k/\rho c$ is called the thermal diffusivity.

Diffusion *(Crank 1979)*

> u: concentration (C);
> k: diffusivity (D);
> C: dimensionless quantity of value 1 in eqn (A.1);
> f: mass source (Q);
> **q**: mass flux.

> Constitutive law: Fick's law, $\mathbf{q} = -D \operatorname{grad} C$.
> Mass conservation law: $-\operatorname{div} \mathbf{q} = \partial u / \partial t + Q$.
> Diffusion equation: $\operatorname{div}(D \operatorname{grad} C) = \partial u / \partial t + Q$.
> For constant parameters and no mass source, $\nabla^2 C = (1/D)(\partial u / \partial t)$.

A.2 Elliptic problems

Our generic field equation is of the form

(A.2)
$$\operatorname{div}(k \operatorname{grad} u) = f.$$

Clearly, time-independent heat conduction and diffusion reduce to elliptic problems of this type.

Electrostatics *(Reitz et al. 1992)*

> u: electrostatic potential (ϕ);
> k: permittivity (ϵ);
> f: charge density (ρ);
> **E**: electric field.

> Constitutive law: $\mathbf{E} = -\operatorname{grad} \phi$.
> Gauss's law: $\operatorname{div}(\epsilon E) = \rho$.
> Field equation: $\operatorname{div}(\epsilon \operatorname{grad} \phi) = -\rho$.
> For constant parameters, $\nabla^2 \phi = -\rho/\epsilon$, Poisson's equation.

Magnetostatics *(Reitz et al. 1992)*

> u: magnetic scalar potential (ϕ);
> k: permeability (μ);
> $f = 0$;
> **H**: magnetic field.

> Constitutive law: $\mathbf{H} = -\operatorname{grad} \phi$, $\operatorname{div}(\mu \mathbf{H}) = 0$.
> Field equation: $\operatorname{div}(\mu \operatorname{grad} \phi) = 0$.
> For constant parameters, $\nabla^2 \phi = 0$, Laplace's equation.

Hydrodynamics *(Lamb 1993)*

> u: velocity potential (ϕ);
> k: dimensionless quantity of value 1 in eqn (A.2);
> $\mathbf{q}$: fluid velocity vector.

> Constitutive law: $\mathbf{q} = \operatorname{grad}\phi$.
> Incompressibility law: $\operatorname{div}\mathbf{q} = 0$.
> Field equation: $\nabla^2 = 0$, Laplace's equation.

A.3   Hyperbolic problems

Our generic field equation is of the form

(A.3)
$$\nabla^2 u = \frac{1}{c^2}\frac{\partial^2 u}{\partial t^2}.$$

Sound waves *(Curle and Davies 1971)*

> u: velocity potential (ϕ);
> c: speed of sound;
> $\mathbf{q}$: fluid velocity vector;
> p: acoustic pressure;
> ρ: fluid density;
> s: condensation.

> Constitutive relations: $\mathbf{q} = \operatorname{grad}\phi$, $p = \rho(\partial\phi/\partial t)$, $\rho = \rho_0(1+s)$.

> Field equations: $\nabla^2\phi = (1/c^2)(\partial^2\phi/\partial t^2)$, $\nabla^2 s = (1/c^2)(\partial^2 s/\partial t^2)$.

Waves on strings and membranes *(Coulson and Jeffrey 1977)*

$$\frac{\partial^2 u}{\partial x^2} = \frac{1}{c^2}\frac{\partial^2 u}{\partial t^2}, \quad \nabla^2 u = \frac{1}{c^2}\frac{\partial^2 u}{\partial t^2}.$$

> u: transverse displacement;
> c: speed of wave.

For waves of the form $u(\mathbf{r}, t) = v(\mathbf{r})e^{\pm i\omega t}$ with angular frequency ω, the hyperbolic equations may be written as the elliptic Helmholtz equation

$$\nabla^2 v + k^2 v = 0,$$

where $k = \omega/c$, the *wavenumber*.

A.4   Initial and boundary conditions

Elliptic equations are associated with steady-state problems and require suitable conditions applied on the boundary, for example Dirichlet, Neumann or Robin conditions; see Section 2.2.

Parabolic and hyperbolic equations are associated with systems that progress with time. As well as suitable boundary conditions, they require initial conditions. Usually we know $u(\mathbf{r}, 0)$ for parabolic problems and $u(\mathbf{r}, 0)$, $(\partial u / \partial t)(\mathbf{r}, 0)$ for hyberbolic problems.

Appendix B   Some integral theorems of the vector calculus

In this appendix, all functions are assumed to satisfy suitable differentiability conditions to ensure that the resulting operations exist.

V is the volume of a three-dimensional region bounded by a surface S, with outward unit normal $\mathbf{n}$.

Gauss's divergence theorem for a vector field $\mathbf{F}$:

$$\int_V \operatorname{div} \mathbf{F} \, dV = \int_S \mathbf{F} \cdot \mathbf{n} \, dS.$$

(i) If $\mathbf{F} = k \operatorname{grad} u$,

$$\int_V \operatorname{div} \mathbf{F} \, dV = \int_S k \frac{\partial u}{\partial n} \, dS.$$

(ii) If $\mathbf{E} = \kappa \operatorname{grad} u$,

$$\int_V \operatorname{div} \mathbf{F} \, dV = \int_S (\kappa \operatorname{grad} u) \cdot \mathbf{n} \, dS.$$

Green's theorem for scalar fields u and v.

First form:

$$\int_V (u \, \nabla^2 v + \operatorname{grad} u \cdot \operatorname{grad} v) \, dV = \int_S u \operatorname{grad} v \cdot \mathbf{n} \, dS.$$

Second form:

$$\int_V (u \, \nabla^2 v - v \, \nabla^2 u) \, dV = \int_S \left(u \frac{\partial v}{\partial n} - v \frac{\partial u}{\partial n} \right) dS.$$

Generalized Green's theorem for scalar fields u and v, with the tensor κ represented by a 3×3 matrix.

First form:

$$\int_V \{ u \operatorname{div}(\kappa \operatorname{grad} v) + \operatorname{grad} u \cdot (\kappa \operatorname{grad} v) \} \, dV = \int_S u (\kappa \operatorname{grad} v) \cdot \mathbf{n} \, dS.$$

Second form, for symmetric κ:

$$\int_V \{u\,\mathrm{div}(\kappa\,\mathrm{grad}\,v) + v\,\mathrm{div}(\kappa\,\mathrm{grad}\,u)\}\,dV = \int_S \{u(\kappa\,\mathrm{grad}\,v) - v(\kappa\,\mathrm{grad}\,u)\}\cdot\mathbf{n}\,dS.$$

The integral theorems stated above may easily be interpreted for a two-dimensional region D bounded by a curve C; for such regions, the operator $\nabla \equiv [\partial/\partial x \;\; \partial/\partial y]^T$, and κ is represented by a 2×2 matrix. In some of the examples in the text, one-dimensional problems are considered, so that $\nabla \equiv [d/dx]$; in this case the boundary integrals are obtained by finding the difference between the values of the integrand at each end of the interval.

Appendix C   A formula for integrating products of area coordinates over a triangle

$$I = \int\int_A L_1^m L_2^n L_3^p \, dx \, dy$$

$$= \int\int_A L_1^m L_2^n (1 - L_1 - L_2)^p \, dx \, dy,$$

using eqn (3.59). It follows from eqns (3.60) and (3.61) that

$$dx \, dy = \begin{vmatrix} x_1 - x_3 & x_2 - x_3 \\ y_1 - y_3 & y_2 - y_3 \end{vmatrix} dL_1 \, dL_2$$

$$= 2A \, dL_1 \, dL_2.$$

Thus

$$I = 2A \int_0^1 L_1^m \, dL_1 \int_0^{1-L_1} L_2^n (1 - L_1 - L_2)^p \, dL_2.$$

Consider

$$I(a, b) = \int_0^\alpha t^a (\alpha - t)^b \, dt$$

$$= \frac{t^{a+1}}{a+1} (\alpha - t)^b \Big|_0^\alpha + \frac{b}{a+1} \int_0^\alpha t^{a+1} (\alpha - t)^{b-1} \, dt$$

$$= \frac{b}{a+1} I(a+1, b-1).$$

Thus

$$I(a, b) = \frac{b(b-1)(b-2)\ldots(2)(1)}{(a+1)(a+2)(a+3)\ldots(a+b)} I(a+b, 0)$$

$$= \frac{a! \, b!}{(a+b)!} I(a+b, 0).$$

Now

$$I(a+b,0) = \int_0^\alpha t^{a+b}\, dt$$

$$= \frac{a^{a+b+1}}{a+b+1},$$

so that

$$I(a,b) = \frac{a!\,b!}{(a+b+1)!}\alpha^{a+b+1}.$$

Therefore

$$I = \frac{2An!p!}{(n+p+1)!} \int_0^1 L_1^m (1-L_1)^{n+p+1}\, dL_1$$

$$= \frac{2A\,n!p!}{(n+p+1)!}\,\frac{m!(n+p+1)!}{\{m+(n+p+1)+1\}!},$$

i.e.

$$I = \frac{2Am!n!p!}{(m+n+p+2)!}.$$

Appendix D   Numerical integration formulae

D.1   One-dimensional Gauss quadrature

$$\int_{-1}^{1} f(\xi)\, d\xi \approx \sum_{g=1}^{G} w_g f(\xi_g),$$

where ξ_g is the coordinate of an integration point and w_g is the corresponding weight; G is the total number of such points. The formula integrates exactly all polynomials of degree $2G - 1$. The coordinates of the integration points and the corresponding weights are given in Table D.1.

D.2   Two-dimensional Gauss quadrature

Rectangular regions:.

$$\int_{-1}^{1}\int_{-1}^{1} f(\xi, \eta)\, d\xi\, d\eta \approx \sum_{g_1=1}^{G_1} \sum_{g_2=1}^{G_2} w_{g_1} w_{g_2} f(\xi_{g_1}, \eta_{g_2}),$$

Table D.1 Coordinates and weights for one-dimensional Gauss quadrature

G	$\pm\xi_g$	w_g
1	0	2
2	0.577 350 269	1
3	0	0.888 888 889
	0.774 596 669	0.555 555 556
4	0.861 136 312	0.347 854 845
	0.339 981 044	0.652 145 155
6	0.932 469 514	0.171 324 492
	0.661 209 386	0.360 761 573
	0.238 619 186	0.467 913 935

Table D.2 Coordinates and weights for Gauss quadrature over a triangle

G	$L_1^{(g)}$	$L_2^{(g)}$	$L_3^{(g)}$	w_g
1	1/3	1/3	1/3	1/2
2	1/2	1/2	0	1/6
	1/2	0	1/2	1/6
	0	1/2	1/2	1/6
4	1/3	1/3	1/3	−9/32
	3/5	1/5	1/5	25/96
	1/5	3/5	1/5	25/96
	1/5	1/5	3/5	25/96

where the coordinates of the integration points and the corresponding weights are given in Table D.1. This formula integrates exactly all polynomials of degree $2G_1 - 1$ in the ξ-direction and $2G_2 - 1$ in the η-direction.

Triangular regions:.

$$\int_0^1 \int_0^{1-L_1} f(L_1, L_2, L_3) \, dL_1 \, dL_2 \approx \sum_{g=1}^G w_g f\left(L_1^{(g)}, L_2^{(g)}, L_3^{(g)}\right),$$

where $\left(L_1^{(g)}, L_2^{(g)}, L_3^{(g)}\right)$ are the coordinates of the integration points, w_g are the corresponding weights and G is the total number of such points. The numerical formulae given in Table D.2 have been chosen in such a way that there is no bias towards any one coordinate (Hammer *et al.* 1956).

D.3 Logarithmic Gauss quadrature

$$\int_0^1 f(\xi) \ln \xi \, d\xi \approx -\sum_{g=1}^G w_g f(\xi_g),$$

where ξ_g is the coordinate of an integration point and w_g is the corresponding weight; G is the total number of such points. The coordinates of the integration points and the corresponding weights are given in Table D.3.

Table D.3 Coordinates and weights for logarithmic Gauss quadrature

G	ξ_g	w_g
2	0.112 008 062	0 718 539 319
	0.602 276 908	0.281 460 681
3	0.063 890 793	0.513 404 552
	0.368 997 064	0.391 980 041
	0.766 880 304	0.094 615 407
4	0.041 448 480	0.383 464 068
	0.245 274 914	0.386 875 318
	0.556 165 454	0.190 435 127
	0.848 982 395	0.039 225 487
6	0.021 634 006	0.238 763 663
	0.129 583 391	0.308 286 573
	0.314 020 450	0.245 317 427
	0.538 657 217	0.142 008 757
	0.756 915 337	0.055 454 622
	0.922 668 851	0.010 168 959

Appendix E   Stehfest's formula and weights for numerical Laplace transform inversion

Stehfest's procedure is as follows (Stehfest 1970a,b).

Given $\bar{f}(\lambda)$, the Laplace transform of $f(t)$, we seek the value $f(\tau)$ for a specific value $t = \tau$. Choose $\lambda_j = j\ln 2/\tau$, $j = 1, 2, \ldots, M$, where M is even; the approximate numerical inversion is given by

$$f(\tau) \approx \frac{\ln 2}{\tau} \sum_{j=1}^{M} w_j \bar{f}(\lambda_j).$$

The weights w_j are given by

$$w_j = (-1)^{M/2+j} \sum_{k=\lfloor 1/2(1+j) \rfloor}^{\min(j,M/2)} \frac{k^{M/2}(2k)!}{(M/2 - k)!\, k!\, (k-1)!(j-k)!(2k-j)!}$$

and are given in Table E.1 (Davies and Crann 2008).

Table E.1 Weights for Stehfest's numerical Laplace transform

$M = 6$	$M = 8$	$M = 10$	$M = 12$	$M = 14$
1	$-1/3$	$1/12$	$-1/60$	$1/360$
-49	$145/3$	$-285/12$	$961/60$	$-461/72$
366	-906	1279	-1247	$18481/20$
-858	$16394/3$	$-46871/3$	$82663/3$	$-6227627/180$
810	$-43130/3$	$505465/6$	$-1579685/6$	$4862890/9$
-270	18730	$-473915/2$	$13241387/10$	$-131950391/30$
	$-35480/3$	$1127735/3$	$-58375583/15$	$189788326/9$
	$8960/3$	$-1020215/3$	$21159859/3$	$-2877521087/45$
		$328125/2$	$-16010673/2$	$2551951591/20$
		$-65625/2$	$11105661/2$	$-2041646257/12$
			$-10777536/5$	$4509824011/30$
			$1796256/5$	$-169184323/2$
				$824366543/30$
				$-117766649/30$

References

Abramowitz, M. and Stegun, A. (1972). *Handbook of Mathematical Functions*. Dover.

Aliabadi, M. H. (2002). *The Boundary Element Method*, Vol. 2. Wiley.

Ames, W. F. (1972). *Non-linear Partial Differential Equations in Engineering*, Vol. II. Academic Press.

Archer, J. S. (1963). Consistent mass matrix for distributed mass systems. *Proc. A.S.C.E.*, **89ST4**, 161–78.

Argyris, J. H. (1955). Energy theorems and structural analysis. *Aircraft Eng.*, reprinted by Butterworths, London, 1960.

Argyris, J. H. (1964). Recent advances in matrix methods of structural analysis. *Prog. Aeron. Sci.*, **4**.

Argyris, J. H. and Kelsey, S. (1960). *Energy Theorems and Structural Analysis*. Butterworths.

Atkinson, B., Card, C. C. M. and Irons, B. M. (1970). Application of the finite element method to creeping flow problems. *Trans. Inst. Chem. Eng.*, **48**, 276–84.

Axelsson, O. and Barker, V. A. (2001). *Finite Element Solution of Boundary-Value Problems: Theory and Computation*. SIAM.

Babuška, I. (1971). Error bounds for finite element methods. *Numer. Math.*, **16**, 322–33.

Babuška, I. (1973). The finite element method with Lagrange multipliers. *Numer. Math.*, **20**, 179–92.

Babuška, I. and Rheinboldt, W. C. (1978). A-posteriori error estimates for the finite element method. *Int. J. Numer. Methods Eng.*, **11**, 1597–1615.

Babuška, I. and Rheinboldt, W. C. (1979). Adaptive approaches and reliability estimates in finite element analysis. *Comput. Methods Appl. Mech. Eng.*, **17/18**, 519–40.

Barnhill, R. E., Birkhoff, G. and Gordon, W. J. (1973). Smooth interpolation in triangles. *J. Approx. Theory*, **8**, 114–28.

Bazeley, G. P., Cheung, Y. K., Irons, B. M. and Zienkiewicz, O. C. (1965). Triangular elements in bending-conforming and non-conforming solutions. *Proc. Conf. Matrix Methods in Struct. Mech.*, Air Force Institute of Technology, Wright Patterson Air Force Base, Ohio, 547–76.

Becker, A. A. (1992). *The Boundary Element Method in Engineering*. McGraw-Hill.

Beer, G. (2001). *Programming the Boundary Element Method: An Introduction for Engineers*. Wiley.

Bettess, P. (1977). Infinite elements. *Int. J. Numer. Methods Eng.*, **11**, 53–64.

Bettess, P. (1992). *Infinite Elements*. Penshaw Press.

Birkhoff, G., Schultz, M. H. and Varga, R. S. (1968). Piecewise Hermite interpolation in one and two variables with applications to partial differential equations. *Numer. Math.*, **11**, 232–56.

Brebbia, C. A. (1978). *The Boundary Element Method for Engineers*. Pentech Press.

Brebbia, C. A. and Dominguez, J. (1977). Boundary element methods for potential problems. *Appl. Math. Model.*, **1**, 372–8.

Brenner, S. C. and Scott, L. R. (1994). *The Mathematical Theory of Finite Element Methods*. Springer.

Broyden, C. G. and Vespucci, M. T. (2004). *Krylov Solvers for Linear Algebraic Systems*. Elsevier.

Burnett, D. S. (1987). *Finite Element Analysis*. Addison-Wesley.

Carslaw, H. S. and Jaeger, J. C. (1986). *Conduction of Heat in Solids*. Oxford Science Publications.

Cheng, A. H.-D. and Cheng, D. T. (2005). Heritage and early history of the boundary element method. *Eng. Anal. Bound. Elements*, **29**, 268–302.

Ciarlet, P. G. (1978). *The Finite Element Method for Elliptic Problems*. North-Holland.

Ciarlet, P. G. and Lions, J. L. (1991). *Handbook of Numerical Analysis*, Vol. II, *Finite Element Methods (Part 1)*. North-Holland.

Clough, R. W. (1960). The finite element in plane stress analysis. *2nd A.S.C.E. Conf. on Electronic Computation*, Pittsburgh, Pennsylvania, 345–78.

Clough, R. W. (1969). Comparison of three-dimensional finite elements. *Proc. A.S.C.E. Symp. on Application of Finite Element Methods in Civil Engineering*, Vanderbilt University, Nashville, Tennessee, 1–26.

Clough, R. W. and Johnson, C. P. (1968). A finite element approximation for the analysis of thin shells. *Int. J. Solids Struct.*, **4**, 43–60.

Connor, J. J. and Brebbia, C. A. (1976). *Finite Element Techniques for Fluid Flow*. Newnes-Butterworths.

Coulson, C. A. and Jeffrey, A. (1977). *Waves: A Mathematical Approach to the Common Types of Wave Motion*. Longman.

Courant, R. (1943). Variational methods for the solution of problems of equilibrium and vibrations. *Bull. Am. Math. Soc.*, **49**, 1–23.

Crandall, S. H. (1956). *Engineering Analysis*. McGraw-Hill.

Crank, J. (1979). *The Mathematics of Diffusion*. Oxford Science Publications.

Curle, N. and Davies, H. J. (1971). *Modern Fluid Dynamics: Compressible Flow*. Van Nostrand Reinhold.

Davies, A. J. and Crann, D. (2008). *A Handbook of Essential Mathematical Formulae*. University of Hertfordshire Press.

Davies, B. and Martin, B. (1979). Numerical inversion of Laplace transforms, a survey and comparison of methods. *J. Comput. Phys.*, **33**, 1–32.

Desai, C. S. and Abel, J. F. (1972). *Introduction to the Finite Element Method: A Numerical Method for Engineering Analysis*. Van Nostrand Rienhold.

Doctors, L. J. (1970). An application of the finite element technique for boundary value problems of potential flow. *Int. J. Numer. Methods Eng.*, **2**, 243–52.

Douglas, J. and Dupont, T. (1974). Galerkin approximations for the two point boundary problem using continuous, piecewise polynomial spaces. *Numer. Math.*, **22**, 99–109.

Elliott, C. M. and Larsson, S. (1995). A finite element model for the time-dependent Joule heating problem. *Math. Commun.*, **64**, 1433–53.

Fausett, L. V. (2007). *Applied Numerical Analysis using MATLAB*. Prentice Hall.

Finlayson, B. A. and Scriven, L. E. (1967). On the search for variational principles. *Int. J. Heat Mass Transfer*, **10**, 799–832.

Fish, J. and Belytschko T. (2007). *A First Course in Finite Elements*. Wiley.

Fletcher, C. A. T. (1984). *Computational Galerkin Methods*. Springer.

Fredholm, I. (1903). Sur une class d'equations functionelles. *Acta Math.*, **27**, 365–90.

Galerkin, B. G. (1915). Series solution of some problems of elastic equilibrium of rods and plates [in Russian]. *Vestn. Inzh. Tech.*, **19**, 897–908.

Gallagher, R. H. (1969). Finite element analysis of plate and shell structures. *Proc. A.S.C.E. Symp. on Application of Finite Element Methods in Civil Engineering*, Vanderbilt University, Nashville, Tennessee, 155–205.

Gallagher, R. H., Padlog, J. and Bijlaard, P. P. (1962). Stress analysis of heated complex shapes. *J. Am. Rocket Soc.*, **32**, 700–7.

Gipson, G. S. (1987). *Boundary Element Fundamentals*. Computational Mechanics Press.

Goldberg, M. A. and Chen, C. S. (1997). *Discrete Projection Methods for Integral Equations*. Computational Mechanics Publications.

Goldberg, M. A. and Chen, C. S. (1999). The method of fundamental solutions for potential, Helmholtz and diffusion problems. In *Boundary and Integral Methods: Numerical and Mathematical Aspects*, Chapter 4. Computational Mechanics Press.

Gordon, W. J. and Hall, C. A. (1973). Transfinite element methods blending-function interpolation over arbitrary curved element domains. *Numer. Math.*, **21**, 109–29.

Grafton, P. E. and Strome, D. R. (1963). Analysis of axisymmetrical shells by the direct stiffness method. *J.A.I.A.A.*, **1**, 2342–7.

Green, G. (1828). *An Essay on the Application of Mathematical Analysis to the Theories of Electricity and Magnetism*. Longman.

Greenstadt, J. (1959). On the reduction of continuous problems to discrete form. *IBM J. Res. Dev.*, **3**, 355–68.

Gurtin, M. E. (1964). Variational principles for linear initial value problems. *Q. Appl. Math.*, **22**, 252–6.

Hadamard, J. (1923). *Lectures on Cauchy's Problem in Linear Partial Differential Equations*. Dover.

Hall, C. A. and Heinrich, J. (1978). A finite element that satisfies natural boundary conditions exactly. *J. Inst. Math. Appl.*, **21**, 237–50.

Hall, W. S. (1993). *The Boundary Element Method*. Kluwer Academic.

Hammer, T. G., Marlowe, O. P. and Stroud, A. H. (1956). Numerical integration over simplexes and cones. *Math. Tables Aids Comput.*, **10**, 130–7.

Hazel, T. G. and Wexler, A. (1972). Variational formulation of the Dirichlet boundary condition. *IEEE Trans. Microwave Theory Technol.*, **MTT20**, 385–90.

Hess, J. L. and Smith, A. M. O. (1964). Calculation of non-lifting potential flow about arbitrary three-dimensional bodies. *J. Ship Res.*, **8**, 22–44.

Hrennikoff, A. (1941). Solution of problems in elasticity by the framework method. *J. Appl. Mech.*, **A8**, 169–75.

Irons, B. M. (1970). A frontal solution program for finite element analysis. *Int. J. Numer. Methods Eng.*, **2**, 5–32.

Irons, B. M. and Draper, J. K. (1965). Inadequacy of nodal connections in a stiffness solution for plate bending. *J.A.I.A.A.*, **3**, 961.

Jameson, A. and Mavriplis, D. (1986). Finite volume solution of the two-dimensional Euler equations on a regular triangular mesh. *AIAA J.*, **24**, 611–8.

Jaswon, M. A. and Symm, G. T. (1977). *Integral Equation Methods in Potential Theory and Elastostatics*. Academic Press.

Jennings, A. and McKeown, J. J. (1992). *Matrix Computation for Engineers and Scientists*. Wiley.

Jones, R. E. (1964). A generalization of the direct stiffness method of structural analysis. *J.A.I.A.A.*, **2**, 821–6.

Kaltan, P. I. (2007). *MATLAB Guide to Finite Elements*. Springer.

Kellog, O. D. (1929). *Foundations of Potential Theory*. Springer.

Koenig, H. A. and Davids, N. (1969). The damped transient behaviour of finite beams and plates. *Int. J. Numer. Methods Eng.*, **1**, 151–62.

Kupradze, O. D. (1965). *Potential Methods in the Theory of Elasticity*. Daniel Davy, New York.

Kwok, W. L., Cheung, V. K. and Delcourt, C. (1977). Application of least squares collocation technique in finite element and finite strip formulation. *Int. J. Numer. Methods Eng.*, **11**, 1391–404.

Kythe, P. (1995). *An Introduction to Boundary Element Methods*. CRC Press.

Kythe, P. K. (1996). *Fundamental Solutions for Differential Operators and Applications*. Birkhäuser.

Lamb, H. (1993). *Hydrodynamics*. Cambridge Mathematical Library.

Lambert, J. D. (2000). *Numerical Methods for Ordinary Differential Systems: The Initial Value Problem*. Wiley.

Liu, G. R. (2003). *Mesh Free Methods*. CRC Press.

Lynn, P. P. and Arya, S. K. (1973). Least squares criterion in the finite element formulation. *Int. J. Numer. Methods Eng.*, **6**, 75–88.

Lynn, P. P. and Arya, S. K. (1974). Finite elements formulated by the weighted discrete least squares method. *Int. J. Numer. Methods Eng.*, **8**, 71–90.

Manzhirov, A. and Polyanin, A. D. (2008). *Handbook of Integral Equations* (2nd edn). Chapman and Hall.

Martin, H. C. (1965). On the derivation of stiffness matrices for the analysis of large deflection and stability problems. *Proc. Conf. Matrix Methods in Struct. Mech.*, Air Force Institute of Technology, Wright Patterson Air Force Base, Ohio, 697–716.

Martin, H. C. and Carey, G. F. (1973). *Introduction to Finite Element Analysis: Theory and Application*. McGraw-Hill.

McHenry, D. (1943). A lattice analogy for the solution of plane stress problems. *J. Instn. Civ. Eng.*, **21**, 59–82.

McMahon, J. (1953). Lower bounds for the electrostatic capacity of a cube. *Proc. R. Irish Acad.*, **55**, 133–67.

Melosh, R. J. (1963). Basis for derivation of matrices for the direct stiffness method. *J.A.I.A.A.*, **1**, 1631–7.

Mikhlin, S. G. (1964). *Variational Methods in Mathematical Physics*. Pergamon.

Mohr, G. A. (1992). *Finite Elements for Solids, Fluids and Optimization*. Oxford Science Publications.

Moridis, G. J. and Reddell, D. L. (1991). The Laplace transform finite element numerical method for the solution of the groundwater equation. *EOS Trans. AGU*, **72**, H22C-4.

Noble, B. (1973). Variational finite element methods for initial value problems. In *The Mathematics of Finite Elements and Applications*, Vol. I (ed. J. Whiteman), Proc. Conf. Brunel University. Academic Press, pp. 14–51.

París, F. and Cañas, J. (1997). *Boundary Element Method: Fundamentals and Applications*. Oxford University Press.

Partridge, P. W., Brebbia, C. A. and Wrobel, L. C. (1992). *The Dual Reciprocity Boundary Element Method*. Elsevier Applied Science.

Phillips, A. T. (2009). The femur as a musculo-skeletal construct: a free boundary condition modelling approach. *Med. Eng. Phys.*, **31**, 673–80.

Polya, G. (1952). Sur une interprétation de la méthod des différences finies qui peut fournir des bornes supériers ou inférieurs. *C. R. Acad. Sci. Paris*, **235**, 995–7.

Popov, P. and Power, H. (2001). An $\mathcal{O}(N)$ Taylor series multipole boundary element method for three-dimensional elasticity problems. *Eng. Anal. Bound. Elements*, **25**, 7–18.

Prager, W. and Synge, J. L. (1947). Approximations in elasticity based on the concept of function space. *Q. Appl. Math.*, **5**, 241–69.

Przemieniecki, J. S. (1968). *Theory of Matrix Structural Analysis*. McGraw-Hill.

Renardy, M. and Rogers, R. C. (1993). *An Introduction to Differential Equations*. Springer.

Reitz, J. R., Milford, F. J. and Christy, R. W. (1992). *Foundations of Electromagnetic Theory*. Addison-Wesley.

Rheinboldt, W. C. (1987). *Methods for Solving Systems of Non-linear Equations* (2nd edn). SIAM.

Ritz, W. (1909). Über eine neue methode zur lösung gewissen variations – problem der mathematischen physik. *J. Reine angew. Math.*, **135**, 1–61.

Rizzo, F. J. (1967). An integral equation approach to boundary value problems. *Q. Appl. Math.*, **25**, 83–95.

Schoenberg, I. J. (1946). Contributions to the problem of approximation of equidistant data by analytic functions. *Q. Appl. Math.*, **4**, 45–9.

Silvester, P. P. and Ferrari, R. L. (1983). *Finite Elements for Electrical Engineers*. Cambridge University Press.

Smith, G. D. (1985). *Numerical Solution of Partial Differential Equations: Finite Difference Methods* (3rd edn). Oxford University Press.

Smith, I. M. and Griffiths, D. V. (2004). *Programming the Finite Element Method* (4th edn). Wiley.

Sokolnikoff, I. S. (1956). *Mathematical Theory of Elasticity*. McGraw-Hill.

Spalding, D. A. (1972). A novel finite difference formulation for differential equations involving both first and second derivatives. *Int. J. Numer. Methods Eng.*, **4**, 551–9.

Stehfest, H. (1970a). Numerical inversion of Laplace transforms. *Commun. ACM*, **13**, 47–9.

Stehfest, H. (1970b). Remarks on algorithm 368[D5] Numerical inversion of Laplace transforms. *Commun. ACM*, **13**, 624.

Strang, G. and Fix, G. J. (1973). *An Analysis of the Finite Element Method*. Prentice-Hall. Now published by Wellesley-Cambridge Press.

Strutt, J. W. (Lord Rayleigh) (1870). On the theory of resonance. *Trans. R. Soc. Lond.*, **19**, 72–118.

Synge, J. L. (1957). *The Hypercircle Method in Mathematical Physics*. Cambridge University Press.

Szabó, B. A. and Lee, G. C. (1969). Derivation of stiffness matrices for problems in plane elasticity by Galerkin's method. *Int. J. Numer. Methods Eng.*, **1**, 301–10.

Tao Jiang, Xin Liu and Zhengzhou Yu (2009). Finite element algorithms for pricing 2-D basket options. *ICISE*, 4881–6.

Taylor, C. and Hood, P. (1973). A numerical solution of the Navier–Stokes equations using the finite element technique. *Comput. Fluids*, **1**, 73–100.

Telles, J. C. F. (1987). A self-adaptive co-ordinate transformation for efficient evaluation of boundary element integrals. *Int. J. Numer. Methods Eng.*, **24**, 959–73.

Topper, J. (2005). Option pricing with finite elements. *Wilmott J.*, **2005-1**, 84–90.

Trefftz, E. (1926). Ein gegenstuck zum Ritz'schen verfahren. *Proc. 2nd. Int. Congr. Appl. Mech.*, Zürich.

Turner, M. J., Clough, R. W., Martin, H. C. and Topp, L. T. (1956). Stiffness and deflection analysis of complex structures. *J. Aeron. Sci.*, **23**, 805–23.

Turner, M. J., Dill, E. H., Martin, H. C. and Melosh, R. J. (1960). Large deflections of structures subjected to heating and external loads. *J. Aeron. Sci.*, **27**, 97–107.

Wade, W. R. (1995). *An Introduction to Analysis*. Pearson Education.

Wait, R. and Mitchell, A. R. (1985). *Finite Element Analysis and Applications*. Wiley.

Weinberger, H. F. (1956). Upper and lower bounds for eigenvalues by finite difference methods. *Commun. Pure Appl. Math.*, **9**, 613–23.

Wilkinson, J. H. (1999). *The Algebraic Eigenvalue Problem*. Oxford University Press.

Wilmott, P., Howison, S. and Dewynne, J. (1995). *The Mathematics of Financial Derivatives*. Cambridge University Press.

Wilson, E. L. and Nickell, R. E. (1966). Application of finite element method to heat conduction analysis. *Nucl. Eng. Des.*, **4**, 1–11.

Wolf, J. P. and Song, C. (1997). *Finite Element Modelling of Unbounded Media*. Wiley.

Wrobel, L. C. (2002). *The Boundary Element Method*, Vol. 1. Wiley.

Zhu, S.-P. (1999). Time-dependent reaction–diffusion problems and the LTDRM approach. In *Boundary Integral Methods: Numerical and Mathematical Aspects* (ed. M. Goldberg). Computational Mechanics Publications, pp. 1–35.

Zienkiewicz, O. C. (1977). *The Finite Element Method* (3rd edn). McGraw-Hill.

Zienkiewicz, O. C. (1995). Origins, milestones and direction of the FEM—a personal view. *Arch. Comput. Methods Eng.*, **2**, 1–48.

Zienkiewicz, O. C. and Cheung, Y. K. (1965). Finite elements in the solution of field problems. *Engineer*, **220**, 507–10.

Zienkiewicz, O. C. and Taylor, R. L. (2000a). *The Finite Element Method*, Vol. 2, *Solid Mechanics* (5th edn). Butterworth-Heinemann.

Zienkiewicz, O. C. and Taylor, R. L. (2000b). *The Finite Element Method*, Vol. 3, *Fluid Dynamics* (5th edn). Butterworth-Heinemann.

Zienkiewicz, O. C., Irons, B. M. and Nath, B. (1966). Natural frequencies of complex, free or submerged structures by the finite element method. In *Proceedings of the Symposium on Vibration in Civil Engineering*. Butterworths, pp. 83–93.

Zienkiewicz, O. C., Heinrich, J. C., Huyakorn, P. S. and Mitchell, A. R. (1977). An upwind finite element scheme for two-dimensional convective transport equations. *Int. J. Numer. Methods Eng.*, **11**, 131–44.

Zienkiewicz, O. C., Kelly, D. W. and Bettess, P. (1977). The coupling of the finite element method and boundary solution procedures. *Int. J. Numer. Methods Eng.*, **11**, 355–75.

Zienkiewicz, O. C., Lyness, J. and Owen, D. R. J. (1977). Three-dimensional magnetic field determination using a scalar potential – a finite element solution. *IEEE Trans. Magn.*, **MAG13**, 1649–56.

Zienkiewicz, O. C., Taylor, R. L. and Zhu, J. Z. (2005). *The Finite Element Method: Its Basis and Fundamentals* (6th edn). Elsevier.

Zienkiewicz, O. C., Watson, M. and King, I. P. (1968). A numerical method of visco-elastic stress analysis. *Int. J. Mech. Sci.*, **10**, 807–27.

Zlamal, M. (1968). On the finite element method. *Numer. Math.*, **12**, 395–409.

Index